# Lecture Notes in Computer Science 7535

Dimitrios M. Thilikos
Gerhard J. Woeginger (Eds.)

# Parameterized and Exact Computation

7th International Symposium, IPEC 2012
Ljubljana, Slovenia, September 12-14, 2012
Proceedings

Volume Editors

Dimitrios M. Thilikos
National and Kapodistrian University of Athens
Department of Mathematics
Panepistimioupolis
15784 Athens, Greece
E-mail: sedthilk@thilikos.info

Gerhard J. Woeginger
Eindhoven University of Technology
Department of Mathematics and Computer Science
P.O. Box 513
5600 MB Eindhoven, The Netherlands
E-mail: gwoegi@win.tue.nl

ISSN 0302-9743 e-ISSN 1611-3349
ISBN 978-3-642-33292-0 e-ISBN 978-3-642-33293-7
DOI 10.1007/978-3-642-33293-7
Springer Heidelberg Dordrecht London New York

Library of Congress Control Number: 2012946191

CR Subject Classification (1998): F.2.1-3, G.1-2, G.2.3, I.3.5, G.4, E.1, I.2.8

LNCS Sublibrary: SL 1 – Theoretical Computer Science and General Issues



*Typesetting:* Camera-ready by author, data conversion by Scientific Publishing Services, Chennai, India

Printed on acid-free paper

Springer is part of Springer Science+Business Media (www.springer.com)

# Preface

This volume contains the 23 papers presented at the *7th International Symposium on Parameterized and Exact Computation*, IPEC 2012 (ipec2012.isoftcloud.gr), held on September 12–14, 2012 as part of the ALGO 2012 (algo12.fri.uni-lj.si) conference in Ljubljana (Slovenia). IPEC is an international symposium series that covers research on all aspects of parameterized and exact algorithms and complexity. The workshop series started in 2004 in Bergen (Norway) as a biennial event, and in 2008 became an annual event. The first four workshops in the series used the five-letter acronym IWPEC (which was bulky and hard to pronounce), whereas from the fifth workshop onwards the catchy four-letter acronym IPEC has been used. Over the years IPEC has become very visible and it has grown into one of the main events for the algorithmics and complexity community.

The IPEC 2012 plenary keynote talks were given by *Andreas Björklund* (Lund University) on "The Path Taken for $k$-Path" and by *Dániel Marx* (MTA SZTAKI) on "Randomized Techniques for Parameterized Algorithms". We had two additional invited tutorial speakers: *Michał Pilipczuk* (University of Bergen) speaking on lower bounds for polynomial kernelization, and *Saket Saurabh* (Chennai) speaking on subexponential parameterized algorithms.

Altogether IPEC 2012 received 37 extended abstracts. Each submission was reviewed by at least three reviewers. The Program Committee thoroughly discussed the submissions in electronic meetings using the EasyChair system, and selected 23 papers for presentation. We expect the full versions of the papers contained in this volume to be submitted for publication in refereed journals.

Many people contributed to the smooth running and the success of IPEC 2012. In particular our thanks go

- to all authors who submitted their current research to IPEC
- to all our reviewers and subreferees whose expertise flowed into the decision process
- to the members of the Program Committee who graciously gave their time and energy
- to the members of the Local Organizing Committee who made the conference possible
- to *Charalampos Tampakopoulos* for his web-hosting services via isoftcloud.gr
- to the EasyChair conference management system for hosting the evaluation process.

July 2012

Dimitrios M. Thilikos
Gerhard J. Woeginger

# Organization

## Program Committee

| | |
|---|---|
| Jianer Chen | Texas A&M University, USA |
| Marek Cygan | University of Warsaw, Poland |
| Henning Fernau | University of Trier, Germany |
| Fedor V. Fomin | University of Bergen, Norway |
| Martin Grohe | Humboldt University Berlin, Germany |
| Daniel Král' | Charles University Prague, Czech Republic |
| Stefan Kratsch | Max-Planck-Institut für Informatik, Saarbrücken, Germany |
| Mikko Koivisto | University of Helsinki, Finland |
| Igor Razgon | University of Leicester, UK |
| Saket Saurabh | Institute of Mathematical Sciences, Chennai, India |
| Dimitrios M. Thilikos | National and Kapodistrian University of Athens, Greece |
| Erik Jan van Leeuwen | Sapienza University of Rome, Italy |
| Magnus Wahlström | Max-Planck-Institut für Informatik, Saarbrücken, Germany |
| Gerhard Woeginger | Eindhoven University of Technology, The Netherlands |

## Organization Committee

Andrej Brodnik
Uroš Čibej
Gašper Fele-Žorž
Matevž Jekovec
Jurij Mihelič
Borut Robič
Andrej Tolić

## External Reviewers

Faisal Abu-Khzam
Mohammadhossein Bateni
Sergio Bermudo
René Van Bevern
Ljiljana Brankovic
Peter Damaschke
Samir Datta
Anuj Dawar
Holger Dell
Pål Drange
Andrew Drucker
Serge Gaspers
Petr Golovach
Fabrizio Grandoni
Sylvain Guillemot
Jiong Guo
Danny Hermelin
Petr Hliněný
Bart Jansen
Iyad Kanj

Petteri Kaski
Pavel Klavík
Łukasz Kowalik
Daniel Lokshtanov
Daniel Marx
Daniel Meister
Matthias Mnich
Jan Obdržálek
Marcin Pilipczuk
Michał Pilipczuk
Marcus Ritt
Noy Rotbart
Ignasi Sau
Pascal Schweitzer
Narges Simjour
Karolina Soltys
Ondřej Suchý
Till Tantau
Yngve Villanger
Ryan Williams

# Table of Contents

# The Path Taken for $k$-Path

Andreas Björklund

Department of Computer Science, Lund University, Sweden
andreas.bjorklund@yahoo.se

**Abstract.** We give a historical account of the parametrized results for the $k$-PATH problem: given a graph $G$ and a positive integer $k$, is there a simple path in $G$ of length $k$. Throughout the years several ingenious approaches have been used, steadily decreasing the run time bound. Moreover, the techniques used have often found lots of other applications. We will revisit some of the old results, as well as cover the state-of-the-art techniques based on algebraic sieves. We will also briefly talk about what is known about counting $k$-paths.

D.M. Thilikos and G.J. Woeginger (Eds.): IPEC 2012, LNCS 7535, p. 1, 2012.

# Randomized Techniques for Parameterized Algorithms*

Dániel Marx

Computer and Automation Research Institute,
Hungarian Academy of Sciences (MTA SZTAKI),
Budapest, Hungary
dmarx@cs.bme.hu

**Abstract.** Since the introduction of the Color Coding technique in 1994 by Alon, Yuster, and Zwick, randomization has been part of the toolkit for proving fixed-parameter tractability results. It seems that randomization is very well suited to parameterized algorithms: if the task is to find a solution of size $k$ and only those random choices need to be correct that are directly related to the solution, then typically we can bound the error probability by a function of $k$. The talk will overview through various concrete examples how randomization appears in fixed-parameter tractability results. We argue that in many cases randomization appears in form of a reduction: it allows us to reduce the problem we are trying to solve to an easier and more structured problem.

* Research supported by the European Research Council (ERC) grant "PARAMTIGHT: Parameterized complexity and the search for tight complexity results," reference 280152.

D.M. Thilikos and G.J. Woeginger (Eds.): IPEC 2012, LNCS 7535, p. 2, 2012.

# Finding a Maximum Induced Degenerate Subgraph Faster Than $2^n$

Marcin Pilipczuk[1,*] and Michał Pilipczuk[2,**]

[1] Institute of Informatics, University of Warsaw, Poland
malcin@mimuw.edu.pl

[2] Department of Informatics, University of Bergen, Norway
michal.pilipczuk@ii.uib.no

**Abstract.** In this paper we study the problem of finding a maximum induced $d$-degenerate subgraph in a given $n$-vertex graph from the point of view of exact algorithms. We show that for any fixed $d$ one can find a maximum induced $d$-degenerate subgraph in randomized $(2-\varepsilon_d)^n n^{\mathcal{O}(1)}$ time, for some constant $\varepsilon_d > 0$ depending only on $d$. Moreover, our algorithm can be used to sample inclusion-wise maximal induced $d$-degenerate subgraphs in such a manner that every such subgraph is output with probability at least $(2-\varepsilon_d)^{-n}$; hence, we prove that their number is bounded by $(2-\varepsilon_d)^n$.

## 1 Introduction

The theory of exact computations studies the design of algorithms for NP-hard problems that compute the answer optimally, however using possibly exponential time. The goal is to limit the exponential blow-up in the best possible running-time guarantee. For some problems, like Independent Set [1], Dominating Set [1, 2], and Bandwidth [3] the research concentrates on achieving better and better constants in the bases of exponents. However, for many important computational tasks designing even a routine faster than trivial brute-force solution or straightforward dynamic program is a challenging combinatorial question; the answer to this question can provide valuable insight into the structure of the problem. Perhaps the most prominent among recent developments in breaking trivial barriers is the algorithm for Hamiltonian Cycle of Björklund [4], but a lot of effort is put also into less fundamental problems, like Maximum Induced Planar Graph [5] or a scheduling problem $1|\text{prec}|\sum C_i$ [6], among many others [7–12]. However, many natural and well-studied problems still lack exact algorithms faster than the trivial ones; the most important examples are TSP, Permanent, Set Cover, #Hamiltonian Cycles and SAT. In particular, hardness of SAT is the starting point for the *Strong Exponential Time Hypothesis* of Impagliazzo and Paturi [13, 14], which is used as an argument that other problems are hard as well [15–18].

* Partially supported by NCN grant N206567140 and Foundation for Polish Science.
** Partially supported by European Research Council (ERC) Grant "Rigorous Theory of Preprocessing", reference 267959.

D.M. Thilikos and G.J. Woeginger (Eds.): IPEC 2012, LNCS 7535, pp. 3–12, 2012.

A group of tasks we are particularly interested in in this paper are the problems that ask for a maximum size induced subgraph belonging to some class $\Pi$. If belonging to $\Pi$ can be recognized in polynomial time, then we have an obvious brute-force solution working in $2^n n^{\mathcal{O}(1)}$ time that iterates through all the subsets of vertices checking which of them induce subgraphs belonging to $\Pi$. Note that the classical INDEPENDENT SET problem can be formulated in this manner for $\Pi$ being the class of edgeless graphs, while if $\Pi$ is the class of forests then we arrive at the MAXIMUM INDUCED FOREST, which is dual to FEEDBACK VERTEX SET. For both these problems algorithms with running time of form $(2-\varepsilon)^n$ for some $\varepsilon > 0$ are known [1, 11, 12]. The list of problems admitting algorithms with similar complexities includes also $\Pi$ being the classes of regular graphs [19], graphs of small treewidth [20], planar graphs [5], 2- or 3-colourable graphs [21], bicliques [22] or graphs excluding a forbidden subgraph [23].

The starting point of our work is the question raised by Fomin et al. in [5]. Having obtained an algorithm finding a maximum induced planar graph in time $\mathcal{O}(1.7347^n)$, they ask whether their result can be extended to graphs of bounded genus or even to $H$-minor-free graphs for fixed $H$. Note that all these graph classes are hereditary and consist of *sparse* graphs, i.e., graphs with the number of edges bounded linearly in the number of vertices. Moreover, for other hereditary sparse classes, such as graphs of bounded treewidth, algorithms with running time $(2-\varepsilon)^n$ for some $\varepsilon > 0$ are also known [20]. Therefore, it is tempting to ask whether the sparseness of the graph class can be used to break the $2^n$ barrier in a more general manner.

In order to formalize this question we study the problem of finding a maximum induced $d$-degenerate graph. Recall that a graph is called $d$-degenerate if each of its subgraphs contains a vertex of degree at most $d$. Every hereditary class of graphs with a number of edges bounded linearly in the number of vertices is $d$-degenerate for some $d$; for example, planar graphs are 5-degenerate, graphs excluding $K_r$ as a minor are $\mathcal{O}(r\sqrt{\log r})$-degenerate, while the class of forests is equivalent to the class of 1-degenerate graphs. However, $d$-degeneracy does not impose any topological constraints; to see this, note that one can turn any graph into a 2-degenerate graph by subdividing every edge. Hence, considering a problem on the class of $d$-degenerate graphs can be useful to examine whether it is just sparseness that makes it more tractable, or one has to add additional restrictions of topological nature [24].

*Our Results and Techniques.* We make a step towards understanding the complexity of finding a maximum induced subgraph from a sparse graph class by breaking the $2^n$-barrier for the problem of finding maximum induced $d$-degenerate subgraph. The main result of this paper is the following algorithmic theorem.

**Theorem 1.** *For any integer $d \geq 1$ there exists a constant $\varepsilon_d > 0$ and a polynomial-time randomized algorithm $\mathcal{A}_d$, which given an $n$-vertex graph $G$ either reports an error, or outputs a subset of vertices inducing a $d$-degenerate subgraph. Moreover, for every inclusion-wise maximal induced $d$-degenerate subgraph, let $X$ be its vertex set, the probability that $\mathcal{A}_d$ outputs $X$ is at least $(2-\varepsilon_d)^{-n}$.*

Let $X_0$ be a set of vertices inducing a maximum $d$-degenerate subgraph. If we run the algorithm $(2-\varepsilon_d)^n$ times, we know that with probability at least 1/2 in one of the runs the set $X_0$ will be found. Hence, outputting the maximum size set among those found by the runs gives the following corollary.

**Corollary 2.** *There exists a randomized algorithm which, given an n-vertex graph G, in* $(2-\varepsilon_d)^n n^{\mathcal{O}(1)}$ *time outputs a set* $X \subseteq V(G)$ *inducing a d-degenerate graph. Moreover, X is maximum with probability at least* $\frac{1}{2}$.

As the total probability that $\mathcal{A}_d$ outputs some set of vertices is bounded by 1, we obtain also the following corollary.

**Corollary 3.** *For any integer* $d \geq 1$ *there exists a constant* $\varepsilon_d > 0$ *such that any n-vertex graph contains at most* $(2-\varepsilon_d)^n$ *inclusion-wise maximal induced d-degenerate subgraphs.*

Let us elaborate briefly on the idea behind the algorithm of Theorem 1. Assume first that $G$ has large average degree, i.e., $|E(G)| > \lambda d|V(G)|$ for some large constant $\lambda$. As $d$-degenerate graphs are sparse, i.e., the number of edges is less than $d$ times the number of vertices, it follows that for any set $X$ inducing a $d$-degenerate graph $G[X]$, only a tiny fraction of edges inside $G$ are in fact inside $G[X]$. Hence, an edge $uv$ chosen uniformly at random can be assumed with high probability to have at least one endpoint outside $X$. We can further choose at random, with probabilities 1/3 each, one of the following decisions: $u \in X$, $v \notin X$ or $u \notin X$, $v \in X$, or $u, v \notin X$. In this manner we fix the status of two vertices of $G$ and, if $\lambda > 4$, the probability that the guess is correct is larger than 1/4. If this randomized step cannot be applied, we know that the average degree in $G$ is at most $\lambda d$ and we can apply more standard branching arguments on vertices of low degrees.

Our algorithm is a polynomial-time routine that outputs an induced $d$-degenerate graph by guessing assignment of consecutive vertices with probabilities slightly better than 1/2. We would like to remark that all but one of the ingredients of the algorithm can be turned into standard, deterministic branching steps. The only truly randomized part is the aforementioned random choice of an edge to perform a guess with enhanced success probability. However, to ease the presentation we choose to present the whole algorithm in a randomized fashion by expressing classical branchings as random choices of the branch.

*Organization.* In Section 2 we settle notation and give preliminary results on degenerate graphs. Section 3 contains the proof of Theorem 1. Section 4 concludes the paper.

## 2 Preliminaries

*Notation.* We use standard graph notation. For a graph $G$, by $V(G)$ and $E(G)$ we denote its vertex and edge sets, respectively. For $v \in V(G)$, its neighborhood $N_G(v)$ is defined as $N_G(v) = \{u : uv \in E(G)\}$. For a set $X \subseteq V(G)$ by $G[X]$ we denote the subgraph of $G$ induced by $X$. For a set $X$ of vertices or edges of $G$, by $G \setminus X$ we denote the graph with the vertices or edges of $X$ removed; in case of vertex removal, we remove also all the incident edges.

*Degenerate Graphs.* For an integer $d \geq 0$, we say that a graph $G$ is $d$-degenerate if every subgraph (equivalently, every induced subgraph) of $G$ contains a vertex of degree at most $d$. Clearly, the class of $d$-degenerate graphs is closed under taking both subgraphs and induced subgraphs. Note that 0-degenerate graphs are independent sets, and the class of 1-degenerate graphs is exactly the class of forests. All planar graphs are 5-degenerate; moreover, every $K_r$-minor-free graph (in particular, any $H$-minor-free graph for $|V(H)| = r$) is $\mathcal{O}(r\sqrt{\log r})$-degenerate [25–27].

The following simple proposition shows that the notion of $d$-degeneracy admits greedy arguments.

**Proposition 4.** *Let $G$ be a graph and $v$ be a vertex of degree at most $d$ in $G$. Then $G$ is $d$-degenerate if and only if $G \setminus v$ is.*

*Proof.* As $G \setminus v$ is a subgraph of $G$, then $d$-degeneracy of $G$ implies $d$-degeneracy of $G \setminus v$. Hence, we only need to justify that if $G \setminus v$ is $d$-degenerate, then so does $G$. Take any $X \subseteq V(G)$. If $v \in X$, then the degree of $v$ in $G[X]$ is at most its degree in $G$, hence it is at most $d$. However, if $v \notin X$ then $G[X]$ is a subgraph of $G \setminus v$ and $G[X]$ contains a vertex of degree at most $d$ as well. As $X$ was chosen arbitrarily, the claim follows. □

Proposition 4 ensures that one can test $d$-degeneracy of a graph by in turn finding a vertex of degree at most $d$, which needs to exist due to the definition, and deleting it. If in this manner we can remove all the vertices of the graph, it is clearly $d$-degenerate. Otherwise we end up with an induced subgraph with minimum degree at least $d + 1$, which is a sufficient proof that the graph is not $d$-degenerate. Note that this procedure can be implemented in polynomial time. As during each deletion we remove at most $d$ edges from the graph, the following proposition is straightforward.

**Proposition 5.** *Any $n$-vertex $d$-degenerate graph has at most $dn$ edges.*

## 3 The Algorithm

In this section we prove Theorem 1. Let us fix $d \geq 1$, an $n$-vertex graph $G$ and an inclusion-wise maximal set $X \subseteq V(G)$ inducing a $d$-degenerate graph.

The behaviour of the algorithm depends on a few constants that may depend on $d$ and whose values influence the final success probability. At the end of this section we propose precise values of these constants and respective values of $\varepsilon_d$ for $1 \leq d \leq 6$. However, as the values of $\varepsilon_d$ are really tiny even for small $d$, when describing the algorithm we prefer to introduce these constants symbolically, and only argue that there exists their evaluation that leads to a $(2 - \varepsilon_d)^{-n}$ lower bound on the probability of successfully sampling $X$.

The algorithm maintains two disjoint sets $A, Z \subseteq V(G)$, consisting of vertices about which we have already made some assumptions: we seek for the set $X$ that contains $A$ and is disjoint from $Z$. Let $Q = V(G) \setminus (A \cup Z)$ be the set of the remaining vertices, whose assignment is not yet decided.

We start with $A = Z = \emptyset$. The description of the algorithm consists of a sequence of *rules*; at each point, the lowest-numbered applicable rule is used. When applying a rule we assign some vertices of $Q$ to the set $A$ or $Z$, depending on some random decision. We say that an application of a rule is *correct* if, assuming that before the application we have $A \subseteq X$ and $Z \cap X = \emptyset$, the vertices assigned to $A$ belong to $X$, and the vertices assigned to $Z$ belong to $V(G) \setminus X$. In other words, a correct application assigns the vertices consistently with the fixed solution $X$.

We start with the randomized rule that is triggered when the graph is dense. Observe that, since $G[X]$ is $d$-degenerate, $G[X \cap Q]$ is $d$-degenerate as well and, by Proposition 5, contains less than $d|X \cap Q|$ edges. Thus, if $|E(G[Q])|/|Q|$ is significantly larger than $d$, then only a tiny fraction of the edges of $G[Q]$ are present in $G[X]$. Hence, an overwhelming fraction of edges of $G[Q]$ has at least one of the endpoints outside $X$, so having sampled an edge of $G[Q]$ uniformly at random with high probability we may assume that there are only three possibilities of the behaviour of its endpoints, instead of four. This observation leads to the following rule. Let $\lambda > 4$ be a constant.

**Rule 1.** If $|E(G[Q])| \geq \lambda d|Q|$, then:

1. choose an edge $uv \in E(G[Q])$ uniformly at random;
2. with probability 1/3 each, make one of the following decisions: either assign $u$ to $A$ and $v$ to $Z$, or assign $u$ to $Z$ and $v$ to $A$, or assign both $u$ and $v$ to $Z$.

**Lemma 6.** *Assume that $A \subseteq X$ and $Z \cap X = \emptyset$ before Rule 1 is applied. Then the application of Rule 1 is correct with probability at least $\frac{\lambda-1}{3\lambda}$.*

*Proof.* As $|E(G[Q])| \geq \lambda d|Q|$, but $|E(G[X \cap Q])| \leq d|X \cap Q| \leq d|Q|$ by Proposition 5, the probability that $uv \notin E(G[X])$ is at least $\frac{\lambda-1}{\lambda}$. Conditional on the assumption $uv \notin E(G[X])$, in the second step of Rule 1 we make a correct decision with probability 1/3. This concludes the proof. □

Note that the bound $\frac{\lambda-1}{3\lambda}$ is larger than 1/4 for $\lambda > 4$.

Equipped with Rule 1, we may focus on the case when $G[Q]$ has small average degree. Let us introduce a constant $\kappa > 2\lambda$ and let $S \subseteq Q$ be the set of vertices having degree less than $\kappa d$ in $G[Q]$. If Rule 1 is not applicable, then $|E(G[Q])| < \lambda d|Q|$. Hence we can infer that $|S| \geq \frac{\kappa-2\lambda}{\kappa}|Q|$, as otherwise by just counting the degrees of vertices in $Q \setminus S$ we could find at least $\frac{1}{2} \cdot \frac{2\lambda}{\kappa}|Q| \cdot \kappa d = \lambda d|Q|$ edges in $G[Q]$. Consider any $v \in S$. Such a vertex $v$ may be of two types: it either has at most $d$ neighbours in $A$, or at least $d+1$ of them. In the first case, we argue that we may perform a good guessing step in the closed neighbourhood of $v$, because the degree of $v$ is bounded and when all the neighbours of $v$ are deleted (assigned to $Z$), then one may greedily assign $v$ to $A$. In the second case, we observe that we cannot assign too many such vertices $v$ to $A$, as otherwise we would obtain a subgraph of $G[A]$ with too high average degree. Let us now proceed to the formal arguments.

**Rule 2.** Assume there exists a vertex $v \in Q$ such that $|N_G(v) \cap Q| < \kappa d$ and $|N_G(v) \cap A| \leq d$. Let $r = |N_G(v) \cap Q|$ and $v_1, v_2, \ldots, v_r$ be an arbitrary ordering of the neighbours of $v$ in $Q$. Let $\gamma = \gamma(r) \geq 1$ be such that

$$\gamma^{-1} + \gamma^{-2} + \ldots + \gamma^{-r-1} = 1.$$

Randomly, make one of the following decisions:

1. for $1 \leq i \leq r$, with probability $\gamma^{-i}$ assign $v_1, v_2, \ldots, v_{i-1}$ to $Z$ and $v_i$ to $A$;
2. with probability $\gamma^{-r-1}$ assign all vertices $v_1, v_2, \ldots, v_r$ to $Z$ and $v$ to $A$.

Note that the choice of $\gamma$ not only ensures that the probabilities of the options in Rule 2 sum up to one, but also that $\gamma(r) \leq \gamma(\lceil \kappa d \rceil - 1) < 2$. We now show a bound on the probability that an application of Rule 2 is correct.

**Lemma 7.** *Assume that $A \subseteq X$ and $Z \cap X = \emptyset$ before Rule 2 is applied. Then exactly one of the decisions considered in Rule 2 leads to a correct application. Moreover, if in the correct decision exactly $i_0$ vertices are assigned to $A \cup Z$, then the probability of choosing the correct one is equal to $\gamma^{-i_0}$.*

*Proof.* Firstly observe that the decisions in Rule 2 contradict each other, so at most one of them can lead to a correct application.

Assume that $(N_G(v) \cap Q) \cap X \neq \emptyset$ and let $v_{i_0}$ be the vertex from $(N_G(v) \cap Q) \cap X$ with the smallest index. Then the decision, which assigns all the vertices of $N_G(v) \cap Q$ with smaller indices to $Z$ and $v_{i_0}$ to $A$ leads to a correct application. Moreover, it assigns exactly $i_0$ vertices to $A \cup Z$ and the probability of choosing it is equal to $\gamma^{-i_0}$.

Assume now that $(N_G(v) \cap Q) \cap X = \emptyset$. We claim that $v \in X$. Assume otherwise; then $v$ has at most $d$ neighbours in $X$, so by Proposition 4 after greedily incorporating it to $X$ we would still have $G[X]$ being a $d$-degenerate graph. This contradicts maximality of $X$. Hence, we infer that the decision which assigns all the neighbours of $v$ from $Q$ to $Z$ and $v$ itself to $A$ leads to a correct application, it assigns exactly $r+1$ vertices to $A \cup Z$ and has probability $\gamma^{-r-1}$. □

We now handle vertices with more than $d$ neighbours in $A$. Intuitively, there can be at most $d|A|$ such vertices assigned to $A$, as otherwise $A$ would have an induced subgraph with too high average degree. Hence, if there is significantly more than $2d|A|$ such vertices in total, then picking one of them at random with probability higher than $1/2$ gives a vertex that needs to be assigned to $Z$. Let us introduce a constant $c > 2$.

**Rule 3.** If there are at least $cd|A|$ vertices in $Q$ that have more than $d$ neighbours in $A$, choose one such vertex uniformly at random and assign it to $Z$.

**Lemma 8.** *Assume that $A \subseteq X$ and $Z \cap X = \emptyset$ before Rule 3 is applied. Then the application of Rule 3 is correct with probability at least* $1 - 1/c$.

*Proof.* Let $P = \{v \in Q : |N_G(v) \cap A| > d\}$. As $|P| \geq cd|A|$, to prove the lemma it suffices to show that $|P \cap X| < d|A|$. Assume otherwise, and consider the set $((P \cap X) \cup A) \subseteq X$. The number of edges of the subgraph of $G[X]$ induced by $(P \cap X) \cup A$ is at least

$$(d+1)|P \cap X| = d|P \cap X| + |P \cap X| \geq d(|P \cap X| + |A|) = d|(P \cap X) \cup A|.$$

This contradicts the assumption that $G[X]$ is $d$-degenerate, due to Proposition 5. □

Note that $1 - 1/c > 1/2$ for $c > 2$.

We now show that if Rules 1, 2 and 3 are not applicable, then $|A \cup Z|$ is large, which means that the algorithm has already made decisions about a significant fraction of the vertices of the graph.

**Lemma 9.** *If Rules 1, 2 and 3 are not applicable, then* $|A \cup Z| > \alpha n$ *for some constant* $\alpha > 0$ *that depends only on the constants* $d$, $\lambda$, $\kappa$ *and* $c$.

*Proof.* As Rule 1 is not applicable, $Q$ contains at most $\frac{2\lambda}{\kappa}|Q|$ vertices of degree at least $\kappa d$ in $G[Q]$. As Rule 2 is not applicable, the remaining vertices have more than $d$ neighbours in $A$. As Rule 3 is not applicable, we have that

$$\frac{\kappa - 2\lambda}{\kappa}|Q| < cd|A| \leq cd|A \cup Z|.$$

As $Q = V(G) \setminus (A \cup Z)$, simple computations show that this is equivalent to

$$\frac{|A \cup Z|}{|V(G)|} > \left(\frac{cd\kappa}{\kappa - 2\lambda} + 1\right)^{-1},$$

and the proof is finished. □

Lemma 9 ensures that at this point the algorithm has already performed enough steps to achieve the desired success probability. Therefore, we may finish by brute-force.

**Rule 4.** If $|A \cup Z| > \alpha n$ for the constant $\alpha$ given by Lemma 9, for each $v \in Q$ independently, assign $v$ to $A$ or $Z$ with probability 1/2 each, and finish the algorithm by outputting the set $A$ if it induces a $d$-degenerate graph, or reporting an error otherwise.

We now summarize the bound on the success probability.

**Lemma 10.** *The algorithm outputs the set* $X$ *with probability at least*

$$\max\left(\sqrt{\frac{3\lambda}{\lambda - 1}}, \gamma(\lceil \kappa d \rceil - 1), \frac{c}{c-1}\right)^{-\alpha n} 2^{-(1-\alpha)n},$$

*which is equal to* $(2 - \varepsilon_d)^n$ *for some* $\varepsilon_d > 0$.

*Proof.* Recall that $\frac{3\lambda}{\lambda-1} < 4$, $\gamma(\lceil \kappa d \rceil - 1) < 2$, $\frac{c}{c-1} < 2$ and $\alpha > 0$, by the choice of the constants and by Lemma 9. Therefore, it suffices to prove that, before Rule 4 is applied, the probability that $A \subseteq X$ and $Z \cap X = \emptyset$ is at least

$$\max\left(\sqrt{\frac{3\lambda}{\lambda-1}}, \gamma(\lceil \kappa d \rceil - 1), \frac{c}{c-1}\right)^{-|A \cup Z|}.$$

However, this is a straightforward corollary of Lemmata 6, 7 and 8. □

This concludes the proof of Theorem 1. In Table 1 we provide a choice of values of the constants for small values of $d$, together with corresponding value of $2-\varepsilon_d$.

**Table 1.** Example values of the constants together with the corresponding success probability

| $d$ | 1 |
|---|---|
| $\lambda$ | 4.0238224 |
| $\kappa$ | 9 |
| $c$ | 2.00197442 |
| $\alpha$ | 0.050203 |
| $2-\varepsilon_d$ | 1.99991 |
| $d$ | 2 |
| $\lambda$ | 4.00009156 |
| $\kappa$ | 17/2 |
| $c$ | 2.00000763 |
| $\alpha$ | 0.01449 |
| $2-\varepsilon_d$ | 1.9999999 |
| $d$ | 3 |
| $\lambda$ | 4.000000357628 |
| $\kappa$ | 25/3 |
| $c$ | 2.0000000298 |
| $\alpha$ | 0.0066225 |
| $2-\varepsilon_d$ | 1.9999999999 |

| $d$ | 4 |
|---|---|
| $\lambda$ | 4.000000001397 |
| $\kappa$ | 33/4 |
| $c$ | 2.0000000001164 |
| $\alpha$ | 0.0037736 |
| $2-\varepsilon_d$ | 1.9999999999996 |
| $d$ | 5 |
| $\lambda$ | 4.000000000005457 |
| $\kappa$ | 41/5 |
| $c$ | 2.0000000000004548 |
| $\alpha$ | 0.0024331 |
| $2-\varepsilon_d$ | 1.999999999999999 |
| $d$ | 6 |
| $\lambda$ | 4.00000000000021316 |
| $\kappa$ | 49/6 |
| $c$ | 2.0000000000000017833 |
| $\alpha$ | 0.0016978 |
| $2-\varepsilon_d$ | 1.9999999999999999997 |

## 4 Conclusions

We have shown that the MAXIMUM $d$-DEGENERATE INDUCED SUBGRAPH problem can be solved in time $(2-\varepsilon_d)^n n^{\mathcal{O}(1)}$ for any fixed $d \geq 1$. There are two natural questions arising from our work. First, can the algorithm be derandomized? Rules 2 and 3 can be easily transformed into appropriate branching rules, but we do not know how to handle Rule 1 without randomization.

Second, our constants $\varepsilon_d$ are really tiny even for small values of $d$. This is mainly caused by two facts: the gain over a straightforward brute-force algorithm in Rule 2 is very small (i.e., $\gamma(\lfloor \kappa d \rfloor)$ is very close to 2) and the algorithm falls back to Rule 4 after processing only a tiny fraction $\alpha$ of the entire graph. Can the running time of the algorithm be significantly improved? Another interesting question would be to investigate, whether the MAXIMUM $d$-DEGENERATE

Induced Subgraph problem can be solved in time $(2-\varepsilon)^n n^{\mathcal{O}(1)}$ for some universal constant $\varepsilon$ that is independent of $d$.

Apart from the above questions, we would like to state here a significantly more challenging goal. Let $\mathcal{G}$ be a polynomially recognizable graph class of bounded degeneracy (i.e., there exists a constant $d$ such that each $G \in \mathcal{G}$ is $d$-degenerate). Can the corresponding Maximum Induced $\mathcal{G}$-Subgraph problem be solved in $(2-\varepsilon_{\mathcal{G}})^n$ time for some constant $\varepsilon_{\mathcal{G}} > 0$ that depends only on the class $\mathcal{G}$? Can we prove some meta-result for such type of problems?

Our Rules 1 and 3 are valid for any such class $\mathcal{G}$; however, this is not true for the greedy step in Rule 2. In particular, we do not know how to handle the Maximum Induced $\mathcal{G}$-Subgraph problem faster than $2^n$ even if the input is assumed to be $d$-degenerate.

**Acknowledgements.** We would like to thank Marek Cygan, Fedor V. Fomin and Pim van 't Hof for helpful discussions.

## References

1. Fomin, F.V., Grandoni, F., Kratsch, D.: A measure & conquer approach for the analysis of exact algorithms. J. ACM 56(5), 1–32 (2009)
2. van Rooij, J.M.M., Nederlof, J., van Dijk, T.C.: Inclusion/Exclusion Meets Measure and Conquer. In: Fiat, A., Sanders, P. (eds.) ESA 2009. LNCS, vol. 5757, pp. 554–565. Springer, Heidelberg (2009)
3. Cygan, M., Pilipczuk, M.: Exact and approximate bandwidth. Theor. Comput. Sci. 411(40-42), 3701–3713 (2010)
4. Björklund, A.: Determinant sums for undirected hamiltonicity. In: 51th Annual IEEE Symposium on Foundations of Computer Science (FOCS), pp. 173–182. IEEE Computer Society (2010)
5. Fomin, F.V., Todinca, I., Villanger, Y.: Exact Algorithm for the Maximum Induced Planar Subgraph Problem. In: Demetrescu, C., Halldórsson, M.M. (eds.) ESA 2011. LNCS, vol. 6942, pp. 287–298. Springer, Heidelberg (2011)
6. Cygan, M., Pilipczuk, M., Pilipczuk, M., Wojtaszczyk, J.O.: Scheduling Partially Ordered Jobs Faster Than $2^n$. In: Demetrescu, C., Halldórsson, M.M. (eds.) ESA 2011. LNCS, vol. 6942, pp. 299–310. Springer, Heidelberg (2011)
7. Cygan, M., Pilipczuk, M., Wojtaszczyk, J.O.: Capacitated Domination Faster Than $O(2^n)$. In: Kaplan, H. (ed.) SWAT 2010. LNCS, vol. 6139, pp. 74–80. Springer, Heidelberg (2010)
8. Binkele-Raible, D., Brankovic, L., Cygan, M., Fernau, H., Kneis, J., Kratsch, D., Langer, A., Liedloff, M., Pilipczuk, M., Rossmanith, P., Wojtaszczyk, J.O.: Breaking the $2^n$-barrier for irredundance: Two lines of attack. J. Discrete Algorithms 9(3), 214–230 (2011)
9. Cygan, M., Pilipczuk, M., Pilipczuk, M., Wojtaszczyk, J.O.: Solving the 2-Disjoint Connected Subgraphs Problem Faster Than $2^n$. In: Fernández-Baca, D. (ed.) LATIN 2012. LNCS, vol. 7256, pp. 195–206. Springer, Heidelberg (2012)
10. Fomin, F.V., Heggernes, P., Kratsch, D., Papadopoulos, C., Villanger, Y.: Enumerating Minimal Subset Feedback Vertex Sets. In: Dehne, F., Iacono, J., Sack, J.-R. (eds.) WADS 2011. LNCS, vol. 6844, pp. 399–410. Springer, Heidelberg (2011)

11. Razgon, I.: Exact Computation of Maximum Induced Forest. In: Arge, L., Freivalds, R. (eds.) SWAT 2006. LNCS, vol. 4059, pp. 160–171. Springer, Heidelberg (2006)
12. Fomin, F.V., Gaspers, S., Pyatkin, A.V., Razgon, I.: On the minimum feedback vertex set problem: Exact and enumeration algorithms. Algorithmica 52(2), 293–307 (2008)
13. Impagliazzo, R., Paturi, R.: On the complexity of k-SAT. J. Comput. Syst. Sci. 62(2), 367–375 (2001)
14. Calabro, C., Impagliazzo, R., Paturi, R.: The Complexity of Satisfiability of Small Depth Circuits. In: Chen, J., Fomin, F.V. (eds.) IWPEC 2009. LNCS, vol. 5917, pp. 75–85. Springer, Heidelberg (2009)
15. Cygan, M., Nederlof, J., Pilipczuk, M., Pilipczuk, M., van Rooij, J.M.M., Wojtaszczyk, J.O.: Solving connectivity problems parameterized by treewidth in single exponential time. In: Ostrovsky, R. (ed.) FOCS, pp. 150–159. IEEE (2011)
16. Lokshtanov, D., Marx, D., Saurabh, S.: Known Algorithms on Graphs of Bounded Treewidth are Probably Optimal. In: Proceedings of the Twenty-Second Annual ACM-SIAM Symposium on Discrete Algorithms (SODA), pp. 777–789 (2011)
17. Pătraşcu, M., Williams, R.: On the possibility of faster SAT algorithms. In: Proceedings of the Twenty-First Annual ACM-SIAM Symposium on Discrete Algorithms (SODA), pp. 1065–1075 (2010)
18. Cygan, M., Dell, H., Lokshtanov, D., Marx, D., Nederlof, J., Okamoto, Y., Paturi, R., Saurabh, S., Wahlström, M.: On problems as hard as CNFSAT. CoRR abs/1112.2275 (2011)
19. Gupta, S., Raman, V., Saurabh, S.: Fast Exponential Algorithms for Maximum $r$-Regular Induced Subgraph Problems. In: Arun-Kumar, S., Garg, N. (eds.) FSTTCS 2006. LNCS, vol. 4337, pp. 139–151. Springer, Heidelberg (2006)
20. Fomin, F.V., Villanger, Y.: Finding induced subgraphs via minimal triangulations. In: Marion, J.Y., Schwentick, T. (eds.) STACS. LIPIcs, vol. 5, pp. 383–394. Schloss Dagstuhl - Leibniz-Zentrum fuer Informatik (2010)
21. Angelsmark, O., Thapper, J.: Partitioning Based Algorithms for Some Colouring Problems. In: Hnich, B., Carlsson, M., Fages, F., Rossi, F. (eds.) CSCLP 2005. LNCS (LNAI), vol. 3978, pp. 44–58. Springer, Heidelberg (2006)
22. Gaspers, S., Kratsch, D., Liedloff, M.: On Independent Sets and Bicliques in Graphs. In: Broersma, H., Erlebach, T., Friedetzky, T., Paulusma, D. (eds.) WG 2008. LNCS, vol. 5344, pp. 171–182. Springer, Heidelberg (2008)
23. Gaspers, S.: Exponential Time Algorithms: Structures, Measures, and Bounds. PhD Thesis, University of Bergen (2008)
24. Cygan, M., Pilipczuk, M., Pilipczuk, M., Wojtaszczyk, J.O.: Kernelization Hardness of Connectivity Problems in $d$-Degenerate Graphs. In: Thilikos, D.M. (ed.) WG 2010. LNCS, vol. 6410, pp. 147–158. Springer, Heidelberg (2010)
25. Kostochka, A.V.: Lower bound of the hadwiger number of graphs by their average degree. Combinatorica 4(4), 307–316 (1984)
26. Thomason, A.: An extremal function for contractions of graphs. Math. Proc. Cambridge Philos. Soc. 95(2), 261–265 (1984)
27. Thomason, A.: The extremal function for complete minors. J. Comb. Theory, Ser. B 81(2), 318–338 (2001)

# The Exponential Time Hypothesis and the Parameterized Clique Problem

Yijia Chen[1], Kord Eickmeyer[2,*], and Jörg Flum[3]

[1] Department of Computer Science, Shanghai Jiaotong University
yijia.chen@cs.sjtu.edu.cn
[2] National Institute of Informatics, Tokyo
eickmeye@nii.ac.jp
[3] Mathematisches Institut, Albert-Ludwigs-Universität Freiburg
joerg.flum@math.uni-freiburg.de

**Abstract.** In parameterized complexity there are three natural definitions of fixed-parameter tractability called strongly uniform, weakly uniform and nonuniform fpt. Similarly, there are three notions of subexponential time, yielding three flavours of the exponential time hypothesis (ETH) stating that 3SAT is not solvable in subexponential time. It is known that ETH implies that $p$-CLIQUE is not fixed-parameter tractable if both are taken to be strongly uniform or both are taken to be uniform, and we extend this to the nonuniform case. We also show that even the containment of weakly uniform subexponential time in nonuniform subexponential time is strict. Furthermore, we deduce from nonuniform ETH that no single exponent $d$ allows for arbitrarily good fpt-approximations of clique.

## 1 Introduction

In parameterized complexity, FPT most commonly denotes the class of *strongly uniformly* fixed-parameter tractable problems, i.e., parameterized problems solvable in time $f(k)\cdot n^{O(1)}$ for some *computable* function $f$. Downey and Fellows also introduced the classes $\mathrm{FPT}_{\mathrm{uni}}$ and $\mathrm{FPT}_{\mathrm{nu}}$ of *uniformly* and *nonuniformly* fixed-parameter tractable problems, where one drops the condition that $f$ be computable or allows for different algorithms for each $k$, respectively. (We give detailed definitions in Section 2.) For example, $p$-CLIQUE $\notin \mathrm{FPT}_{\mathrm{nu}}$, where $p$-CLIQUE denotes the parameterized clique problem, means that for all $d \in \mathbb{N}$ and sufficiently large fixed $k$ determining whether a graph $G$ contains a clique of size $k$ is not in DTIME $\left(n^d\right)$. The obvious inclusions between the classes FPT, $\mathrm{FPT}_{\mathrm{uni}}$, and $\mathrm{FPT}_{\mathrm{nu}}$ can be shown to be strict [1].

In classical complexity, the subexponential time classes

$$\mathrm{DTIME}\left(2^{o^{\mathrm{eff}}(n)}\right), \quad \mathrm{DTIME}\left(2^{o(n)}\right), \quad \text{and} \quad \bigcap_{\varepsilon>0}\mathrm{DTIME}\left(2^{\varepsilon\cdot n}\right), \qquad (1)$$

* This work was supported by a fellowship of the second author within the FIT-Programme of the German Academic Exchange Service (DAAD).

D.M. Thilikos and G.J. Woeginger (Eds.): IPEC 2012, LNCS 7535, pp. 13–24, 2012.

have been considered. In particular, there are the corresponding versions of the exponential time hypothesis, namely the *strongly uniform exponential time hypothesis* ETH, the *uniform exponential time hypothesis* $\mathrm{ETH}_{\mathrm{uni}}$, and the *nonuniform exponential time hypothesis* $\mathrm{ETH}_{\mathrm{nu}}$, which are the statements that 3SAT is not in these respective classes, where $n$ denotes the number of variables of the input formula.[1] By results due to Impagliazzo et al. [3] we know that these three statements are equivalent to the ones obtained by replacing 3SAT by the clique problem CLIQUE; then, $n$ denotes the number of vertices of the corresponding graph.

Furthermore, it is known [4] that ETH implies $p$-CLIQUE $\notin$ FPT, where $p$-CLIQUE denotes the parameterized clique problem. As pointed out, for example in [5], there is a correspondence between subexponential algorithms having a running time $2^{o(n)}$ and $\mathrm{FPT}_{\mathrm{uni}}$ similar to that between running time $2^{o^{\mathrm{eff}}(n)}$ and FPT. Therefore, it is not surprising that $\mathrm{ETH}_{\mathrm{uni}}$ implies $p$-CLIQUE $\notin \mathrm{FPT}_{\mathrm{uni}}$ (we include a proof in Section 4).

The first main result of this paper shows that $\mathrm{ETH}_{\mathrm{nu}}$ implies that $p$-CLIQUE $\notin \mathrm{FPT}_{\mathrm{nu}}$. So, putting the three results together, we see that:

(i) if ETH holds, then $p$-CLIQUE $\notin$ FPT;
(ii) if $\mathrm{ETH}_{\mathrm{uni}}$ holds, then $p$-CLIQUE $\notin \mathrm{FPT}_{\mathrm{uni}}$;
(iii) if $\mathrm{ETH}_{\mathrm{nu}}$ holds, then $p$-CLIQUE $\notin \mathrm{FPT}_{\mathrm{nu}}$.

One of the most important complexity classes of (apparently) intractable parameterized problems is the class W[1], the class of problems (strongly uniformly) fpt-reducible to $p$-CLIQUE. By replacing (strongly uniformly) fpt-reducible by uniformly fpt-reducible and nonuniformly fpt-reducible, we get the classes $\mathrm{W}[1]_{\mathrm{uni}}$ and $\mathrm{W}[1]_{\mathrm{nu}}$, respectively. So, the previous results can be seen as providing some evidence that FPT $\neq$ W[1], $\mathrm{FPT}_{\mathrm{uni}} \neq \mathrm{W}[1]_{\mathrm{uni}}$, and $\mathrm{FPT}_{\mathrm{nu}} \neq \mathrm{W}[1]_{\mathrm{nu}}$. Note that the strongest separation, $\mathrm{FPT}_{\mathrm{nu}} \neq \mathrm{W}[1]_{\mathrm{nu}}$, obtained in this paper under $\mathrm{ETH}_{\mathrm{nu}}$, plays a role in [6] (see the comment following Theorem 1.1 of that paper). In a recent paper [7] on the history of Parameterized Complexity Theory, Downey remarks that the question $\mathrm{FPT}_{\mathrm{nu}} \neq \mathrm{W}[1]_{\mathrm{nu}}$ is a central issue of the theory.

We get the implication (iii) by using a family of algorithms witnessing $p$-CLIQUE $\in \mathrm{FPT}_{\mathrm{nu}}$ to obtain an algorithm showing $p$-CLIQUE is in "$\mathrm{FPT}_{\mathrm{uni}}$ for positive instances." Once we have this single algorithm for $p$-CLIQUE, we adapt the techniques we used in [8] to show (i), to get that the clique problem CLIQUE is in $\bigcap_{\varepsilon>0} \mathrm{DTIME}\,(2^{\varepsilon\cdot n})$, that is, the failure of $\mathrm{ETH}_{\mathrm{nu}}$.

The basic idea of parameterized approximability is explained in [9] as follows: "Suppose we have a problem that is hard to approximate. Can we at least approximate it efficiently for instances for which the optimum is small? The classical theory of inapproximability does not seem to help answering this question, because usually the hardness proofs require fairly large solutions." In [9] and [10] the framework of parameterized approximability was introduced. In particular, the concept of an fpt approximation algorithm with a given approximation ratio was coined and some (in)approximability results were proven.

[1] We should mention that in [2] a different statement is called nonuniform ETH.

Our second main result is a nonapproximability result for $p$-CLIQUE: under the assumption $\text{ETH}_{\text{nu}}$, we show that for every $d \in \mathbb{N}$ there is a $\rho > 1$ such that $p$-CLIQUE has no parameterized approximation algorithm with approximation ratio $\rho$ and running time $f(k) \cdot n^d$ for some function $f : \mathbb{N} \to \mathbb{N}$.

The content of the different sections is the following. We recall concepts and fix notation in Section 2. In Section 3, we derive a result on the clique problem used to obtain both main results (in Section 4 and Section 5). Then, in Section 6, we show that the results relating the complexity of $p$-CLIQUE and the different variants of ETH are of a more general nature. Finally, in Section 7, we show that the second and third time class in (1) are distinct.

## 2 Some Preliminaries

Let $\mathbb{N}$ denote the positive natural numbers. If a function $f : \mathbb{N} \to \mathbb{N}$ is nondecreasing and unbounded, then $f^{-1}$ denotes the function $f^{-1} : \mathbb{N} \to \mathbb{N}$ with

$$f^{-1}(n) := \begin{cases} \max\{i \in \mathbb{N} \mid f(i) \leq n\}, & \text{if } n \geq f(1) \\ 1, & \text{otherwise.} \end{cases}$$

Then, $f(f^{-1}(n)) \leq n$ for all $n \geq f(1)$ and the function $f^{-1}$ is nondecreasing and unbounded.

Let $f, g : \mathbb{N} \to \mathbb{N}$ be functions. Then $f \in o^{\text{eff}}(g)$ (often written as $f(n) \in o^{\text{eff}}(g(n))$) if there is a computable function $h : \mathbb{N} \to \mathbb{N}$ such that for all $\ell \geq 1$ and $n \geq h(\ell)$, we have $f(n) \leq g(n)/\ell$.

We denote the alphabet $\{0, 1\}$ by $\Sigma$. The length of a string $x \in \Sigma^*$ is denoted by $|x|$. We identify problems with subsets $Q$ of $\Sigma^*$. If $x \in Q$ we say that $x$ is a *positive* instance of the problem $Q$. Clearly, as done mostly, we present concrete problems in a verbal, hence uncodified form. For example, we introduce the problem CLIQUE in the form:

> CLIQUE
> *Instance:* A graph $G$ and $k \in \mathbb{N}$.
> *Problem:* Does $G$ have a clique of size $k$?

A graph $G$ is given by its vertex set $V(G)$ and its edge set $E(G)$. By $|G|$ we denote the length of a string naturally encoding $G$. The cardinality or size of a set $S$ is also denoted by $|S|$.

If $\mathbb{A}$ is an algorithm and $\mathbb{A}$ halts on input $x$, then we denote by $t_{\mathbb{A}}(x)$ the number of steps of $\mathbb{A}$ on input $x$; if $\mathbb{A}$ does not halt on $x$, then $t_{\mathbb{A}}(x) := \infty$.

We view *parameterized problems* as pairs $(Q, \kappa)$ consisting of a classical problem $Q \subseteq \Sigma^*$ and a *parameterization* $\kappa : \Sigma^* \to \mathbb{N}$, which is required to be polynomial time computable. We will present parameterized problems in the form we do for $p$-CLIQUE:

| $p$-CLIQUE |
|---|
| *Instance:* A graph $G$ and $k \in \mathbb{N}$. |
| *Parameter:* $k$. |
| *Problem:* Does $G$ have a clique of size $k$? |

A parameterized problem $(Q, \kappa)$ is *(strongly uniformly) fixed-parameter tractable* or, in FPT, if there is an algorithm $\mathbb{A}$ deciding $Q$, a natural number $d$, and a computable function $f : \mathbb{N} \to \mathbb{N}$ such that $t_{\mathbb{A}}(x) \leq f(\kappa(x)) \cdot |x|^d$ for all $x \in \Sigma^*$.

If in this definition we do not require the computability of $f$, then $(Q, \kappa)$ is *uniformly fixed-parameter tractable* or, in $\text{FPT}_{\text{uni}}$. Finally, $(Q, \kappa)$ is *nonuniformly fixed-parameter tractable* or, in $\text{FPT}_{\text{nu}}$, if there is a natural number $d$ and a function $f : \mathbb{N} \to \mathbb{N}$ such that for every $k \in \mathbb{N}$ there is an algorithm $\mathbb{A}_k$ deciding the set $\{x \in Q \mid \kappa(x) = k\}$ with $t_{\mathbb{A}_k}(x) \leq f(\kappa(x)) \cdot |x|^d$ for all $x \in \Sigma^*$.

In Section 6 we assume that the reader is familiar with the notion of (strongly uniform) fpt-reduction, with the classes of the W-hierarchy, and for $t, d \in \mathbb{N}$ with the weighted satisfiability problem $p$-WSAT$(\Gamma_{t,d})$ (e.g., see [11]). We write $(Q, \kappa) \leq^{\text{fpt}} (Q', \kappa')$ if there is an fpt-reduction from $(Q, \kappa)$ to $(Q', \kappa')$.

## 3 Going from Nonuniform to Uniform on Positive Instances

In this section we show how to get a single algorithm detecting cliques of size $k$ in time $f(k) \cdot |G|^{o(k)}$ on positive instances from the existence of such algorithms of running time $O(|G|^{e_\ell + k/\ell})$ for each pair of natural numbers $k, \ell$. We now state this assumption formally; and we will later show it to be unlikely because it implies that $\text{ETH}_{\text{nu}}$ is not true.

**Definition 1.** *We say that* CLIQUE *satisfies* $(*)$ *if*

> for every $\ell \in \mathbb{N}$ there is an $e_\ell \in \mathbb{N}$ such that for every $k \in \mathbb{N}$ there is a constant $a_{\ell,k} \in \mathbb{N}$ and an algorithm $\mathbb{A}_{\ell,k}$ which on every graph $G$ which contains a clique of size $k$ outputs such a clique in time
>
> $$a_{\ell,k} \cdot |G|^{e_\ell + k/\ell}.$$
>
> The behaviour of $\mathbb{A}_{\ell,k}$ on graphs without a clique of size $k$ or on inputs not encoding graphs may be arbitrary.

By a standard self-reduction argument we have:

**Lemma 2.** *If* $p$-CLIQUE $\in \text{FPT}_{\text{nu}}$*, then* CLIQUE *satisfies* $(*)$.

We now use the algorithms in $(*)$ to obtain a single algorithm with a guaranteed running time on positive instances.

**Lemma 3.** *If* CLIQUE *satisfies* $(*)$, *then there is an algorithm* $\mathbb{A}$ *deciding* CLIQUE *and there is a function* $f : \mathbb{N} \to \mathbb{N}$ *such that*

$$t_{\mathbb{A}}(G, k) \leq f(k) \cdot |G|^{o(k)}$$

*for every* positive *instance* $(G, k)$ *of* CLIQUE.

*Proof.* We let $\mathbb{C}$ be any algorithm which on input $(G, k)$ decides in time $|G|^{O(k)}$ whether $G$ contains a clique of size $k$, e.g., by brute force. Let $\{M_1, M_2, \ldots\}$ be any recursive enumeration of all Turing machines. By standard arguments we may assume that, given inputs $t$, $i$, and $x$, we can simulate $t$ steps of machine $M_i$ on input $x$ in time polynomial in $i$, $t$ and $|x|$.

We define the algorithm $\mathbb{A}$ as follows:

$\mathbb{A}$ // $G = (V(G), E(G))$ a graph and $k \in \mathbb{N}$

*1.* do the following in parallel:
*2.* simulate $\mathbb{C}$ on $(G, k)$ and
*3.* simulate $M_i$ on $G$ for $i = 1, \ldots, |G|$.
*4.* **if** the simulation of $\mathbb{C}$ accepts **then** accept
*5.* **if** the simulation of $\mathbb{C}$ rejects **then** never halt
*6.* **if** one of the machines $M_i$ finds a clique of size $k$ **then** accept.

Obviously, this algorithm will accept an input $(G, k)$ if and only if $G$ contains a clique of size $k$. We now turn to the claimed running time. Let $\ell \geq 1$, and let $e_\ell$ be the corresponding constant from assumption $(*)$. For $k > \ell(\ell+1)e_\ell$, the running time of $\mathbb{A}_{\ell,k}$ is bounded by

$$a_{\ell,k} \cdot |G|^{\frac{k}{\ell-1}},$$

and for all but finitely many instances $G$, the algorithm $\mathbb{A}_{\ell,k}$ will be among the ones simulated by $\mathbb{A}$. For such instances $G$, the running time of $\mathbb{A}$ is bounded by

$$\begin{aligned} &((\#\text{ machines to be simulated in parallel}) \cdot (\#\text{ of steps}) \cdot |G|)^{O(1)}, \\ \leq &\left(|G| \cdot a_{\ell,k} \cdot |G|^{\frac{k}{\ell-1}} \cdot |G|\right)^{O(1)} \\ \leq &c_k \cdot |G|^{\frac{d\cdot(k+1)}{\ell-1}} \end{aligned}$$

for suitable constants $c_k$ and $d$, the latter one not depending on $\ell$, $k$ or $G$.

## 4 $\text{ETH}_{\text{nu}}$ and the Complexity of $p$-Clique

In this section we show our first main result, namely:

**Theorem 4.** *If* $\text{ETH}_{\text{nu}}$ *holds, then* $p$-CLIQUE $\notin \text{FPT}_{\text{nu}}$.

To obtain this result we prove the following chain of implications

$$\text{(a)} \Rightarrow \text{(b)}_{\text{pos}} \Rightarrow \text{(c)}_{\text{pos}} \Rightarrow \text{(d)},$$

where

**(a)** $p$-CLIQUE $\in \text{FPT}_{\text{nu}}$;

**(b)$_{\text{pos}}$** There is an algorithm $\mathbb{A}$ deciding CLIQUE such that for all *positive* instances $(G, k)$ of CLIQUE and some function $f : \mathbb{N} \to \mathbb{N}$ we have

$$t_{\mathbb{A}}(G, k) \leq f(k) \cdot |G|^{o(k)}.$$

**(c)$_{\text{pos}}$** There is an algorithm $\mathbb{A}$ deciding CLIQUE such that for all *positive* instances $(G, k)$ of CLIQUE, where $G$ has vertex set $V(G)$, we have

$$t_{\mathbb{A}}(G, k) \leq 2^{o(|V(G)|)}.$$

**(d)** $\text{ETH}_{\text{nu}}$ does not hold.

Note that $\neg$(d) $\Rightarrow$ $\neg$(a) is the claim of Theorem 4.

The implication (a) $\Rightarrow$ (b)$_{\text{pos}}$ was shown in the previous section (Lemma 2 and Lemma 3). We turn to the implication (b)$_{\text{pos}}$ $\Rightarrow$ (c)$_{\text{pos}}$. Let (b) and (c) be the statements obtained from (b)$_{\text{pos}}$ and (c)$_{\text{pos}}$, respectively, by deleting the restriction to positive instances. Note that (c) is equivalent to the failure of $\text{ETH}_{\text{uni}}$. Furthermore, we let (b)$_{\text{eff}}$ be the statement (b) with the additional requirement that the function $f$ is computable and let (c)$_{\text{eff}}$ be the statement obtained from (c) by replacing $2^{o(|V(G)|)}$ by $2^{o^{\text{eff}}(|V(G)|)}$. Again note that (c)$_{\text{eff}}$ is equivalent to the failure of ETH.

**Lemma 5.** *(1)* (b)$_{\text{eff}}$ *implies* (c)$_{\text{eff}}$*;*
*(2)* (b) *implies* (c)*;*
*(3)* (b)$_{\text{pos}}$ *implies* (c)$_{\text{pos}}$.

Part (1) was shown as Theorem 27 in [8] (and previously in [4]). We argue similarly to get parts (2) and (3). In particular, we use the following lemma stated and proved in [8] as Lemma 28. Its proof uses the fact that a clique in a graph $G$ can be viewed as an "amalgamation of local cliques" of subgraphs of $G$.

**Lemma 6.** *There is an algorithm* $\mathbb{D}$ *that assigns to every graph* $G = (V, E)$ *and* $k, m \leq |V|$ *in time polynomial in* $|V| \cdot 2^m$ *a graph* $G' = (V', E')$ *with* $|V'| \leq |V|^2 \cdot 2^m$ *such that*

$$G \text{ has a clique of size } k \iff G' \text{ has a clique of size } \lceil |V|/m \rceil. \tag{2}$$

*Proof (of Lemma 5 (3)).* The proof of Lemma 5 (2) is obtained by the obvious modification and is left to the reader.

Let the algorithm $\mathbb{D}$ be as in Lemma 6. Assuming (b)$_{\text{pos}}$ there is an algorithm $\mathbb{A}$ deciding CLIQUE such that for all positive instances $(G, k)$ of CLIQUE we have $t_{\mathbb{A}}(G, k) \leq f(k) \cdot |G|^{o(k)}$ and hence,

$$t_{\mathbb{A}}(G, k) \leq f(k) \cdot |V(G)|^{o(k)} \tag{3}$$

for some $f : \mathbb{N} \to \mathbb{N}$. We consider the following algorithm deciding CLIQUE:

```
𝔹     // G a graph and k ∈ ℕ

1. Do in parallel for every m ≤ |V(G)| the following
2.        simulate 𝔻 on (G, k, m) and let G' be its output
3.        simulate 𝔸 on (G', ⌈|V(G)|/m⌉)
4.        if 𝔸 accepts for some m then accept
5.                  else reject.
```

Let $(G, k)$ be a positive instance of CLIQUE and $n := |V(G)|$. Without loss of generality, we can assume that $f$ is nondecreasing and unbounded. For

$$m := \max\left\{\left\lceil \frac{n}{f^{-1}(n)} \right\rceil, \lceil \log n \rceil\right\}$$

we have $m \geq \log n$ and $m \in o(n)$ and, by (3),

$$t_{\mathbb{A}}(G', \lceil |V(G)|/m \rceil) \leq f(\lceil n/m \rceil) \cdot (n^2 \cdot 2^m)^{o(n/m)} = 2^{o(n)}.$$

Thus, the running time for Line 2 to Line 5 is bounded by $2^{o(n)}$. Therefore

$$t_{\mathbb{B}}(G, k) \leq O(n \cdot 2^{o(n)}) \leq 2^{o(n)}. \qquad \square$$

We already remarked that (c) is equivalent to the failure of $\text{ETH}_{\text{uni}}$. Thus, part (2) of the previous lemma yields:

**Corollary 7.** *If* $\text{ETH}_{\text{uni}}$, *then* $p$-CLIQUE $\notin \text{FPT}_{\text{uni}}$.

*Proof of* $(\text{c})_{\text{pos}} \Rightarrow$ (d): For every $\epsilon > 0$ there is an $n_0$ such that for graphs with $|V(G)| > n_0$ the running time of the algorithm asserted by $(\text{c})_{\text{pos}}$ is bounded by $2^{\epsilon|V(G)|}$ on positive instances. For graphs with at least $n_0$ vertices we let the algorithm run for at most this many steps and reject if it does not hold within this time bound. For smaller graphs we use brute force.

For later purposes we remark:

**Corollary 8.** *If* CLIQUE *satisfies* $(*)$, *then* $\text{ETH}_{\text{nu}}$ *does not hold.*

*Proof.* If CLIQUE satisfies $(*)$, then $(\text{b})_{\text{pos}}$ holds by Lemma 3. We have shown that $(\text{b})_{\text{pos}}$ implies (d), thus, $\text{ETH}_{\text{nu}}$ does not hold.

## 5 $\text{ETH}_{\text{nu}}$ and the Parameterized Approximability of $p$-Clique

Let $\rho > 1$ be a real number. As in [9], we say that an algorithm $\mathbb{A}$ is an *fpt*$_{uni}$ *parameterized approximation algorithm for* $p$-CLIQUE *with approximation ratio* $\rho$ if

(i) $t_{\mathbb{A}}(G, k) \leq f(k) \cdot |V(G)|^{O(1)}$ for all instances $(G, k)$ of $p$-CLIQUE and some function $f : \mathbb{N} \to \mathbb{N}$;

(ii) for all positive instances $(G, k)$ of $p$-CLIQUE the algorithm $\mathbb{A}$ outputs a clique of size at least $k/\rho$; otherwise, the output of $\mathbb{A}$ can be arbitrary.

If $d \in \mathbb{N}$ and we get $t_{\mathbb{A}}(G, k) \leq f(k) \cdot |V(G)|^d$ in (i), then we say that $\mathbb{A}$ is an *fpt$_{uni}$ parameterized approximation algorithm for p-*CLIQUE *with approximation ratio $\rho$ and exponent $d$.*

Now we can state the main result of this section:

**Theorem 9.** *If* $\text{ETH}_{\text{nu}}$ *holds, then for every $d \in \mathbb{N}$ there is a $\rho > 1$ such that p-*CLIQUE *has no fpt$_{uni}$ parameterized approximation algorithm with approximation ratio $\rho$ and exponent $d$.*

The key observation which, together with Corollary 8, will yield this theorem is contained in the following lemma.

**Lemma 10.** *Assume that p-*CLIQUE *has an fpt$_{uni}$ parameterized approximation algorithm with approximation ratio $\rho > 1$ and exponent $d \geq 2$. Then, for every rational number $r$ with $0 < r \leq \frac{1}{\log \rho}$, there is an algorithm $\mathbb{B}$ deciding* CLIQUE *such that for some function $g : \mathbb{N} \to \mathbb{N}$ and every instance $(G, k)$ of* CLIQUE

$$t_{\mathbb{B}}(G, k) \leq g(k) \cdot |V(G)|^{r+2+d\cdot\lceil k/r \rceil}.$$

*Proof.* The main idea is as follows: we assume the existence of an $\text{fpt}_{\text{uni}}$ parameterized approximation algorithm $\mathbb{A}$ for $p$-CLIQUE. Given an instance $(G, k)$ of CLIQUE we stretch it by passing to an equivalent "product instance" $(G', k')$. By applying $\mathbb{A}$ to $(G', k')$ we can decide whether $(G, k) \in$ CLIQUE.

For a graph $G = (V, E)$ we let $\omega(G)$ be the size of a maximum clique in $G$. Furthermore, for every $m \in \mathbb{N}$ with $m \geq 1$ we denote by $G^m$ the graph $(V(G^m), E(G^m))$, where

$$\begin{aligned} V(G^m) &:= V^m = \{(v_1, \ldots, v_m) \mid v_1, \ldots, v_m \in V\} \\ E(G^m) &:= \Big\{\{(u_1, \ldots, u_m), (v_1, \ldots, v_m)\} \;\Big|\; \{u_1, \ldots, u_m, v_1, \ldots, v_m\} \\ &\qquad \text{is a clique in } G \text{ and } (u_1, \ldots, u_m) \neq (v_1, \ldots, v_m)\Big\}. \end{aligned}$$

One easily verifies that

$$\omega(G^m) = \omega(G)^m. \tag{4}$$

Now we let $\mathbb{A}$ be an $\text{fpt}_{\text{uni}}$ parameterized approximation algorithm for $p$-CLIQUE with approximation ratio $\rho > 1$ and exponent $d \geq 2$, say, with running time bounded by $f(k) \cdot |V(G)|^d$. Let $r$ be a rational number with $0 < r \leq \frac{1}{\log \rho}$. Then, $\rho \leq \sqrt[r]{2}$ and for every $k \in \mathbb{N}$ with $k \geq 2$ we get

$$\left(\frac{k}{k-1}\right)^{\lceil k/r \rceil} > \rho. \tag{5}$$

We let $\mathbb{B}$ be the following algorithm:

```
𝔹      // G a graph and k ∈ ℕ
1. if k = 1 or k < r then decide whether G has a clique of size k by
   brute force
2.     else simulate 𝔸 on (G^⌈k/r⌉, k^⌈k/r⌉)
3.         if 𝔸 outputs a clique of G^⌈k/r⌉ of size k^⌈k/r⌉/ρ
4.             then accept else reject.
```

The algorithm $\mathbb{B}$ decides CLIQUE: Clearly, the answer is correct if $k = 1$ or $k < r$. So assume that $k \geq 2$ and $k \geq r$. If $G$ has no clique of size $k$, that is, $\omega(G) \leq k - 1$, then, by (4), $\omega(G^{\lceil k/r \rceil}) \leq (k-1)^{\lceil k/r \rceil}$. By (5),

$$\frac{k^{\lceil k/r \rceil}}{\rho} > (k-1)^{\lceil k/r \rceil};$$

thus, compare Line 3 and Line 4, the algorithm $\mathbb{B}$ rejects $(G, k)$. If $\omega(G) \geq k$ and hence, $\omega(G^{\lceil k/r \rceil}) \geq k^{\lceil k/r \rceil}$, then the approximation algorithm $\mathbb{A}$ outputs a clique of $G^{\lceil k/r \rceil}$ of size $k^{\lceil k/r \rceil}/\rho$; thus $\mathbb{B}$ accepts $(G, k)$.

Moreover, on every instance $(G, k)$ with $G = (V, E)$ the running time of $\mathbb{B}$ is bounded by

$$|V|^{r+2} + |V|^{2\cdot\lceil k/r \rceil + 2} + f\left(k^{\lceil k/r \rceil}\right) \cdot \left|V\left(G^{\lceil k/r \rceil}\right)\right|^{d} \leq g(k) \cdot |V|^{r+2+d\cdot\lceil k/r \rceil},$$

for a suitable $g : \mathbb{N} \to \mathbb{N}$.

Setting $r := 1/\log \rho$ in the previous lemma, we get:

**Corollary 11.** *If there is an fpt$_{uni}$ parameterized approximation algorithm for $p$-*CLIQUE *with approximation ratio $\rho \geq 1$ and exponent $d \geq 2$, then there exists $e \in \mathbb{N}$ and an algorithm $\mathbb{B}$ deciding* CLIQUE *with $t_{\mathbb{B}}(G, k) \leq g(k) \cdot |V(G)|^{e+d\cdot k\cdot\log \rho}$*

*Proof of Theorem 9*: By contradiction, assume that for some $d \geq 2$ and all $\rho > 1$ the problem $p$-CLIQUE has an fpt$_{\text{uni}}$ parameterized approximation algorithm with approximation ratio $\rho$ and exponent $d$.

If $\ell \in \mathbb{N}$, then $d \cdot \ell \leq \frac{1}{\log \rho}$ for suitable $\rho > 1$. Thus, by Lemma 10, there is an algorithm $\mathbb{A}_\ell$ deciding CLIQUE such that for some $e_\ell \in \mathbb{N}$ and some function $g : \mathbb{N} \to \mathbb{N}$ and every instance $(G, k)$

$$t_{\mathbb{A}_\ell}(G, k) \leq g(k) \cdot |V(G)|^{e_\ell + k/\ell}.$$

Fix $k \in \mathbb{N}$. Then, again using the self-reducibility of CLIQUE, there is an algorithm $\mathbb{A}_{\ell,k}$ which on every graph $G$ outputs a clique of size $k$, if one exists, in time

$$O\left(|G|^{e_\ell + 1 + k/\ell}\right).$$

Thus, CLIQUE satisfies $(*)$ (the property introduced in Definition 1). Therefore, ETH$_{\text{nu}}$ does not hold by Corollary 8.

## 6 Some Extensions and Generalisations

Some results of Section 3 and of Section 4 can be stated more succinctly and in a more general form in the framework of parameterized complexity theory. We do this in this section, at the same time getting some open questions.

The class $\mathrm{FPT}_{\mathrm{nu}}$ is closed under fpt-reductions, that is,

$$\text{if } (Q,\kappa) \leq^{\mathrm{fpt}} (Q',\kappa') \text{ and } (Q',\kappa') \in \mathrm{FPT}_{\mathrm{nu}}, \text{ then } (Q,\kappa) \in \mathrm{FPT}_{\mathrm{nu}}. \tag{6}$$

Thus, for every W[1]-complete problem $(Q,\kappa)$ (complete under fpt-reductions), we have

$$(Q,\kappa) \in \mathrm{FPT}_{\mathrm{nu}} \iff \mathrm{W}[1] \subseteq \mathrm{FPT}_{\mathrm{nu}}. \tag{7}$$

Denote by $\mathrm{FPT}^{+}_{\mathrm{uni}}$ the class of problems $(Q,\kappa)$ such that there is an algorithm deciding $Q$ and with running time $h(\kappa(x)) \cdot |x|^{O(1)}$ for $x \in Q$, that is, for positive instances $x$ of $Q$. The class $\mathrm{FPT}^{+}_{\mathrm{uni}}$ is closed under fpt-reductions, too. So, again we have for every W[1]-complete problem $(Q,\kappa)$,

$$(Q,\kappa) \in \mathrm{FPT}^{+}_{\mathrm{uni}} \iff \mathrm{W}[1] \subseteq \mathrm{FPT}^{+}_{\mathrm{uni}}. \tag{8}$$

**Corollary 12.** *For every* W[1]*-complete problem* $(Q,\kappa)$,

$$(Q,\kappa) \in \mathrm{FPT}_{\mathrm{nu}} \quad \textit{implies} \quad (Q,\kappa) \in \mathrm{FPT}^{+}_{\mathrm{uni}}.$$

*Proof.* By Lemma 2 and Lemma 3, we know that the implication holds for the W[1]-complete problem $p$-CLIQUE. Now, the claim follows by (7) and (8).

It is not clear whether the previous implication holds for all problems $(Q,\kappa) \in$ W[1] (and not only for the complete ones). Of course, it does if FPT = W[1]. The proof of Lemma 3 makes essential use of a self-reducibility property of $p$-CLIQUE. For $t, d \in \mathbb{N}$ the weighted satisfiability problem $p$-WSAT$(\Gamma_{t,d})$ has this self-reducibility property, too. So, along the lines of Lemma 3, one gets (we leave the details to the reader):

**Lemma 13.** *Let* $t, d \in \mathbb{N}$. *Then*

$$p\text{-WSAT}(\Gamma_{t,d}) \in \mathrm{FPT}_{\mathrm{nu}} \quad \textit{implies} \quad p\text{-WSAT}(\Gamma_{t,d}) \in \mathrm{FPT}^{+}_{\mathrm{uni}}.$$

And thus, we get the extension of Corollary 12 to all levels of the W-hierarchy:

**Proposition 14.** *Let* $t \in \mathbb{N}$. *For every* W[$t$]*-complete problem* $(Q,\kappa)$,

$$(Q,\kappa) \in \mathrm{FPT}_{\mathrm{nu}} \quad \textit{implies} \quad (Q,\kappa) \in \mathrm{FPT}^{+}_{\mathrm{uni}}.$$

After Theorem 4, we have considered two further properties of the clique problem there denoted by $(\mathrm{b})_{\mathrm{pos}}$ and $(\mathrm{c})_{\mathrm{pos}}$. One could also define these properties for arbitrary parameterized problems (even though, there are some subtle points as the terms $2^{o(|V(G)|)}$ and $2^{o(|G|)}$ may be distinct). More importantly, these properties are not closed under fpt-reductions. So somehow one has to check whether other implications of Section 4 survive problem by problem. We do that here for the most prominent W[2]-complete problem, the parameterized dominating set problem $p$-DS:

> *p*-DS
> *Instance:* A graph $G$ and $k \in \mathbb{N}$.
> *Parameter:* $k$.
> *Problem:* Does $G$ have a dominating set of size $k$?

We denote by DS the underlying classical problem. In [8, Theorem 29] we have shown:

*If* DS *can be decided in time* $f(k) \cdot |V(G)|^{o^{\text{eff}}(k)}$ *for some computable* $f : \mathbb{N} \to \mathbb{N}$, *then* DS *can be decided in time* $2^{o^{\text{eff}}(|V(G)|)}$.

The reader should compare this result with the following one in the spirit of this paper.

**Theorem 15.** *If there is an algorithm* $\mathbb{A}$ *deciding* DS *such that for all positive instances* $(G,k)$ *of* DS *we have*

$$t_{\mathbb{A}}(G,k) \leq f(k) \cdot |G|^{o(k)}$$

*for some function* $f : \mathbb{N} \to \mathbb{N}$, *then there is an algorithm* $\mathbb{B}$ *deciding* DS *such that for all positive instances* $(G,k)$ *of* DS *we have*

$$t_{\mathbb{B}}(G,k) \leq 2^{o(|V(G)|)}.$$

The proof of the corresponding result for CLIQUE, namely the implication $(b)_{\text{pos}} \Rightarrow (c)_{\text{pos}}$, was based on Lemma 6 which used the fact that a clique in a graph can be viewed as an "amalgamation of local cliques" of subgraphs. As dominating sets are not necessarily an "amalgamation of local dominating sets," in [8] we took a detour via the weighted satisfiability problem for propositional formulas in CNF. As an inspection of the exposition in [8] shows, it can be adapted to a proof of Theorem 15.

## 7 An Example

We believe that the three statements ETH, $\text{ETH}_{\text{uni}}$, and $\text{ETH}_{\text{nu}}$ are true and hence equivalent. Here we consider the "underlying" complexity classes (see (1)). Clearly,

$$\text{DTIME}\left(2^{o^{\text{eff}}(n)}\right) \subseteq \text{DTIME}\left(2^{o(n)}\right) \subseteq \bigcap_{\varepsilon>0}\text{DTIME}\left(2^{\varepsilon \cdot n}\right) \quad (9)$$

To the best of our knowledge it is open whether the first inclusion is strict. Here we show the strictness of the second inclusion in (9). We remark that in [8, Proposition 5] we proved that the first class, that is, the effective version of the second one, coincides with an effective version of the third class.

For $m \in \mathbb{N}$ let $1^m$ be the string in $\Sigma^*$ consisting of $m$ ones. Recall that $\Sigma = \{0,1\}$. For a Turing machine $\mathbb{M}$ we denote by $\text{enc}(\mathbb{M})$ a string in $\Sigma^*$ reasonably encoding the Turing machine $\mathbb{M}$. Furthermore, $|\mathbb{M}|$ denotes the length of $\text{enc}(\mathbb{M})$, $|\mathbb{M}| = |\text{enc}(\mathbb{M})|$.

**Theorem 16.** *The problem*

| |
|---|
| EXP-HALT<br>*Instance:* A Turing machine $\mathbb{M}$, $x \in \Sigma^*$, and $1^m$ with $m \in \mathbb{N}$.<br>*Problem:* Does $\mathbb{M}$ accept $x$ in time $2^{\lfloor m/(\lvert\mathbb{M}\rvert + \lvert x\rvert)\rfloor}$? |

*is in* $\bigcap_{\varepsilon>0} \mathrm{DTIME}\,(2^{\varepsilon \cdot n}) \setminus \mathrm{DTIME}\,(2^{o(n)})$.

Due to space limitations we cannot present a proof in this extended abstract.

## References


1. Downey, R., Fellows, M.: Fixed-parameter tractability and completeness iii: some structural aspects of the w hierarchy. In: Ambos-Spies, K., Homer, S., Schöning, U. (eds.) Complexity Theory, New York, NY, USA, pp. 191–225. Cambridge University Press (1993)
2. Ganian, R., Hlinený, P., Langer, A., Obdrzálek, J., Rossmanith, P., Sikdar, S.: Lower bounds on the complexity of $\mathrm{MSO}_1$ model-checking. In: Proc. STACS 2012, pp. 326–337 (2012)
3. Impagliazzo, R., Paturi, R., Zane, F.: Which problems have strongly exponential complexity? Journal of Computer and System Sciences 63, 512–530 (2001)
4. Chen, J., Huang, X., Kanj, I.A., Xia, G.: Linear fpt reductions and computational lower bounds. In: Proc. of STOC 2004, pp. 212–221 (2004)
5. Flum, J., Grohe, M.: Parametrized complexity and subexponential time (column: Computational complexity). Bulletin of the EATCS 84, 71–100 (2004)
6. Grohe, M.: The complexity of homomorphism and constraint satisfaction problems seen from the other side. J. ACM 54(1) (2007)
7. Downey, R.: The Birth and Early Years of Parameterized Complexity. In: Bodlaender, H.L., Downey, R., Fomin, F.V., Marx, D. (eds.) Fellows Festschrift 2012. LNCS, vol. 7370, pp. 17–38. Springer, Heidelberg (2012)
8. Chen, Y., Flum, J.: On miniaturized problems in parameterized complexity theory. Theoretical Computer Science 351(3), 314–336 (2006)
9. Chen, Y., Grohe, M., Grüber, M.: On Parameterized Approximability. In: Bodlaender, H.L., Langston, M.A. (eds.) IWPEC 2006. LNCS, vol. 4169, pp. 109–120. Springer, Heidelberg (2006)
10. Marx, D.: Parameterized complexity and approximation algorithms. The Computer Journal 51, 60–78 (2008)
11. Flum, J., Grohe, M.: Parameterized Complexity Theory. Springer, Heidelberg (2006)

# New Results on Polynomial Inapproximability and Fixed Parameter Approximability of EDGE DOMINATING SET*

Bruno Escoffier[1], Jérôme Monnot[1], Vangelis Th. Paschos[1,2], and Mingyu Xiao[3]

[1] PSL Research University, Université Paris-Dauphine, LAMSADE, CNRS UMR 7243, France
{escoffier,monnot,paschos}@lamsade.dauphine.fr
[2] Institut Universitaire de France
[3] School of Computer Science and Engineering, University of Electronic Science and Technology of China, China
myxiao@gmail.com

**Abstract.** An edge dominating set in a graph $G = (V, E)$ is a subset $S$ of edges such that each edge in $E - S$ is adjacent to at least one edge in $S$. The EDGE DOMINATING SET problem, to find an edge dominating set of minimum size, is a basic and important NP-hard problem that has been extensively studied in approximation algorithms and parameterized complexity. In this paper, we present improved hardness results and parameterized approximation algorithms for EDGE DOMINATING SET. More precisely, we first show that it is NP-hard to approximate EDGE DOMINATING SET in polynomial time within a factor better than 1.18. Next, we give a parameterized approximation schema (with respect to the standard parameter) for the problem and, finally, we develop an $O^*(1.821^\tau)$-time exact algorithm where $\tau$ is the size of a minimum vertex cover of $G$.

## 1 Introduction

As one of the basic problems in Garey and Johnson's work on NP-completeness [17], EDGE DOMINATING SET has received high attention in history. It is NP-hard even in planar or bipartite graphs of maximum degree 3 [26]. Due to its theoretical and practical interests, many algorithms have been developed in order to tackle it. There is a simple 2-approximation algorithm for EDGE DOMINATING SET in unweighted graphs. It is not hard to verify that any maximal matching in the graph is an edge dominating set of size at most double of the minimum size. Carr et al. [7] proved a $(2 + \frac{1}{10})$-approximation algorithm for WEIGHTED EDGE DOMINATING SET (the generalization of EDGE DOMINATING SET where weights are assigned to the edges of the input graph and the objective becomes

* Research partially supported by the French Agency for Research under the DEFIS program TODO, ANR-09-EMER-010 and the National Natural Science Foundation of China under the Grant 60903007.

D.M. Thilikos and G.J. Woeginger (Eds.): IPEC 2012, LNCS 7535, pp. 25–36, 2012.

to determine a minimum total-weight edge dominating set), the ratio of which was later improved to 2 by Fujito and Nagamochi [16]. Improved results have also been obtained in sparse graphs [6] and in dense graphs [22]. However, providing an approximation algorithm with ratio (strictly) smaller than 2, or proving that such algorithm does not exist (under some likely complexity hypothesis) still remains as an open problem. Chlebik and Chlebikova [9] proved that it is NP-hard to approximate it within any factor better than $\frac{7}{6}$. Assuming the unique game conjecture (UGC), [22] showed some inapproximability results on dense instances, a corollary of which is that for every $\varepsilon > 0$ EDGE DOMINATING SET is inapproximable within ratio $3/2 - \varepsilon$ (under UGC).

In terms of parameterized complexity, EDGE DOMINATING SET, with parameter $k$ being the size of the solution, is fixed-parameter tractable (FPT). Fernau [14] gave an $O^*(2.6181^k)$-time algorithm that has been subsequently improved by Fomin *et al.* [15] downto $O^*(2.4181^k)$ and by Binkele-Raible and Fernau [1] downto $O^*(2.3819^k)$. Currently, the best result is the $O^*(2.3147^k)$-time algorithm by Xiao *et al.* [23]. When the graph is restricted to be of maximum degree 3, the result can be further improved to $O^*(2.1479^k)$ [24]. There is also a long list of contributions to exact algorithms for EDGE DOMINATING SET, such as the $O^*(1.4423^{|V|})$-time algorithm by Raman *et al.* [20], the $O^*(1.4082^{|V|})$-time algorithm by Fomin et al. [15], the $O^*(1.3226^{|V|})$-time algorithm by Rooij and Bodlaender [21], and finally the $O^*(1.3160^{|V|})$-time algorithm by Xiao and Nagamochi [25].

In this paper, we study parameterized approximation for EDGE DOMINATING SET. A parameterized approximation algorithm is a technique combining parameterization and approximation for getting approximation algorithms with fixed-parameter running time. In this way, we may be able to achieve approximation ratios unachievable (or yet unachieved) in polynomial time via fixed-parameter running times that are smaller than the running times of exact algorithms. We may also be able to use this technique to handle W[1]-hard problems which unlikely have fixed-parameter tractable algorithms. The interested reader can be referred to [4,11,19] for more about this issue. Let the parameter $k$ be the size of the solution to our problem. In the FPT framework, we want to design algorithms with running time $f(k)|I|^{O(1)}$ that decide whether there is a solution of size at most $k$ or not, where $f$ is a computable function. In approximation algorithms, we are interested in designing polynomial-time algorithms to find a solution of size $g(k)$, where $g$ is a computable function. In parameterized approximation, we wish to design algorithms with running time $f(k)|I|^{O(1)}$ that either find an approximate solution of size $g(k)$ or report that there is no solution of size $k$. Clearly, any fixed-parameter tractable problem allows parameterized approximation algorithms for any computable function $g$. However, this may not hold for W[1]-hard problems. For example, the dominating set problem (find a set $S$ of $k$ vertices in graph $G = (E, V)$ such that each vertex in $V - S$ is adjacent to at least one vertex in $S$) does not allow parameterized approximation algorithms for $g(k)$ of the form $k + c$ with fixed constant $c$ [11]. For EDGE DOMINATING SET, we are interested in designing parameterized approximation algorithms, which produce edge dominating sets of size at most $(1+\varepsilon)k$ (or assert that there is no solution of size $k$) in $f(k,\varepsilon)|I|^{O(1)}$ time for some computable function $f$. Of course, the goal is to find such an algorithm for a function $f$ which is smaller than the $O^*(2.3147^k)$-time (exact) FPT algorithm by Xiao et al. [23]. This issue has already been considered for other FPT problems, in particular for the MIN VERTEX COVER problem. In [2,3,13] several parameterized approximation algorithms running faster than (exact) FPT algorithms and achieving ratios better than the ratio 2 (achievable in polynomial time) are given. Note that [3,13] ask as open question if similar results can be achieved for EDGE DOMINATING SET.

The remaining parts of this paper are organized as follows. In Section 2, we give an improved hardness result for EDGE DOMINATING SET by showing that it is not $5\sqrt{5}-10+\varepsilon < 1.18$ approximable in polynomial time unless P=NP. In Sections 3 and 4 we tackle parameterized approximation algorithms, answering positively to the open question in [3]. More precisely, in Section 3, we first give a simple algorithm to present the basic ideas, and then improve this algorithm in Section 4. We conclude the article in Section 5 by devising a parameterized algorithm for EDGE DOMINATING SET where the parameter is the vertex cover number of the graph. Due to lengthe limits, results are presented here without proofs that can be found in [12].

## 2 An Improved Polynomial-Time Lower Bound

In this section, we give some new hardness results for EDGE DOMINATING SET, which are based on a reduction preserving approximation from the famous MIN VERTEX COVER problem (find a minimum subset $S$ of vertices in a graph such that each edge has at least one endpoint in $S$) to EDGE DOMINATING SET.

Before, recall some existing results between MIN VERTEX COVER and EDGE DOMINATING SET. The first two are rather folklore: there exist two simple approximation preserving reductions between MIN VERTEX COVER and EDGE DOMINATING SET transforming a polynomial-time $\rho$-approximation algorithm for one of them into a polynomial-time $2\rho$-approximation algorithm for the other one. Let $G=(V,E)$ be a simple graph and let $M^* \subseteq E$ and $C^* \subseteq V$ be a minimum edge dominating set and a minimum vertex cover of $G$, respectively. We will use $\tau = |C^*|$ to denote the size of a minimum vertex cover of $G$. Since, it is well known that $M^*$ can be supposed to be a maximal matching, we get $\tau = |C^*| \geq |M^*|$. Also $V(M^*)$, the set of endpoints of $M^*$, forms a vertex cover of $G$ and then $2|M^*| \geq \tau$. Thus, $\tau \geq |M^*| \geq \frac{\tau}{2}$. Now, from any $\rho$-approximation algorithm for MIN VERTEX COVER given by $V'$, we can polynomially find an edge dominating set $E'$ by taking at most one arbitrary edge incident to each vertex of $V'$. Thus, using the above expression for $\tau$, we get $|E'| \leq |V'| \leq \rho \times \tau \leq 2\rho|M^*|$. Conversely, from any $\rho$-approximation algorithm for EDGE DOMINATING SET given by $M'$, we can construct a vertex cover $V' = V(M')$ of $G$ by taking the endpoints of $M^*$. Hence, using expression for $\tau$, we deduce: $|V'| = 2|M'| \leq 2\rho|M^*| \leq 2\rho \times \tau$.

In Theorem 1 just below, we improve the expansion $2\rho$ of the reduction to $2\rho - 1$. Dealing with weighted versions of these two problems, it is proved in [5] that weighted MIN VERTEX COVER can be approximated as well as weighted EDGE DOMINATING SET.

**Theorem 1.** *For any $\rho \geq 1$, if there is a polynomial-time $\rho$-approximation algorithm for* EDGE DOMINATING SET, *then there exists a polynomial-time $(2\rho - 1)$-approximation algorithm for* MIN VERTEX COVER.

In order to prove Theorem 1, we show that for each instance $G = (V, E)$ of MIN VERTEX COVER, we can construct at most $|V|$ instances $G_i = (V_i, E_i)$ (where $|V_i| \leq 3|V|$) of EDGE DOMINATING SET such that a $(2\rho - 1)$-approximation solution to $G$ can be found in polynomial time based on a $\rho$-approximation solution to each $G_i$. For each positive integer $1 \leq i \leq |V|$, the graph $G_i = (V_i, E_i)$ is a graph constructed from $G$ in the following way: $V_i = V \cup \{a_j, a'_j : j \in \{1, \ldots, i\}\}$ and $E_i = E \cup F_i \cup H_i$, where $F_i = \{(a_j, a'_j) : j \in \{1, \ldots, i\}\}$ and $H_i = \{(v, a_j) : v \in V, j \in \{1, \ldots, i\}\}$. Informally, $G_i$ contains a copy of $G$, an induced matching $F_i$ and a complete bipartite graph between the vertices of $G$ and the left part of the induced matching $F_i$. It is NP-hard to approximate MIN VERTEX COVER within any factor smaller than $10\sqrt{5} - 21$ by a result of Dinur and Safra [10]. By this result and Theorem 1, we get the following corollary.

**Corollary 1.** *For any $\varepsilon > 0$,* EDGE DOMINATING SET *is not $(5\sqrt{5} - 10 + \varepsilon)$-approximable in polynomial time unless $P = NP$.*

Note that under UGC, since MIN VERTEX COVER cannot be approximated to within $2 - \varepsilon$ for any $\varepsilon > 0$ [18], we get that for any $\varepsilon > 0$, EDGE DOMINATING SET is not $(3/2 - \varepsilon)$-approximable in polynomial time, which is the same lower bound recently achieved in [22].

## 3 A Simple Parameterized Approximation Schema

In this section, we design a simple parameterized approximation schema for EDGE DOMINATING SET. As mentioned in Introduction, this algorithm contains the basic idea upon which the improved algorithms in Section 4 is built.

### 3.1 CONSTRAINED EDGE DOMINATING SET

First of all, we introduce a CONSTRAINED EDGE DOMINATING SET problem and present some properties for it. Given a graph $G = (V, E)$ and a prescribed subset $V_1 \subseteq V$ of non-isolated vertices, an edge dominating set $M$ is called a constrained edge dominating set of $G$, if $V_1 \subseteq V(M)$. In the CONSTRAINED EDGE DOMINATING SET problem, we are asked to find a constrained edge dominating set of minimum size. CONSTRAINED EDGE DOMINATING SET is a natural generation of EDGE DOMINATING SET where $V_1 = \emptyset$. We show a simple approximation algorithm for CONSTRAINED EDGE DOMINATING SET.

**Lemma 1.** *For an instance $(G, V_1)$ of* CONSTRAINED EDGE DOMINATING SET, *let $M_1$ be a maximum matching in the induced graph $G[V_1]$, $M_2$ be a maximum* *matching in the induced graph* $G[V - V_1]$*, and* $M_3$ *be a set of* $|V_1 - V(M_1)|$ *edges such that each edge in* $M_3$ *is incident on a different vertex in* $V_1 - V(M_1)$*. Edge set* $M' = M_1 \cup M_2 \cup M_3$ *is a constrained edge dominating set with size* $|M'| \leq (2 - \rho_1)\nu$*, where* $\nu$ *is the size of a minimum constrained edge dominating set* $M^*$ *and* $\rho_1\nu$ *is the number of edges in* $M^*$ *with both endpoints in* $V_1$*, for some* $\rho_1$*.*

Note that Lemma 1 is a special case of Lemma 3 in the next section (but we prefer to give a proof of both lemmas for readability).

Lemma 1 implies a 2-approximation algorithm for CONSTRAINED EDGE DOMINATING SET and a possible way to design a parameterized approximation algorithm for EDGE DOMINATING SET. Note that we can first find a vertex set $V_1$ such that $V_1 \subseteq V(M^*)$ for some minimum edge dominating set $M^*$ of $G$ and then use the algorithm in Lemma 1 to get an approximation algorithm for EDGE DOMINATING SET. The approximation ratio is related to the size of $V_1$: the larger the set $V_1$, the better the ratio.

### 3.2 A Parameterized Approximation Schema for EDGE DOMINATING SET

As already mentioned in introduction, deciding whether a graph contains an edge dominating set of size $k$ can be done in $O^*(2.3147^k)$ time by the parameterized algorithm presented in [23]. Here we design a parameterized approximation algorithm for it. It is based on the following fact:

*Suppose that there are a set* $V_1$ *and an edge dominating set* $M$ *such that* $V_1 \subseteq V(M)$*,* $|M| \leq k$ *and* $|V_1| = k + \rho' k$*. Then the number of edges in* $M$ *that have both endpoints in* $V_1$ *is at least* $\rho' k$*.*

Indeed, if there were $\alpha < \rho' k$ edges in $M$ with both endpoints in $V_1$, then the number of vertices in $V_1$ would be at most $2\alpha + (|M| - \alpha) \leq |M| + \alpha < k + \rho' k = |V_1|$, a contradiction. Putting together the above emphasized fact and Lemma 1 and taking $M = M^*$, one can see that the computed edge set $M'$ is of size at most $(2 - \rho')k$.

Then, our goal is to find such a large set $V_1$. As in several articles devising FPT algorithms for EDGE DOMINATING SET, we can use the fact that $V(M^*)$ for a minimum edge dominating set $M^*$ is a vertex cover of $G$. For each edge in the graph, at least one endpoint of it is in $V(M^*)$. Then, we can use a branching algorithm to construct a set $V_1$ of size up to $k + \rho' k$ such that $V_1$ is part of the vertex set of a minimum edge dominating set $V(M^*)$ in $G$. We iteratively select an edge $(a, b)$ in the current graph and branch into two branches by including either $a$ or $b$ into $V_1$ and deleting it from the graph until the size of $V_1$ becomes $k + \rho' k$ or the remaining graph has no edge. This process produces at most $2^{k+\rho' k}$ vertex sets $V_1$ of size at most $k + \rho' k$ in $O^*(2^{k+\rho' k})$ time and at least one of them is contained in $V(M^*)$. For each of the vertex sets $V_1$, we use the algorithm in Lemma 1 to compute $M'$ and return a smallest one. The returned edge set is an edge dominating set of size at most $(2 - \rho')k$ if $|M^*| \leq k$ (note that if in a leaf of the search tree we have a set $V_1 \subseteq V(M^*)$ with $|V_1| < k + \rho' k$, this means that the remaining graph is empty and the output solution is then optimal by Lemma 1). By taking $\rho' = 1 - \rho$, we deduce the following result.

**Lemma 2.** *For any $\rho > 0$, there exists a $(1+\rho)$-approximation algorithm to* $k$-EDGE DOMINATING SET *running in $O^*(2^{(2-\rho)k})$ time for $0 \leq \rho \leq 1$.*

When $\rho = 0$, Lemma 2 implies that $k$-EDGE DOMINATING SET can be solved in $O^*(4^k)$ time, which is far away from the current best parameterized algorithm of running time $O^*(2.3147^k)$. To reduce the gap, we will improve the running time bound of our parameterized approximation schema in the next section.

# 4 Improved Parameterized Approximation Schemata

In the algorithm presented in Section 3.2, in order to search $V_1$ we may need to branch on each edge. One way to reduce the running time is to reduce the number of branchings in the algorithm. This approach has been used for (exact) FPT algorithms to obtain improved running times. We will use some of these improved branchings, but we need to combine them with approximability. We first deal with these approximation properties in Section 4.1 and then present the improved parameterized approximation algorithm in Section 4.2.

## 4.1 More Approximation Algorithms for CONSTRAINED EDGE DOMINATING SET

Given a graph $G = (V, E)$. We consider a partition $(V_1, V_2, V_3)$ of the vertex set $V$ such that:

- Each connected component of the induced graph $G[V_2]$ is a clique.
- There is no edge between a vertex in $V_2$ and a vertex in $V_3$.

Once the set $V_1$ is given, we can find in linear time the set of connected components of $G[V - V_1]$ which are cliques and which constitute $V_2$. Let us now give more properties of our problems based on this partition.

We consider an instance $(G = (V, E), V_1)$ of CONSTRAINED EDGE DOMINATING SET. Let $M^*$ be a minimum constrained edge dominating set of $(G = (V, E), V_1)$ and $\nu = |M^*|$.

We denote by $\alpha_1$ (resp., $\alpha_2$, $\alpha_3$) the number of edges in $M^*$ with both endpoints in $V_1 \cup V_2$ (resp., with one endpoint in $V_1$ and one in $V_3$, both endpoints in $V_3$). This partitions the edge set $M^*$ into three sets, hence, $\nu = \alpha_1 + \alpha_2 + \alpha_3$.

Moreover, since the connected components of $G[V_2]$ are cliques and $V(M^*)$ is a vertex cover of $G$, we know that $V(M^*)$ contains at least $|C_i| - 1$ vertices in each clique $C_i$ of $G[V_2]$. Assume that there are $p$ cliques $C_1, \cdots, C_p$ in $G[V_2]$ among which $q$ cliques $Q_1, \cdots, Q_q$ are such that $V(Q_i) \subseteq V(M^*)$. Then $V(M^*) \cap V_2 = |V_2| - p + q$. In other words, we have:

$$2\alpha_1 + \alpha_2 = |V(M^*) \cap (V_1 \cap V_2)| = |V_1| + |V_2| - p + q \quad (1)$$

We are ready now to specify an approximation algorithm for CONSTRAINED EDGE DOMINATING SET (Algorithm ApproxPoly1 in Figure 1), which is a generation of the algorithm in Lemma 1.

**Input**: A graph $G = (V = V_1 \cup V_2 \cup V_3, E)$ with the above partition of $V$.
**Output**: An edge dominating set $M$ such that $V_1 \subseteq V(M)$.

1. Add a vertex $c'_i$ to each clique $C_i$ in $G[V_2]$, to create a clique of size $|C_i|+1$. Let $V'_2 = \{c'_1, \cdots, c'_p\}$ be the set of added vertices.
2. Compute a maximum matching $M_1$ in $G[V_1 \cup V_2 \cup V'_2]$.
3. While there is an edge $e = (u, c'_i)$ in $M_1$ with $c'_i \in V'_2$ and there exists a neighbor $w$ of $u$ not saturated by $M_1$, replace $e$ with $(u, w)$ in $M_1$.
4. Let $M'_1$ be the set of edges in $M_1$ with an endpoint in $V'_2$.
5. Compute a maximum matching $M_2$ in $G[V_3]$.
6. For each unsaturated vertex in $V_1$, select an arbitrary edge incident on it. Let $M_3$ be the set of such edges.
7. Output $M = M_1 \cup M_2 \cup M_3 - M'_1$.

**Fig. 1.** Algorithm ApproxPoly1

**Lemma 3.** *Edge set $M$ =ApproxPoly1$(G)$ is a constrained edge dominating set of $(G, V_1)$ with size $|M| \leq (2 - \rho_1)\nu$, where $\rho_1\nu = \alpha_1$ is the number of edges in $M^*$ with both endpoints in $V_1 \cup V_2$.*

Note that Lemma 1 is a special case of Lemma 3 where the vertex set $V_2$ is an empty set. Lemma 3 shows that we do not need to branch on each clique component in $G[V - V_1]$ in order to search the vertex set of a constrained edge dominating set.

To improve the running time of our parameterized approximation schema, we also need to consider a particular case of the graph where in the partition $(V_1, V_2, V_3)$ each connected component of $G[V_3]$ is a path of length 2.

Let $N$ be the number of these paths in $G[V_3]$. Considering a minimum constrained edge dominating set $M^*$, we denote by:

- $N_1$ the set of paths in $G[V_3]$ such that there is an edge in $M^*$ between a vertex in $V_1$ and the central vertex of the path; set $n_1 = |N_1|$;
- $N_2$ the set of paths in $G[V_3]$ such that there is an edge of the path in $M^*$; set $n_2 = |N_2|$;
- $N_3$ the set of remaining paths in $G[V_3]$; set $n_3 = |N_3|$.

Observe that some paths of $G[V_3]$ may be counted twice (once with $N_1$ and once with $N_2$); so, $N \leq n_1 + n_2 + n_3$. Note that for each of the $n_3$ remaining paths, $M^*$ has to take two edges (between $V_1$ and the endpoints of the path) to cover the edges of the path. In other words, $\alpha_2 \geq 2n_3 + n_1$. Moreover, by definition, $n_2 = \alpha_3$.

Consider Algorithm ApproxPoly2 (Figure 2) on an instance $(G, V_1)$ of CONSTRAINED EDGE DOMINATING SET.

**Input**: A graph $G = (V = V_1 \cup V_2 \cup V_3, E)$, where each component in $G[V_3]$ is a path of length 2.
**Output**: An edge dominating set $M$ such that $V_1 \subseteq V(M)$.

1. Add a vertex $c'_i$ to each clique $C_i$ in $G[V_2]$, to create a clique of size $|C_i|+1$. Let $V'_2 = \{c'_1, \cdots, c'_p\}$ be the set of added vertices.
2. Compute a maximum matching $M_1$ in $G[V_1 \cup V_2 \cup V'_2 \cup V'_3]$, where $V'_3$ is the set of central vertices of paths in $G[V_3]$.
3. While there is an edge $e = (u, c'_i)$ in $M_1$ such that $c'_i \in V'_2$ and there exists a neighbor $w$ of $u$ not saturated by $M_1$, then replace $e$ with $(u, w)$ in $M_1$.
4. Let $M'_1$ be the set of edges in $M_1$ with an endpoint in $V'_2$.
5. For each path where the central vertex is not saturated by $M_1$, take one edge in this path.
   Let $M_2$ be this set of edges.
6. For each unsaturated vertex in $V_1$, select an arbitrary edge. Let $M_3$ be the set of such edges.
7. Output $M = M_1 \cup M_2 \cup M_3 - M'_1$.

**Fig. 2.** Algorithm `ApproxPoly2`

The following lemma holds.

**Lemma 4.** *Edge set $M$ =`ApproxPoly2`$(G)$ is a constrained edge dominating set of $(G, V_1)$ with size $|M| \leq \nu + n_3$.*

### 4.2 An Improved Parameterized Approximation Schema

Now we are able to give the improved parameterized approximation schema `ApproxFPT` for $k$-EDGE DOMINATING SET as well as $k$-CONSTRAINED EDGE DOMINATING SET. As explained earlier, the principle is to search the vertex set $V_1$ by using some 'good' branchings. Then, in each leaf of our search tree, we will use the approximation algorithms devised in Section 4.1 (either directly, or after some other steps).

We consider a $k$-constrained edge dominating set $(G, V_1)$ with partition $I = (V_1, V_2, V_3)$ of the vertex set. Let $t = |V_1| + |V_2| - p$ (where $p$ is the number of cliques in $G[V_2]$). When $t \geq (2-\rho)k$ $(0 \leq \rho \leq 1)$, there are at least $(1-\rho)k$ edges in any optimal solution $M^*$ with both endpoints in $V_1 \cup V_2$. Therefore, Lemma 3 implies that a $(1+\rho)$-approximation solution to $k$-CONSTRAINED EDGE DOMINATING SET can be found in polynomial time, if $t \geq (2-\rho)k$. We will use a branch-and-search method to move vertices from $V_3$ to $V_1 \cup V_2$ and therefore to increase the parameter $t$. Note that for each vertex $v \in V_3$, it is either in $V(M^*)$ or not. For the second case, all neighbors of $v$ should be in $V(M^*)$ since $V(M^*)$ is a vertex cover of the graph. Then, we can branch on $v$ by either moving $v$ into $V_1$ (this means $v \in V(M^*)$) or by moving the neighbor set $N(v)$ of $v$ in $G[V_3]$ into $V_1$ (this means $v \notin V(M^*)$) and moving all newly created clique components in $G[V_3]$ into $V_2$. When $v$ is a vertex of degree $\geq 3$ in $G[V_3]$, we can branch with recurrence:

$$C(t) \leq C(t+1) + C(t+3) \quad (2)$$

where $C(t)$ is the worst size of the search tree in the algorithm when the current value of $|V_1| + |V_2| - p$ is $t$. When the maximum degree of $G[V_3]$ is at most 2, we may only get $C(t) \leq C(t+1) + C(t+2)$, by branching on a maximum degree vertex. In fact, there are some techniques to branch on a component $H$ in $G[V_3]$ with a recurrence not worse than (2), if $H$ is not a path of length 2 [21,25,23].

For a path $p_1p_2p_3p_4\dots$ of length at least 3, we can branch on $p_3$ by including it into $V_1$ or including its neighbors $p_2$ and $p_4$ into $V_1$. For the first branch, we will also move a clique component $p_1p_2$ into $V_2$. Then we can get:

$$C(t) \leq C(t+2) + C(t+2) \quad (3)$$

which is better than (2).

For a cycle of length at least 5, we branch on an arbitrary vertex in the cycle and then branch on the generated paths in each branch and finally we can get a recurrence not worse than (2). For a cycle $c_1c_2c_3c_4$ of length 4, we can also branch with (3) by including either $\{c_1, c_3\}$ or $\{c_2, c_4\}$ into $V_1$. For the details about the proof of this fact, reader is referred to [21,25,23].

It turns out that only for a component of path of length 2 in $G[V_3]$ we cannot branch with a recurrence as good as (2). We will call a branching with recurrence at least as (2) a *good branching*.

The main steps of the improved parameterized approximation schema, called ApproxFPT, are listed in Figure 3.

Let $\rho^\star \simeq 0.21$ be the number such that $1.466 = 1.619^{(1-\rho^\star)}$. Then the following holds.

**Theorem 2.** *For any $\rho$ with $0 \leq \rho \leq 1$, ApproxFPT is a $(1+\rho)$-approximation algorithm running in time $O^*(2.374^{(1-\rho)k})$ if $\rho \leq \rho^\star$ and in time $O^*(1.466^{(2-\rho)k})$ if $\rho \geq \rho^\star$.*

## 5 Parametrization by the Vertex Cover Number

Since the size of any vertex cover in a graph is at least the size of any matching in this graph, any parameterized algorithm for EDGE DOMINATING SET working in $O(f(k)|I|^{O(1)})$ time also works in $O(f(\tau)|I|^{O(1)})$ time, where $\tau$ is the size of the minimum vertex cover of the graph. Hence, it is possible to solve EDGE DOMINATING SET within time $O^*(2.3147^\tau)$ by using the algorithm in [23]. In this section we show that this result can be improved down to $O^*(1.821^\tau)$.

To this aim, let us consider the algorithm $\text{FPT}_\tau$ presented in Figure 4, which outputs a minimum edge dominating set in graph $G$. Let $\alpha \simeq 0.2864$ be such that $2.3147^{(1-\alpha)} = \left(\frac{1}{\alpha^\alpha(1-\alpha)^{1-\alpha}}\right)$.

**Theorem 3.** *$\text{FPT}_\tau(G)$ computes a minimum edge dominate set in $O^*(1.821^\tau)$ time.*

**Input**: A graph $G = (V = V_1 \cup V_2 \cup V_3, E)$, an integer $k > 0$ and a real number $0 \leq \rho \leq 1$.
**Output**: A $(1+\rho)$-approximation solution $M$ to $k$-CONSTRAINED EDGE DOMINATING SET such that $V_1 \subseteq V(M)$.

1. **While** $t < (2-\rho)k$ and there is a connected component of $V_3$ which is not a 2-path, **do** a good branching.
2. **If** $t \geq (2-\rho)k$, compute `ApproxPoly1`$(G)$.
3. **Elseif** $\rho \geq 1/2$, compute `ApproxPoly2`$(G)$.
4. **Elseif** $t \geq (1-\rho)k$, **do**
   (a) **While** $t \leq 2(1-\rho)k$ and $V_3 \neq \emptyset$, **do** branch on a 2-path in $G[V_3]$ by including either its central vertex or its two endpoints into $V_1$;
   (b) Compute `ApproxPoly2`$(G)$.
5. **Elseif** $N \geq (1-\rho)k$, then compute `ApproxPoly2`$(G)$.
6. **Elseif** $N \leq 2(1-\rho)k/3$, branch into $2^N$ branches by considering the $2^N$ subsets of paths. For each subset $S$, include the central vertex of paths in $S$ into $V_1$, include the two endpoints of the paths not in $S$ into $V_1$, and compute an optimal solution (now $V_3 = \emptyset$).
7. **Else** consider any subset $S$ of the set of the $N$ paths in $G[V_3]$ with size $|S|$ at most $(1-\rho)k - N$. For each such subset $S$, include the two extremities of the paths in $S$ in $V_1$, and compute `ApproxPoly2`$(G)$.
8. If an optimal solution among all the leaves in the search tree is of size at most $(1+\rho)k$, then return it. Else report that there is no solution of size at most $k$.

**Fig. 3.** Algorithm `ApproxFPT`

**Input**: A graph $G = (V, E)$.
**Output**: A minimum edge dominate set.

1. Compute a minimum vertex cover $V^*$ of $G$ by using the algorithm in [8], and let $S^* = V \setminus V^*$.
2. For $k = 1$ to $(1-\alpha)\tau$ determine whether there exists an edge dominating set of size at most $k$ by using the algorithm in [23]. If any, output the minimum edge dominating set and quit.
3. Otherwise, for each subset $V_1$ of $V^*$ of size at most $\alpha\tau$:
   (a) Let $V_2 = V^* \setminus V_1$, $S_1 = N(V_1) \cap S^*$, and $S_2 = S^* \setminus S_1$;
   (b) Compute a maximum matching $M(V_1)$ in $G[V_2 \cup S_1]$;
   (c) For each vertex in $V_2 \cup S_1$ unsaturated by $M(V_1)$, take one edge incident to this vertex. Together with $M(V_1)$, this gives a set $M'(V_1)$ of edges.
4. Output a minimum edge dominating set computed in Step 3 (note that some of the edge sets $M'(V_1)$ are not edge dominating sets).

**Fig. 4.** Algorithm $\text{FPT}_\tau$

## 6 Conclusion

We provide in this article new insights on the approximability of EDGE DOMINATING SET. Our parameterized approximation algorithm first apply some steps of a branching algorithm, and then exploit the specificity of obtained instances to get an approximate solution on them. This is rather different from the notions of *fidelity preserving transformation* recently introduced in [13] where informally the instance is *first* reduced in an approximate way (and then an (exact) FPT algorithm is applied). In particular, our approximation algorithm relies on the branching steps; this is not the case in the approach of [13] and applying this latter approach for EDGE DOMINATING SET is an interesting open question mentioned in [13]. Moreover, our algorithm has complexity $O^*(\gamma_\rho^k)$ for a ratio $\rho$ where $\gamma_1 = 2.374$ (exact algorithm) and $\gamma_2 = 1.466$. Since achieving a ratio 2 is polynomial, one could hope to find approximation algorithms where $\gamma_\rho \to 1$ when $\rho \to 2$, which we leave as open question.

## References

1. Binkele-Raible, D., Fernau, H.: Enumerate and Measure: Improving Parameter Budget Management. In: Raman, V., Saurabh, S. (eds.) IPEC 2010. LNCS, vol. 6478, pp. 38–49. Springer, Heidelberg (2010)
2. Bourgeois, N., Escoffier, B., Paschos, V.T.: Approximation of max independent set, min vertex cover and related problems by moderately exponential algorithms. Discrete Applied Mathematics 159(17), 1954–1970 (2011)
3. Brankovic, L., Fernau, H.: Combining Two Worlds: Parameterised Approximation for Vertex Cover. In: Cheong, O., Chwa, K.-Y., Park, K. (eds.) ISAAC 2010. LNCS, vol. 6506, pp. 390–402. Springer, Heidelberg (2010)
4. Cai, L., Huang, X.: Fixed-Parameter Approximation: Conceptual Framework and Approximability Results. In: Bodlaender, H.L., Langston, M.A. (eds.) IWPEC 2006. LNCS, vol. 4169, pp. 96–108. Springer, Heidelberg (2006)
5. Carr, R.D., Fujito, T., Konjevod, G., Parekh, O.: A $2\frac{1}{10}$-Approximation Algorithm for a Generalization of the Weighted Edge-Dominating Set Problem. J. Comb. Optim. 5(3), 317–326 (2001)
6. Cardinal, J., Langerman, S., Levy, E.: Improved approximation bounds for edge dominating set in dense graphs. Theoretical Computer Science 410(8-10), 949–957 (2009)
7. Carr, R., Fujito, T., Konjevod, G., Parekh, O.: A $2\frac{1}{10}$-approximation algorithm for a generalization of the weighted edge-dominating set problem. Journal of Combinatorial Optimization 5, 317–326 (2001)
8. Chen, J., Kanj, I.A., Xia, G.: Improved upper bounds for vertex cover. Theoretical Computer Science 411(40-42), 3736–3756 (2010)
9. Chlebik, M., Chlebikova, J.: Approximation hardness of edge dominating set problems. Journal of Combinatorial Optimization 11(3), 279–290 (2006)
10. Dinur, I., Safra, M.: The importance of being biased. In: Proc. STOC 2002, pp. 33–42 (2002)
11. Downey, R.G., Fellows, M.R., McCartin, C., Rosamond, F.A.: Parameterized approximation of dominating set problems. Inf. Process. Lett. 109(1), 68–70 (2008)

12. Escoffier, B., Monnot, J., Paschos, V.T., Xiao, M.: New results on polynomial inapproximability and fixed parameter approximability of edge dominating set (manuscript, 2012)
13. Fellows, M.R., Kulik, A., Rosamond, F., Shachnai, H.: Parameterized Approximation via Fidelity Preserving Transformations. In: Czumaj, A., Mehlhorn, K., Pitts, A., Wattenhofer, R. (eds.) ICALP 2012, Part I. LNCS, vol. 7391, pp. 351–362. Springer, Heidelberg (2012)
14. Fernau, H.: Edge Dominating Set: Efficient Enumeration-Based Exact Algorithms. In: Bodlaender, H.L., Langston, M.A. (eds.) IWPEC 2006. LNCS, vol. 4169, pp. 142–153. Springer, Heidelberg (2006)
15. Fomin, F., Gaspers, S., Saurabh, S., Stepanov, A.: On two techniques of combining branching and treewidth. Algorithmica 54(2), 181–207 (2009)
16. Fujito, T., Nagamochi, H.: A 2-approximation algorithm for the minimum weight edge dominating set problem. Discrete Appl. Math. 118, 199–207 (2002)
17. Garey, M.R., Johnson, D.S.: Computers and intractability: A guide to the theory of NP-completeness. Freeman, San Francisco (1979)
18. Khot, S., Regev, O.: Vertex cover might be hard to approximate to within 2-epsilon. J. Comput. Syst. Sci. 74(3), 335–349 (2008)
19. Marx, D.: Parameterized complexity and approximation algorithms. The Computer Journal 51(1), 60–78 (2008)
20. Raman, V., Saurabh, S., Sikdar, S.: Efficient exact algorithms through enumerating maximal independent sets and other techniques. Theory of Computing Systems 42(3), 563–587 (2007)
21. van Rooij, J.M.M., Bodlaender, H.L.: Exact Algorithms for Edge Domination. In: Grohe, M., Niedermeier, R. (eds.) IWPEC 2008. LNCS, vol. 5018, pp. 214–225. Springer, Heidelberg (2008)
22. Schmied, R., Viehmann, C.: Approximating edge dominating set in dense graphs. Theoretical Computer Science 414(1), 92–99 (2012)
23. Xiao, M., Kloks, T., Poon, S.-H.: New Parameterized Algorithms for the Edge Dominating Set Problem. In: Murlak, F., Sankowski, P. (eds.) MFCS 2011. LNCS, vol. 6907, pp. 604–615. Springer, Heidelberg (2011)
24. Xiao, M., Nagamochi, H.: Parameterized Edge Dominating Set in Cubic Graphs (Extended Abstract). In: Atallah, M., Li, X.-Y., Zhu, B. (eds.) FAW-AAIM 2011. LNCS, vol. 6681, pp. 100–112. Springer, Heidelberg (2011)
25. Xiao, M., Nagamochi, H.: A Refined Exact Algorithm for Edge Dominating Set. In: Agrawal, M., Cooper, S.B., Li, A. (eds.) TAMC 2012. LNCS, vol. 7287, pp. 360–372. Springer, Heidelberg (2012)
26. Yannakakis, M., Gavril, F.: Edge dominating sets in graphs. SIAM J. Appl. Math. 38(3), 364–372 (1980)

# A New Algorithm for Parameterized MAX-SAT*

Ivan Bliznets and Alexander Golovnev

St. Petersburg University of the Russian Academy of Sciences, St. Petersburg, Russia

**Abstract.** We show how to check whether at least $k$ clauses of an input formula in CNF can be satisfied in time $O^*(1.358^k)$. This improves the bound $O^*(1.370^k)$ proved by Chen and Kanj more than 10 years ago. Though the presented algorithm is based on standard splitting techniques its core are new simplification rules that often allow to reduce the size of case analysis. Our improvement is based on a simple algorithm for a special case of MAX-SAT where each variable appears at most 3 times.

**Keywords:** exact algorithms, maximum satisfiability, parameterized algorithms, satisfiability.

## 1 Introduction

### 1.1 Problem Statement

Maximum Satisfiability (MAX-SAT) is a well known NP-hard problem where for a given boolean formula in conjunctive normal form one is asked to find the maximum number of clauses that can be simultaneously satisfied. In the parameterized version of MAX-SAT the question is to check whether it is possible to find an assignment that satisfies at least $k$ clauses. The best known upper bound $O^*(1.370^k)$ for this problem was given in 2002 by Chen and Kanj [1]. The previously known bounds are listed in the following table.

| Bound | Authors | Year |
|---|---|---|
| $O^*(1.618^k)$ | Mahajan, Raman [2] | 1999 |
| $O^*(1.3995^k)$ | Niedermeier, Rossmanith [3] | 1999 |
| $O^*(1.3803^k)$ | Bansal, Raman [4] | 1999 |
| $O^*(1.3695^k)$ | Chen, Kanj [1] | 2002 |

In this paper, we present an algorithm with the running time $O^*(1.358^k)$ for parameterized MAX-SAT and $O^*(1.273^k)$ for parameterized $(n,3)$-MAX-SAT. $(n,3)$-MAX-SAT is a special case of MAX-SAT where each variable appears at most three times.

An alternative way to parametrize MAX-SAT is to ask whether at least $\lceil \frac{m}{2} \rceil + k'$ clauses can be satisfied ([2], [5], [6]). This is called parametrization above

* Research is partially supported by Federal Target Program “Scientific and scientific-pedagogical personnel of the innovative Russia” 2009–2013 and Russian Foundation for Basic Research.

D.M. Thilikos and G.J. Woeginger (Eds.): IPEC 2012, LNCS 7535, pp. 37–48, 2012.

guaranteed values since one can always satisfy at least $\lceil \frac{m}{2} \rceil$ clauses (indeed, the expected number of clauses satisfied by a random assignment is $\frac{m}{2}$). It is shown in [2] that an upper bound $\phi^k$ for the parametrization considered in this paper implies an upper bound $\phi^{6k'}$ for an alternative parametrization. We do not known any better results for this parametrization.

### 1.2 General Setting

**Literals and Formulas.** Throughout the paper, $n$ denotes the number of variables, $m$ denotes the number of clauses, and $k$ denotes the number of clauses one is asked to satisfy. By MAX-SAT($F$) we denote the maximum number of clauses in $F$ that can be satisfied simultaneously. MAX-SAT($F, k$) = true iff MAX-SAT($F$) $\geq k$. The constants true and false are denoted just by 1 and 0. Let $\#_F(l)$ be the number of occurrences of a literal $l$. By $F[l]$ we denote a formula obtained from $F$ by removing all occurrences of $\bar{l}$ and deleting all clauses containing $l$. By $F[x = \bar{y}]$ we denote a formula obtained by replacing $x$ and $\bar{x}$ by $\bar{y}$ and $y$, respectively. We say that a variable *has degree* $p$ if it occurs in the formula exactly $p$ times. Also we say that a variable $x$ is of *type* $(a, b)$ if the literal $x$ occurs $a$ times and the literal $\bar{x}$ occurs $b$ times. We say that a variable $x$ is a $(k, 1)$*-singleton* ($(k, 1)$*-non-singleton*) if it is of type $(k, 1)$ and the only negation is contained (is not contained) in a unit clause. Unit clause is a clause of length 1. A literal $l$ is called *pure* if the literal $\bar{l}$ does not appear in the formula. A literal $y$ *dominates* a literal $x$ if all clauses containing $x$ contain also $y$. Two literals are called *inconsistent* if one of them is a negation of the second. A literal $y$ is a *neighbor* of a literal $x$ if they appear in a clause together. We use "..." to indicate the rest of a clause. E.g., $(x \vee \bar{y} \vee \dots)$ is a clause containing literals $x, \bar{y}$ and probably something else.

**Branchings.** Instance of a problem is a pair $(F, k)$. The question is whether it is possible to satisfy at least $k$ clauses in a formula $F$. For $q > 1$, we say that there exists a branching $(a_1, \dots, a_q)$ if we can quickly construct formulas $F_1, F_2, \dots, F_q$ such that the answer for the original problem can be found from the answers to the problems $(F_1, k - a_1), (F_2, k - a_2), \dots, (F_q, k - a_q)$. If $l$ is a literal of $F$, then clearly there exists a branching $(F[l], k - \#_F(l)), (F[\bar{l}], k - \#_F(\bar{l}))$. It is well-known that if an algorithm on each stage uses only branchings from the set

$$(a_{1,1}, \dots, a_{1,q_1}), (a_{2,1}, \dots, a_{2,q_2}), \dots, (a_{t,1}, \dots, a_{t,q_t}),$$

where $a_{i,1} \leq a_{i,2} \leq \dots \leq a_{i,q_i}$, for $1 \leq i \leq t$, then its running time is $O^*(c^k)$ where $c$ is the largest positive root of a polynomial

$$p(X) = \prod_{j=1}^{t} (X^{a_{j,q_j}} - \sum_{i=1}^{q_j} X^{a_{j,q_j} - a_{j,i}}).$$

For branching $(a_1, \dots, a_q)$, where $a_1 \leq a_2 \leq \dots \leq a_q$, we denote by $\tau(a_1, a_2, \dots, a_q)$ the unique positive root of a polynomial $X^{a_q} - (X^{a_q - a_1} + X^{a_q - a_2} + \dots + X^{a_q - a_q})$. $\tau(a_1, a_2, \dots, a_q)$ is called *a branching number.*

We say that a branching $(b_1, b_2, \ldots, b_q)$ *is dominated* by branching a $(a_1, a_2, \ldots, a_q)$ iff for every $i, a_i \geq b_i$.

### 1.3 The Main Idea of the Algorithm

A straightforward branching on a variable of high degree immediately gives a good branching number. As it is common with branching algorithms, the main bottleneck is when a formula consists of variables of low degree only. It is easy to see that variables of degree at most 2 can be eliminated from the formula. Consider a variable $x$ of degree 3: $(x \vee A)(x \vee B)(\bar{x} \vee C)$, where $A, B, C$ are disjunctions of literals. If $A$ or $B$ consists of just one literal, then we can replace $(x \vee A)(x \vee B)(\bar{x} \vee C)$ with $(\bar{A} \vee B \vee C)(A \vee C)$. If $A$ and $B$ are long, then we can branch according to the following "resolution-like" rule:

- replace $(x \vee A)(x \vee B)(\bar{x} \vee C)$ by $(A \vee C)(B \vee C)$;
- set to 0 all literals from $A, B, C$.

More formally the correctness of these steps is shown in Simplification Rule 5 and Branching Rule 2.

For solving MAX-SAT restricted to instances consisting of (3,1)- and (4,1)-singletons we use the algorithm for the Minimum Set Cover problem by van Rooij and Bodlaender [7]. The running time of the algorithm is estimated in Theorem 3.

### 1.4 Organization of the Paper

In Section 3 we present a very simple algorithm for $(n, 3)$-MAX-SAT. Its analysis is based on tricks mentioned above and contains no case analysis at all. In Section 4 we show that the presented rules can be used to simplify a case analysis of branching on a variable of degree 3. In Section 5 we improve the upper bound for Parameterized MAX-SAT.

## 2 Preliminaries: Simplification and Branching Rules

The following simplification rule is straightforward so we state it without a proof.

**Simplification Rule 1.** *A literal $l$ can be assigned the value 1 if $l$ is a pure literal or number of unit clauses $(l)$ is not smaller than number of clauses containing $\bar{l}$.*

**Simplification Rule 2.** *A variable of degree $\leq 2$ can be eliminated.*

**Correctness:** If $l$ is a pure literal, then we can set $l = 1$. Otherwise, $F = G \wedge (l \vee A) \wedge (\bar{l} \vee B)$. It is easy to see that MAX-SAT$(F, k)$ = MAX-SAT$(F \wedge (A \vee B), k-1)$. □

**Simplification Rule 3.** *Pairs of clauses $(x)$ and $(\bar{x})$ can be removed.*

**Correctness:** Clearly, MAX-SAT$(F \vee (x) \vee (\bar{x}), k)$ = MAX-SAT$(F, k-1)$. It does not matter whether the variable $x$ appears in $F$ or not. □

**Simplification Rule 4.** *If two variables $x$ and $y$ of degree 3 appear together in 3 clauses, then all these 3 clauses can be satisfied by assigning $x$ and $y$.*

**Correctness:** One can satisfy 2 clauses by assigning $x$ the remaining clause can be satisfied by assigning $y$. □

**Simplification Rule 5.** *Let $x$ be a variable of degree 3: $F = G \wedge (x \vee A) \wedge (x \vee B) \wedge (\bar{x} \vee C)$. If $A$ or $B$ has length $< 2$ then we can reduce the problem.*

**Correctness:** Wlog, assume that $A$ has length $< 2$. If the length of $A$ equals to 1 then $A$ is a single literal. It is easy to see that

$$\text{MAX-SAT}(F, k) = \text{MAX-SAT}(G \wedge (\bar{A} \vee B \vee C) \wedge (A \vee C), k-1).$$

If $A$ is empty we can set $x = 1$. The parameter is reduced by 2. □

*Remark 1.* It is easy to see that all Simplification Rules can be applied in polynomial time and decrease $k$ at least by one. Note that some of the simplifications rules make several clauses of a formula satisfied while others may replace existing clauses with new clauses and reduce the parameter (like SR2 and SR5). Since as a result of applying a rule the number of satisfied clauses increases we usually say that applying a simplification rule satisfies some clauses.

**Branching Rule 1.** *For any literal $l$, one can branch as $(F[l], k - \#_F(l)), (F[\bar{l}], k - \#_F(\bar{l}))$.*

**Branching Rule 2.** *Let $x$ be a variable of degree 3: $F = G \wedge (x \vee A) \wedge (x \vee B) \wedge (\bar{x} \vee C)$. Then there is a branching:*

- $(G \wedge (A \vee C) \wedge (B \vee C), k-1)$
- $(G', k-2)$*, where $G'$ is obtained from $G$ by assigning all literals from $A, B, C$ to 0.*

**Correctness:** Let $R = (A \vee C) \wedge (B \vee C)$. It is a simple observation that if an optimal assignment satisfies $s$ clauses from $R$, where $s = 1, 2$, then we can satisfy $s+1$ clauses from $F - G$ but cannot satisfy $s+2$. However, if an optimal assignment does not satisfy any clause from $R$ we can still satisfy two clauses from $F - G$ by setting $x = 1$. □

**Corollary 1.** *If $A \cup B \cup C$ contains inconsistent literals, then one can consider only the first branch. It means that one can reduce $(F, k)$ to $(G \wedge (A \vee C) \wedge (B \vee C), k-1)$.*

*Remark 2.* We write $BR2(x)$ if we apply Branching Rule 2 to a variable $x$. Branching on a variable means applying Branching Rule 1. We write $SRi(x)$ for $1 \leq i \leq 5$, if we apply Simplification Rule $i$ to a variable $x$.

**Lemma 1 (Kulikov, Kutzkov [8]).** *If a literal $y$ dominates a literal $x$, then one can branch as*

- $x = 1, y = 0$;
- $x = 0$.

*Proof.* If in some assignment the literals $x$ and $y$ both have the value 1, then flipping the value of $x$ cannot decrease the number of satisfied clauses. Indeed, all clauses that can be satisfied by $x = 1$ are also satisfied by $y = 1$. □

**Lemma 2.** *Let $x$ be a $(t, 1)$-non-singleton variable. Then branching on $x$ is a $(t, 2)$-branching.*

*Proof.* Let $y$ be a neighbor of $\bar{x}$. In the branch $x = 1$ we satisfy at least $t$ clauses. In the branch $x = 0$ we can set $y = 0$ and satisfy at least 2 clauses. The lemma follows from Lemma 1, in this case we use literals $y, \bar{x}$ instead of $y, x$. □

**Lemma 3.** *If $F$ contains a variable $x$ of degree $\geq 6$, then branching number on $x$ is at most $\tau(1, 5)$.*

*Proof.* This follows from the fact that $\tau(1, 5) > \tau(2, 4) > \tau(3, 3)$. □

## 3 Solving $(n, 3)$-MAX-SAT in $1.2721^k$ Time

By $(n, 3)$-MAX-SAT we denote MAX-SAT restricted to instances in which each variable appears in at most 3 clauses. In this section we give a simple algorithm for $(n, 3)$-MAX-SAT. The running time of the algorithm is $1.2721^k$. Note that the previous known upper bound for $(n, 3)$-MAX-SAT w.r.t. $k$ is $1.3247^k$ and it follows from proof of Chen and Kanj for the general MAX-SAT. Throughout this section we assume that $F$ is an $(n, 3)$-MAX-SAT formula.

**Lemma 4.** *Let $x$ be a variable of degree 3: $F = G \wedge (x \vee A) \wedge (x \vee B) \wedge (\bar{x} \vee C)$ and rules SR1-4 are not applicable to $F$. If $A$ has length $< 2$, then we have $(2, 4)$-branching and the resulting formulas are $(n, 3)$-MAX-SAT formulas.*

*Proof.* If $A$ has length 0, then we can set $x = 1$. Otherwise, by Simplification Rule 5 we eliminate one clause and get a new formula $F' = G \wedge (\bar{A} \vee B \vee C) \wedge (A \vee C)$. Variables of degree 4 in the formula $F'$ can appear only in $A$ and $C$. Branching on the variable $A$ gives $(n, 3)$-MAX-SAT formulas in both branches. $A$ has degree 4, so the branching gives at least $(1, 3)$-branching (note that $\tau(2, 2) < \tau(1, 3)$). As one clause is already satisfied, the resulting branching is at least $(2, 4)$. □

**Lemma 5 (Bliznets [9]).** *If each variable of $F$ appears once negatively and twice positively and all negative literals occur in unit clauses, then MAX-SAT($F$) can be computed in polynomial time.*

*Proof.* Construct a graph $G_F = (V, E)$ in the following way. Introduce a vertex for each clause consisting of positive literals, introduce an edge between two vertices if the corresponding two clauses share a variable. Then MAX-SAT$(F) = n + \nu(G_F)$, where $\nu(G_F)$ is the size of a maximum matching in $G_F$. □

**Algorithm 1.** $(n, 3)$-MAX-SAT-ALG — solving $(n, 3)$-MAX-SAT in time $1.2721^k$.

**Input:** $F$ — instance of $(n, 3)$-MAX-SAT.
**Parameter:** $k$ — number of clauses asked to satisfy.
**Output:** *true*, if $k$ clauses can be satisfied simultaneously; *false* otherwise.

1: apply Simplification Rules 1–4
2: **if** all negations are singletons **then**
3:     return answer (use Lemma 5).
4: choose $x$, s.t. $\bar{x}$ is not a singleton: $(x \vee A)(x \vee B)(\bar{x} \vee C), |C| > 0, |A| \leq |B|$
5: **if** $|A| \leq 1$ **then**
6:     use Lemma 4 for branching
7: **if** $|A| \geq 2$ **then**
8:     branch BR $1(x)$

**Theorem 1.** *Algorithm* $(n, 3)$-MAX-SAT-ALG *solves* $(n, 3)$-*MAX-SAT in time* $1.2721^k$.

*Proof.* By Lemma 5 the running time of step 3 is polynomial. Step 6 gives $(2, 4)$-branching by Lemma 4. Branching at step 8 gives at least $(4, 2)$-branching. Indeed, $|C| > 0$ and Lemma 1 implies that in case $x = 0$ we can satisfy at least two clauses: one is $(\bar{x} \vee C)$ and one more with a literal $\bar{t}$ where $t$ is some literal from $C$. By SR4, there are at least 4 clauses containing variables from $A$. In the branch $x = 1$ two clauses are satisfied by $x$ and there exist variables $y, z$ that appear one or two times in the new formula. There are at least 2 clauses that contain variable $y$ or $z$. Hence, SR2 applied to variables $y, z$ from $A$ satisfies two new clauses. Thus, branching on $x$ gives $(4, 2)$-branching. The running time of the algorithm is $\max(\tau(2, 4), \tau(3, 3))^k = \tau(2, 4)^k < 1.2721^k$. □

## 4 Removing Variables of Degree 3

In this section we show that if a formula contains a variable of degree 3 then we can either decrease $k$ or find a good branching. Suppose that $x$ occurs three times in $F$ and no simplification rules are applicable to $F$. Assume that $F$ contains clauses $(x \vee A), (x \vee B), (\bar{x} \vee C)$ where $A, B, C$ are disjunctions of literals. We consider only cases where $|A|, |B| \geq 2$ because otherwise we can apply SR5($x$) or assign 1 to $x$.

**Definition 1.** *Denote by* $LN(!A_1, \ldots, !A_k)$ *the set of all clauses of* $F$ *containing a negation of some literal from* $A_1 \cup \cdots \cup A_k$

Throughout this section we assume that $A \cup B \cup C$ does not contain inconsistent literals. Otherwise, by Corollary 1 the formula can be simplified.

**Lemma 6.** *Let* $y, z$ *be literals such that* $y, z \in A \cup B \cup C$, $\bar{y}$ *occurs two or more times and dominates* $\bar{z}$ *(call this situation* a first special case of domination*). Then there is a* $(2, 4)$-*branching.*

*Proof.* Recall that $\bar{y}, \bar{z}$ appear in some clauses from $LN(!A, !B, !C)$. Consider the branching $F[y], F[\bar{y}]$. In the second case two clauses are satisfied with $\bar{y}$ and $z$ can be substituted by 1, because it becomes a pure literal. $z = 1$ satisfies at least one of the following clauses: $(x \vee A), (x \vee B), (\bar{x} \vee C)$. After that we use SR2($x$) since $x$ appears at most two times. This satisfies at least 4 clauses. In the formula $F[y]$ we use SR2($x$) and that is why the parameter is decreased by 2. □

**Lemma 7.** *Let $y, z$ be literals from $A \cup B \cup C$ such that $\bar{y}$ occurs once and dominates $\bar{z}$ (call this situation* a second special case of domination*). Then there is a* (3, 3)*-branching.*

*Proof.* Like in the previous lemma in $F[\bar{y}]$ we can assign 1 to $z$. A literal $z$ appears at least two times because $\bar{z}$ occurs once in the formula. So $z = 1$ satisfies two more clauses($z$ and $\bar{z}$ do not appear in one clause). This satisfies at least 3 clauses. In $F[y]$ — we satisfy 2 clauses containing $y$ and one using SR2($x$) because after assigning $y = 1$ we have one or two clauses containing the variable $x$. We get a $(3, 3)-$branching. □

**Lemma 8.** *If $|LN(!A, !B, !C)| < 3$ then we have a special case of domination or $A = B = y \vee z$. In the former case we have a good branching ((2, 4)-, (3, 3)-, (1, 6)-branching or better), while in the latter case the parameter can be decreased.*

*Proof.* We know that $|A|, |B| \geq 2$. So $A \cup B$ contains more than two literals or equals $y \vee z$. In the former case three literals should occur in two clauses, so this is a domination. In the latter case we can replace clauses with the variable $x$ by the clause $(y \vee z \vee C)$ and decrease the parameter by two. □

Now we can assume that we only work with formulas where $|LN(!A, !B, !C)| > 2$. If $|LN(!A, !B, !C)| > 3$ using BR2($x$) we immediately get (1, 6)-branching. So for the rest of this section $|LN(!A, !B, !C)| = 3$.

**Lemma 9.** *If $|A \cup B \cup C| > 3$ then we have a special case of domination and hence one of the Lemmas 6, 7 is applicable. So there is a (2, 4)- or (3, 3)-branching.*

*Proof.* At least 4 negations of the literals from $|A \cup B \cup C|$ should be placed in 3 clauses and it is impossible without special case of domination. □

From the previous lemma we conclude that it is enough to consider formulas where $|A \cup B \cup C| \leq 3$.

**Lemma 10.** *If $min\{|A|, |B|\} \geq 3$ then either there is a special case of domination or the parameter can be decreased.*

*Proof.* If $|A \cup B \cup C| > 3$ we have a special case of domination because of Lemma 9. Otherwise from $|A \cup B \cup C| \leq 3$ and $min\{|A|, |B|\} \geq 3$ it follows that $|A| = |B| = 3$ and $x \vee A = x \vee B = x \vee y_1 \vee y_2 \vee y_3$. So, we can replace $x \vee A, x \vee B, \bar{x} \vee C$ by $A \vee C$ and decrease the parameter by 2. □

Now wlog we can assume that $x \vee A = x \vee y \vee z$.

**Lemma 11.** *If we have an instance $(F, k)$ and for all variables $x$ that appear three times the following holds:*

- *$x$ occurs in clauses $x \vee A_x, x \vee B_x, \bar{x} \vee C_x$*
- *$LN(!A_x, !B_x, !C_x) = 3$*

*then we can decrease the parameter or apply one of the following branchings:* $(3,3), (2,4)$ *or better.*

*Proof.* Suppose that all previous lemmas are not applicable otherwise we are done. So we can choose a variable $x$ that occurs three times and $A_x = y \vee z$. Note that if $\bar{y}$ occurs three times then we have a case of domination and in this situation a good branching ((2, 4)-, (3, 3)-, (1, 6)-branching or better) exists. Consider two cases: $\bar{y}$ appears exactly once or twice.

**Case 1: $\bar{y}$ appears exactly once.**

**Case 1.1: $\bar{y}$ has a neighbor.**
$F$ contains the following clauses:

$$(x \vee y \vee z), \quad (x \vee \ldots), \quad (\bar{x} \vee \ldots), \quad (\bar{y} \vee w \vee \ldots).$$

In $F[\bar{y}]$ by Lemma 1 we can assign 0 to $w$ and use SR5($x$) or SR2($x$). So, we decrease the parameter by 3. In $F[y]$ we satisfy at least two clauses and using SR2($x$) we decrease the parameter by 1. We obtain (3, 3)-branching. So in all other cases we must have a clause $(\bar{y})$.

**Case 1.2: $y$ appears more than 2 times and there is an occurrence outside a variable $x$.**
After $F[y]$ and SR2($x$) we satisfy at least 4 clauses. In $F[\bar{y}]$ using SR5($x$) we decrease the parameter by 2.

**Case 1.3: a literal $y$ appears in all clauses with a variable $x$.**
We have the following clauses:

$$(x \vee y \vee z), \quad (x \vee y \vee \ldots), \quad (\bar{x} \vee y \vee \ldots), \quad (\bar{y}).$$

The variable $y$ does not occur in the rest of the formula, otherwise we can treat it as a case 1.2. So, it is enough to consider $y = 0$. Because an assignment with $x = 1, y = 0$ is not worse than the same assignment with $x = y = 1$ and $x = 0, y = 1$. It means that we can satisfy one clause and one variable without branching.

**Case 1.4: $y$ occurs exactly twice**
Using symmetry we can conclude that the clause with $\bar{x}$ does not contain any other literals. Otherwise we have case 1.1. Again using symmetry ideas we may conclude that either $\bar{z}$ appears twice or $\bar{z}$ appears once and $z$ appears exactly twice and there is a clause $(\bar{z})$ . Consider these two subcases separately.

**Case 1.4.1: $\bar{z}$ appears once and $z$ appears exactly twice**

$$(x \vee y \vee z), \quad (x \vee \dots), \quad (\bar{x}), \quad (\bar{y}), \quad (\bar{z}).$$

In $F[x]$ using SR2($y$), SR2($z$) we decrease the parameter by 4. In $F[\bar{x}]$ it is easy to see that we can assume $y = \bar{z}$. So we obtain $(4, 4)$-branching.

**Case 1.4.2: $\bar{z}$ appears twice.**

In this case we have the following clauses:

$$(x \vee y \vee z), \quad (x \vee \dots), \quad (\bar{x}), \quad (\bar{y}), \quad (\bar{z} \vee \dots), \quad (\bar{z} \vee \dots).$$

$F[z]$ and then SR2($x$), SR2($y$) decrease the parameter by 3. In $F[\bar{z}]$ we can assume that $x = \bar{y}$. We get a $(3, 4)$-branching.

**Case 2: $\bar{y}$ appears exactly twice.**

Using symmetry we conclude that $\bar{z}$ also appears twice otherwise we have the situation described in case 1. So, we have the following family of clauses:

$$(x \vee y \vee z), \quad (x \vee B), \quad (\bar{x} \vee \dots), \quad (\bar{y} \vee \dots), \quad (\bar{y} \vee \dots).$$

Assume $y \in B$ (the case $z \in B$ is similar). $F[y]$ and then $x = 0$ removes 3 clauses. In $F[\bar{y}]$ we use SR5($x$) and this removes 3 clauses. If $y, z$ do not appear in $B$ we have $|B| < 2$ or $|A \cup B| \geq 4$ and we get a special domination case, considered before.

□

**Theorem 2.** *If $x$ occurs exactly 3 times in $F$, then either the parameter can be decreased or there is a $(1, 6)$-, $(2, 4)$- or $(3, 3)$-branching.*

## 5 Solving MAX-SAT in $1.358^k$

In this section we present a simple algorithm that improves the upper bound for Parameterized MAX-SAT (Algorithm MAX-SAT-Alg). The main bottleneck of the analysis is when all variables are $(1, 3)$-singletons or $(1, 4)$-singletons. We consider this case separately.

We reduce an instance of this restricted MAX SAT to the instance of Minimum Set Cover. The Minimum Set Cover is, given a universe $U$ and a collection $\mathcal{S}$ of subsets of $U$, to find the minimum cardinality of a subset $S' \subset \mathcal{S}$ which covers $U$: $\bigcup_{S_i \in S'} S_i = U$. For $e \in U$, $f(e)$ (frequency of $e$) denotes the number of subsets of $\mathcal{S}$ in which $e$ is contained.

It can be shown that algorithm for the Minimum Dominating Set given by van Rooij and Bodlaender [7] in fact solves also the Minimum Set Cover in time $1.28759^{k(U,\mathcal{S})}$, where $k(U, \mathcal{S}) = \sum_{e \in U} v(f(e)) + \sum_{S_i \in \mathcal{S}} w(|S_i|)$, and $v, w$ are weight functions. The maximum value of $v$ is 0.595723 and the maximum value of $w$ is 1. We note that for sets of cardinality $\leq 4$ the maximum value of $w$ is 0.866888 (see the end of section 3 in [7]). Therefore, we can use the following lemma due to van Rooij and Bodlaender [7].

**Theorem 3.** *Algorithm* MSC *solves the Minimum Set Cover problem where the cardinality of each set in $S$ is at most 4 in time* $O^*(1.28759^{0.595723|U|+0.866888|\mathcal{S}|}) = O^*(1.29^{0.6|U|+0.9|\mathcal{S}|})$.

The following theorem was proved by Lieberherr and Specker [10]. Later Yannakakis [11] gave a simple proof of this bound by the probabilistic method.

**Theorem 4.** *If any three clauses in $F$ are satisfiable, then at least $\frac{2m}{3}$ clauses are simultaneously satisfiable.*

Theorem 3 is used for instances with $m < 1.5k$, while Theorem 4 is used for instances with $m \geq 1.5k$. We are now ready to prove an upper bound.

**Theorem 5.** *Algorithm* MAX-SAT-ALG *solves MAX-SAT in time* $O^*(1.3579^k)$.

*Proof.* Below we show that in each case the algorithm branches with branching number at most $\tau(5, 10, 1) < 1.3579$, so the running time of the algorithm is $O^*(1.3579^k)$.

- Step 3. If $deg(x) \geq 6$ then by Lemma 3 we get $(1, 5)$-branching. $\tau(1, 5) \approx 1.3248 < 1.3579$.
- Step 5. If $deg(x) = 3$ then by Theorem 2 we get $(1, 6)$-branching. $\tau(1, 6) \approx 1.2852 < 1.3579$.
- Step 7. A $(3, 2)$-variable gives $\tau(3, 2) \approx 1.3248 < 1.3579$. By Corollary 2, branching on $(4, 1)$-non-singleton or $(3, 1)$-non-singleton gives at least $\tau(3, 2) \approx 1.3248 < 1.3579$.
- Step 10. $x$ is a $(2, 2)$-variable, $y$ is a $(1, 4)$-singleton, neighbor of $x$ and literal $y$ does not dominate $x, \bar{x}$ simultaneously. Branching on $y$ gives $\tau(4, 1)$ and the next iteration in branch $y = 1$ has a variable of degree 3 or smaller. The overall branching number is smaller than $\tau(4 + 1, 4 + 6, 1) = \tau(5, 10, 1) < 1.3579$.
- Step 12. $x$ is a $(2, 2)$-variable. Neighbors of $x$ are variables of degree 4 or $x, \bar{x}$ are dominated by $y$. So, both $F[x]$ and $F[\bar{x}]$ contain a variable of degree 3. By Theorem 2, a variable of degree 3 gives $(1, 6)$,- $(2, 4)$- or $(3, 3)$-branching. So the possible branchings are $\tau(2+1, 2+6, 2+1, 2+6), \tau(2+2, 2+3, 2+2, 2+3), \tau(2+3, 2+3, 2+3, 2+3)$ the worst case among them is $\tau(3, 8, 3, 8) \approx 1.3480 < 1.3579$.

In the following we assume that all variables are $(3, 1)$- or $(4, 1)$-singletons.

- Step 14. Now all variables are singletons. It means that we can satisfy $n$ clauses by setting all variables to 0. If $k \leq n$ this solves the problem.
- Step 16. We assume that each variable occurs 3 or 4 times positively and once negatively in a unit clause. Note that all clauses are either negative singletons or positive clauses (all variables in positive clauses occur only positively). We claim that for such a formula there always exists an optimal assignment satisfying all positive clauses. Indeed, if some positive clause is not satisfied then by flipping any of its variables we can only increase the number of

**Algorithm 2.** MAX-SAT-ALG — solving MAX-SAT in time $1.3579^k$.

**Input:** $F$ — instance of MAX-SAT.
**Parameter:** $k$ — number of clauses asked to satisfy.
**Output:** 1, if $k$ clauses can be satisfied simultaneously; 0 otherwise.

1: apply Simplification Rules 1–5.
2: **if** there is $x$, s.t. $deg(x) \geq 6$ **then**
3:     branch on $x$
4: **if** there is $x$ of degree 3 **then**
5:     branch on $x$ according to Theorem 2
    {Now we have only variables of degree 4 and 5.}
6: **if** $F$ contains a variable $x$ of type $(3,2)$, $(3,1)$-non-singleton or $(4,1)$-non-singleton **then**
7:     branch on $x$
    {Now we have only singletons and $(2,2)$-variables.}
8: **if** $F$ contains a variable $x$ of type $(2,2)$ **then**
9:     **if** $x$ has a neighbor $(4,1)$-singleton $y$ and $x, \bar{x}$ are not simultaneously dominated by $y$ **then**
10:         branch on $y$
11:     **else**
12:         branch on $x$
    {Now all variables are $(3,1)$-singletons or $(4,1)$-singletons.}
13: **if** $k \leq n$ **then**
14:     **return** 1
15: **if** $m < 1.5k$ **then**
16:     **return** $k \leq \text{MSC}(F)$
17: **if** there is a clause of length 2: $(x \vee y)$ **then**
18:     branch as $F[x, y]; F[x = \bar{y}]$.
19: **else**
20:     **return** 1

satisfied clauses. It means that we want to assign 1 to the minimal number of variables to satisfy all positive clauses. It is the Minimum Set Cover problem. We construct an instance of the Minimum Set Cover problem in the following way. $U$ is the set of positive clauses ($|U| = m - n$). $\mathcal{S}$ contains $n$ sets. Set $S_i \in \mathcal{S}$ consists of positive clauses, which contains a variable $x_i$. Now we would like to cover $U$ by the minimal number of sets from $\mathcal{S}$. If $t$ is the minimal number of sets required to cover $U$, then the maximum number of satisfied clauses is $m - t$. We can compare this number to $k$ and return the result. By Theorem 3, the algorithm for Minimum Set Cover for sets of cardinality $\leq 4$ has running time $T(F) = O^*(1.29^{(0.6|U|+0.9|\mathcal{S}|)}) = O^*(1.29^{(0.6(m-n)+0.9n)})$. We know that $k > n$ and $m < 1.5k$. Thus $T(F) \approx 1.3574^k < 1.3579^k$.

- Step 18. We know that the formula contains clauses $(\bar{x})$ and $(\bar{y})$. If there is a clause $(x \vee y)$, then some optimal solution satisfies clause $(x \vee y)$. Indeed, if it does not, we can just set $x = 1$ and the number of satisfied clauses does not decrease. So, we can branch as $x = y = 1$ and $x = \bar{y}$. In the first branch we satisfy at least 3 clauses, because $x$ is a $(3,1)-$ or $(4,1)-$variable. In the second branch we satisfy clause $(x \vee y)$ and by Simplification Rule 3 we satisfy one of the clauses $(x)$ and $(\bar{x})$. We obtain $(2, 3)$-branching.

- Step 20. Now we have a formula with $m \geq 1.5k$ clauses. $F$ does not contain clauses of length 2. Therefore, every triple of clauses is satisfiable. By Theorem 4 there is an assignment, which satisfies at least $\frac{2m}{3} \geq k$ clauses. □

**Acknowledgments.** We are greatful to Konstantin Kutzkov for fruitful discussions and suggesting Branching Rule 2 . We would like to thank our supervisor Alexander S. Kulikov for help in writing this paper and valuable comments. Also we thank anonymous reviewers who helped improve the paper.

## References

1. Chen, J., Kanj, I.A.: Improved Exact Algorithms for MAX-SAT. In: Rajsbaum, S. (ed.) LATIN 2002. LNCS, vol. 2286, pp. 341–355. Springer, Heidelberg (2002)
2. Mahajan, M., Raman, V.: Parameterizing above guaranteed values: MaxSat and MaxCut. J. Algorithms 31(2), 335–354 (1999)
3. Niedermeier, R., Rossmanith, P.: New Upper Bounds for MaxSat. In: Wiedermann, J., Van Emde Boas, P., Nielsen, M. (eds.) ICALP 1999. LNCS, vol. 1644, pp. 575–584. Springer, Heidelberg (1999)
4. Bansal, N., Raman, V.: Upper Bounds for MaxSat: Further Improved. In: Aggarwal, A.K., Pandu Rangan, C. (eds.) ISAAC 1999. LNCS, vol. 1741, pp. 247–258. Springer, Heidelberg (1999)
5. Alon, N., Gutin, G., Kim, E., Szeider, S., Yeo, A.: Solving MAX-r-SAT Above a Tight Lower Bound. Algorithmica 61, 638–655 (2011)
6. Crowston, R., Gutin, G., Jones, M., Yeo, A.: A New Lower Bound on the Maximum Number of Satisfied Clauses in Max-SAT and Its Algorithmic Applications. Algorithmica 64(1), 56–68 (2012)
7. van Rooij, J.M., Bodlaender, H.L.: Exact algorithms for dominating set. Discrete Applied Mathematics 159(17), 2147–2164 (2011)
8. Kulikov, A., Kutzkov, K.: New Bounds for MAX-SAT by Clause Learning. In: Diekert, V., Volkov, M.V., Voronkov, A. (eds.) CSR 2007. LNCS, vol. 4649, pp. 194–204. Springer, Heidelberg (2007)
9. Bliznets: A New Upper Bound for $(n, 3)$-MAX-SAT. Zapiski Nauchnikh Seminarov POMI, 5–14 (2012)
10. Lieberherr, K.J., Specker, E.: Complexity of Partial Satisfaction. J. ACM 28, 411–421 (1981)
11. Yannakakis, M.: On the approximation of maximum satisfiability. In: SODA 1992, pp. 1–9 (1992)

# Restricted and Swap Common Superstring: A Parameterized View

Paola Bonizzoni[2], Riccardo Dondi[1], Giancarlo Mauri[2], and Italo Zoppis[2]

[1] DSLCS, Università degli Studi di Bergamo, Bergamo, Italy
[2] DISCo, Università degli Studi di Milano-Bicocca, Milano, Italy
{bonizzoni,mauri,zoppis}@disco.unimib.it, riccardo.dondi@unibg.it

**Abstract.** In several areas, in particular in bioinformatics and in AI planning, Shortest Common Superstring problem (SCS) and variants thereof have been successfully applied. In this paper we consider two variants of SCS recently introduced (Restricted Common Superstring, RCS) and (Swapped Common Superstring, SWCS). In RCS we are given a set $S$ of strings and a multiset, and we look for an ordering $\mathcal{M}_o$ of $\mathcal{M}$ such that the number of input strings which are substrings of $\mathcal{M}_o$ is maximized. In SWCS we are given a set $S$ of strings and a text $\mathcal{T}$, and we look for a swap ordering $\mathcal{T}_o$ of $\mathcal{T}$ (an ordering of $\mathcal{T}$ obtained by swapping only some pairs of adjacent characters) such that the number of input strings which are substrings of $\mathcal{T}_o$ is maximized. In this paper we investigate the parameterized complexity of the two problems. We give two fixed-parameter algorithms, where the parameter is the size of the solution, for SWCS and $\ell$-RCS (the RCS problem restricted to strings of length bounded by a parameter $\ell$). Furthermore, we complement these results by showing that SWCS and $\ell$-RCS do not admit a polynomial kernel.

## 1 Introduction

In several areas, such as bioinformatics [11] and data compression [15], the Shortest Common Superstring problem (SCS) has been successfully applied for strings comparison. For example, in bioinformatics, SCS aims to reconstruct the original string from a set of different fragments of that string. Recently, some variants of the SCS problem have been proposed to deal with problems in bioinformatics and AI planning [10,6]. In such variants, a set of strings is given and we are asked to rearrange a given multiset of characters or a given text in order to maximize the number of strings which are substrings of the resulting text. This can be the case, when the strings represent proteins and only the (multi)-set of amino acids is given (or an ordering which may be affected by some errors), and we want to infer the right ordering of such amino-acids that contains the given strings, or at least a fraction of them. Another application of the above problem is AI planning, where a set of tasks which have to be accomplished is given, and we want to compute a plan that achieves as many goals as possible. Usually, the plan corresponds to compute a SCS of strings representing the tasks. However,

D.M. Thilikos and G.J. Woeginger (Eds.): IPEC 2012, LNCS 7535, pp. 49–60, 2012.

in practice we may have some constraints on the given tasks, hence the plan we want to compute is a SCS with some restrictions [10].

Two combinatorial problems recently introduced in this context are the Restricted Common Superstring (RCS) problem and the Swapped Common Superstring (SWCS) problem. RCS is the more general problem: we are given a set $S$ of $n$ strings over an alphabet $\Sigma$ and a multiset $\mathcal{M}$ over $\Sigma$, and we look for an ordering $\mathcal{M}_o$ of $\mathcal{M}$ such that the number of input strings which are substrings of $\mathcal{M}_o$ is maximized.

In the SWCS problem we are given a set $S$ of $n$ strings over an alphabet $\Sigma$ and a text $\mathcal{T}$ over $\Sigma$, and we look for a swap ordering $\mathcal{T}_o$ of $\mathcal{T}$ (an ordering of $\mathcal{T}$ obtained by swapping only some pairs of adjacent characters) such that the number of input strings which are substrings of $\mathcal{T}_o$ is maximized.

The complexity of the SCS problem has been extensively studied in the past [2,14,16]: the problem is known to be APX-hard [2], even for equal length strings over binary alphabet [17].

The RCS problem is known to be NP-hard, even in the restricted case that the input strings are defined over a binary alphabet or have length bounded by 2 [6]. Furthermore, in [6] it is shown that the problem is not approximable within factor $O(n^{1-\varepsilon})$, with a reduction from Maximum Independent Set. It is easy to see that the reduction is also a parameterized reduction [9,12], thus implying the W[1]-hardness of RCS when parameterized by the size of the solution. The SWCS is known to be NP-hard [10]. However, it is shown in [10] that a relaxed version of the problem, where each occurrence of a string in the swap ordering $\mathcal{T}_o$ is counted, is polynomial time solvable.

For both problems we investigate the parameterized complexity, under some natural parameterizations (for more details on parameterized complexity see [9,12]). The ultimate goal of our investigations is to provide a multivariate analysis of the complexity of the two problems [7,13]. We consider as natural parameters, the size of the solution, that is the number of input strings which are substrings of the computed solution, and the maximum length of the input strings.

In Section 2, we provide some preliminary definitions and we formally state RCS and SWCS. In Section 3 we give two fixed-parameter algorithms for $\ell$-RCS (the RCS problem restricted to strings of length bounded by a parameter $\ell$) and SWCS, when the two problems are parameterized by the size of the solution. We complement these two positive results with two negative results, that is we show in Section 4 that $\ell$-RCS and SWCS do not admit a polynomial kernel. *Kernelization* is a well-known technique in parameterized complexity [9,12]. The goal of kernelization is to preprocess (in polynomial time) an instance of a given problem, so that the resulting instance, called *kernel*, has size depending only on the considered parameter. Recently, the kernelization complexity has been widely investigated [3,5,8,4], and different techniques have been introduced to prove that a problem, although fixed-parameter tractable, does not admit a *polynomial size kernel*.

Some of the proofs are omitted due to space limitation.

## 2 Preliminaries

In this section, we introduce some basic definitions. Given a string $s$ over an alphabet $\Sigma$, denote by $|s|$ the length of $s$. The $i$-th symbol of $s$ is denoted by $s[i]$. For two positions $i, j$ in $s$, with $1 \le i \le j \le |s|$, denote by $s[i, j]$ the substring of $s$ that starts at position $i$ and ends at position $j$. Given a set $S$ of strings, for each $s \in S$ define $incl(s) = \{s' \in S : s' \textit{ is a proper substring of } s\}$. Given a string $s$ and a substring $s_x$ of $s$, we say that $s_x$ is *covered* by $s$. Furthermore, if $s[i, j] = s_x$, we say that $s[i, j]$ is an *occurrence* of $s_x$ in $s$.

Given a multiset $\mathcal{M}$ over alphabet $\Sigma$ and a symbol $\sigma \in \Sigma$, we denote by $occ_{\mathcal{M}}(\sigma)$ the number of occurrences of $\sigma$ in $\mathcal{M}$. Given a multiset $\mathcal{M}$ over an alphabet $\Sigma$, we define an *ordering* $\mathcal{M}_o$ of $\mathcal{M}$, as a string over $\Sigma$ containing exactly $occ_{\mathcal{M}}(\sigma)$ occurrences for each $\sigma \in \Sigma$. Now, we are able to define the first problem we are interested in.

*Problem 1.* [6] RCS
**Input:** a set $S = \{s_1, \ldots, s_n\}$ of strings over alphabet $\Sigma$, a multiset $\mathcal{M}$ over $\Sigma$.
**Output:** an ordering $\mathcal{M}_o$ of $\mathcal{M}$ that maximizes the number of strings in $S$ that are substrings of $\mathcal{M}_o$.

We will consider the restriction of RCS, denoted by $\ell$-RCS, where the strings in $S$ have length bounded by a parameter $\ell$.

Before giving the formal definition of the second problem we are interested in, we need to introduce the definition of *swap ordering*. Given a text $\mathcal{T} = t_1 t_2 \ldots t_m$, where each $t_i$, $1 \le i \le m$, is a character in $\Sigma$, a text $\mathcal{T}_o = t_{\pi(1)} t_{\pi(2)} \cdots t_{\pi(m)}$ is called a *swap ordering of* $\mathcal{T}$ if it is induced by a permutation $\pi : \{1, \ldots, m\} \to \{1, \ldots, m\}$ such that: (1) if $\pi(i) = j$, then $\pi(j) = i$, (2) for all i, $\pi(i) \in \{i-1, i, i+1\}$, (3) if $\pi(i) \neq i$ then $t_{\pi(i)} \neq t_i$. It follows that a swap ordering $\mathcal{T}_o$ of $\mathcal{T}$ is obtained by swapping only some pairs of adjacent distinct characters of $\mathcal{T}$. Notice that the swaps must be *consistent swaps*, that is if $\mathcal{T}_o$ is a swap ordering of $\mathcal{T}$ obtained by swapping characters in positions $p_1$, $p_2$ and characters in positions $p_3$, $p_4$, with $p_1 + 1 = p_2 \le p_3 = p_4 - 1$, then these swaps are *consistent*, if $p_2 < p_3$ (see Example 1).

*Example 1.* Consider the text:

$$\mathcal{T} = abxcyz$$

The text $\mathcal{T}_o = abcxyz$ is a swap ordering of $\mathcal{T}$ obtained by swapping the characters $x$ and $c$ of $\mathcal{T}$. The text $\mathcal{T}'_o = axcbyz$ is not a swap ordering of $\mathcal{T}$, since it requires two non-consistent swaps: first a swap between characters in positions 2 and 3 of $\mathcal{T}$, then a swap between characters in position 3 and 4 of the resulting text.

Now, we are ready to define the SWCS problem.

*Problem 2.* [10] SWCS
**Input:** a set $S = \{s_1, \ldots, s_n\}$ of strings over alphabet $\Sigma$, a text $\mathcal{T}$.
**Output:** a swap ordering $\mathcal{T}_o$ of the text $\mathcal{T}$ that maximizes the number of strings in $S$ that are substrings of $\mathcal{T}_o$.

Assume that $S^* \subseteq S$ is a set of input strings covered by a solution of RCS or SWCS. Then a string $s \in S^*$ is called a *maximal string* of $S^*$ if there is no string $s' \in S^*$ such that $s \in incl(s')$.

**Kernelization Complexity**

In Section 4 we will prove some lower bounds on the *kernelization complexity* of RCS and SWCS, so we introduce here some preliminary notions.

Let $\Delta$ be a finite alphabet and denote with $\Delta^*$ the set of all finite length strings over $\Delta$. Let $\Pi \subseteq \Delta^* \times \mathbb{N}$ be a parameterized problem, and let $1 \notin \Delta$. The *derived classical problem* $\Pi^C$ associated with $\Pi$ is $\{x1^k : (x, k) \in \Pi\}$. In [5], it is introduced the following definition of a class of reductions that can be used to prove kernel lower bounds.

**Definition 1.** *[5] Consider two parameterized problems $\Pi_1$ and $\Pi_2$. Then, $\Pi_1$ is* polynomial time and parameter reducible *to $\Pi_2$, when there exists a function $f : \{0,1\}^* \times \mathbb{N} \rightarrow \{0,1\}^* \times \mathbb{N}$ computable in polynomial time and a polynomial $p : \mathbb{N} \rightarrow \mathbb{N}$ such that for each $x_1 \in \{0,1\}^*$ and $k_1 \in \mathbb{N}$, denoted $(x_2, k_2) = f(x_1, k_1)$, then $(x_1, k_1) \in \Pi_1$ holds iff $(x_2, k_2) \in \Pi_2$, and $k_2 \leq p(k_1)$. Such a function $f$ is a* Polynomial Parameter Transformation (PPT) *from $\Pi_1$ to $\Pi_2$.*

The fundamental result proved in [5] shows that a PPT can be applied to prove kernel lower bounds:

**Theorem 1.** *[5] Let $\Pi_1$ and $\Pi_2$ be two parameterized problems whose derived classical problems $\Pi_1^c$ and $\Pi_2^c$ respectively, are NP-complete. If there exists a PPT from $\Pi_1$ to $\Pi_2$, then, if $\Pi_2$ has a polynomial kernel, it follows that $\Pi_1$ has a polynomial kernel.*

## 3 Fixed-Parameter Algorithms for $\ell$-RCS and SWCS

In this section we give two fixed-parameter algorithms for $\ell$-RCS and SWCS, both based on the color-coding technique [1]. First, we recall the basic definition of perfect hash functions.

**Definition 2.** *Let $I$ be a set, a family $F$ of hash functions from $I$ to $\{1, \ldots, k\}$ is called* perfect *if for any subset $I' \subseteq I$ consisting of $k$ elements, there exists a function $f \in F$ which is injective on $I'$.*

A perfect family $F$ of hash functions from $I$ to $\{1, \ldots, k\}$, having size $O(\log |I| 2^{O(k)})$ can be constructed in time $O(2^{O(k)} |I| \log |I|)$ [1].

### 3.1 A Fixed-Parameter Algorithm for SWCS

First, we present a fixed-parameter algorithm for SWCS, where the parameter $k$ is the number of covered input strings. The algorithm is based on the color-coding technique [1] and it is inspired by the polynomial time algorithm given in [10] for a variant of the SWCS problem, where each occurrence of an input string in the solution $\mathcal{T}_o$ contributes to the solution (hence each covered input string can contribute more than once to the value of a solution).

First, we introduce some notation. Given a string $s$, two positions $i, j$, $1 \leq i \leq j \leq |\mathcal{T}|$, in the text $\mathcal{T}$, with $i = j - |s| + 1$, and two values $b_1, b_2 \in \{0, 1\}$, define $sw(s, i, j, b_1, b_2) = 1$ if there is a swap ordering $\mathcal{T}_o[i, j]$ of $\mathcal{T}[i, j]$ such that:

1. $\mathcal{T}_o[i, j]$ is an occurrence of $s$;
2. if $b_1 = 1$ ($b_1 = 0$ respectively), $\mathcal{T}_o[i, j]$ is obtained by swapping (not swapping respectively) the characters in positions $i - 1$ and $i$ of $\mathcal{T}$;
3. if $b_2 = 1$ ($b_2 = 0$ respectively), $\mathcal{T}_o[i, j]$ is obtained by swapping (not swapping respectively) characters in positions $j$ and $j + 1$ of $\mathcal{T}$.

In any other case $sw(s, i, j, b_1, b_2) = 0$. Notice that if $i = 1$ ($j = |\mathcal{T}|$ respectively), then it must hold $b_i = 0$ ($b_j = 0$ respectively).

Let $F : \{s_1, \ldots, s_n\} \to \{l_1, \ldots, l_k\}$ be a family of perfect hash functions. Fix a function $f \in F$ such that each string of $S$ covered by a solution of SWCS is assigned a unique label in $\{l_1, \ldots, l_k\}$.

Before giving the details, we present the high-level idea of the algorithm. We design a dynamic programming algorithm that, given a position $i$, considers (if it exists) the maximal substring $s_j$ that is a suffix of a swap ordering $\mathcal{T}_o[1, i]$ of $\mathcal{T}[1, i]$. Hence $\mathcal{T}_o[i - |s_j| + 1, i]$ covers $s_j$, and all the input strings that are substrings of $s_j$. Notice that a non-maximal input string may be covered in different positions of $\mathcal{T}_o$. In this case, we assume that each non-maximal input string is covered by its leftmost occurrence in $\mathcal{T}_o$. Any maximal substring $s_h \in S \setminus \{s_j\}$ covered by $\mathcal{T}_o[1, i]$ either does not overlap with $s_j$ (Case 1, Case 2 and Case 3 of the recurrence), or it overlaps with $s_j$ (Case 4 of the recurrence). In the latter case $s_j$ and $s_h$ must be identical in the overlapping positions. In the former case, we consider three possible cases (Case 1, Case 2 and Case 3 of the recurrence), since, depending on the occurrence of string $s_h$ in $\mathcal{T}_o$, we have to check that swaps are possible (see Example 2).

*Example 2.* Let $(\mathcal{T}, S)$ be an instance of SWCS, defined as follows:

$\mathcal{T} = abxcyz$ $\qquad S = \{s_1 = abx, s_2 = xyz\}$

Notice that $sw(s_1, 1, 3, 0, 0) = 1$, and $sw(s_1, i, j, b_1, b_2) = 0$ in any other case; $sw(s_2, 4, 6, 1, 0) = 1$, and $sw(s_2, i, j, b_1, b_2) = 0$ in any other case. Now, if $s_2$ is the rightmost input string that occurs in $\mathcal{T}_o$, this implies a swap of the characters $x$ and $c$ of $\mathcal{T}$. Then, $s_1$ cannot be covered by $\mathcal{T}_o$ (this condition is tested in Case 2 of the recurrence).

Let $(\mathcal{T}, S)$ be an instance of SWCS, defined as follows:

$\mathcal{T} = abxcyde$ $\qquad S = \{s_1 = abc, s_2 = cde\}$

Notice that $sw(s_1, 1, 3, 0, 1) = 1$, and $sw(s_1, i, j, b_1, b_2) = 0$ in any other case; $sw(s_2, 5, 7, 1, 0) = 1$, and $sw(s_2, i, j, b_1, b_2) = 0$ in any other case. If $s_2$ is the rightmost input string that occurs in $\mathcal{T}_o$, this implies a swap of the characters $c$ and $y$ of $\mathcal{T}$. Then, $s_1$ cannot be covered by $\mathcal{T}_o$, since it would require a swap between characters $x$ and $c$. Indeed the two swaps are not consistent, since $c$ has already been swapped with $y$ (the inconsistency of these swaps is tested in Case 3 of the recurrence).

Now, we give the formal description of the algorithm. Define $D[i, j, L, b]$, where $L \subseteq \{l_1, \ldots, l_k\}$, $1 \leq i \leq |\mathcal{T}|$, $1 \leq j \leq |S|$, and $b \in \{0, 1\}$, as follows:

- $D[i, j, L, b] = 1$ if there is a swap ordering $\mathcal{T}_o[1, i]$ of $\mathcal{T}[1, i]$ such that:
  1. $\mathcal{T}_o[1, i]$ covers a set $S^*$ of strings uniquely labeled by the set $L$
  2. $s_j$ is a maximal string in $S^*$ and it is a suffix of $\mathcal{T}_o[1, i]$
  3. if $b = 1$ (if $b = 0$ respectively) $\mathcal{T}_o$ is obtained by swapping (not swapping respectively) the characters of $\mathcal{T}$ in positions $i, i + 1$
- else $D[i, j, L, b] = 0$.

Now, we can define the recurrence to compute $D[i, j, L, b]$. We assume that each entry $D[i, j, L, b]$ is initialized to 0. $D[i, j, L, b]$ is the maximum, with $1 \leq y \leq i$, $1 \leq h \leq |S|$ and $b'\{0, 1\}$, of the following values:

**Case 1.** $D[y, h, L', b'] \wedge sw(s_j, i - |s_j| + 1, i, b_x, b)$, where $y < i - |s_j| - 1$, $1 \leq h \leq n$, $L' \subseteq L \setminus \{f(s_j)\}$, $L' \supseteq L \setminus (\{f(s_j)\} \cup \{f(s_p) : s_p \in incl(s_j)\})$, and $b_x \in \{0, 1\}$;
**Case 2.** $D[y, h, L', b'] \wedge sw(s_j, y + 1, i, b', b)$, where $y = i - |s_j|$, $1 \leq h \leq n$, $L' \subseteq L \setminus \{f(s_j)\}$, $L' \supseteq L \setminus (\{f(s_j)\} \cup \{f(s_p) : s_p \in incl(s_j)\})$;
**Case 3.** $D[y, h, L', b'] \wedge sw(s_j, y + 2, i, b_x, b)$, where $y = i - |s_j| - 1$, $1 \leq h \leq n$, $L' \subseteq L \setminus \{f(s_j)\}$, $L' \supseteq L \setminus (\{f(s_j)\} \cup \{f(s_p) : s_p \in incl(s_j)\})$, and $b' + b_x \leq 1$;
**Case 4.** $D[y, h, L', b'] \wedge sw(s_j[y - i + |s_j| + 1, |s_j|], y + 1, i, b', b)$, where $i - |s_j| < y \leq i - 1$, $1 \leq h \leq n$, the leftmost $y - i + |s_j|$ characters of $s_j$ are identical to the rightmost $y - i + |s_j|$ characters of $s_j$, $L' \subseteq L \setminus \{f(s_j)\}$, and $L' \supseteq L \setminus (\{f(s_j)\} \cup \{f(s_p) : s_p \in incl(s_j[y - i + |s_j| + 1, |s_j|])\})$.

For the basic case, it holds: $D[i, j, L', b] = 1$, for each position $i$ in the text $\mathcal{T}$, such that there is a swap ordering $\mathcal{T}_o[1, i]$ of $\mathcal{T}[1, i]$ where $s_j$ is covered by $\mathcal{T}_o[1, i]$, $L' = \{f(s_j)\} \cup \{f(s_p) : s_p \in incl(s_j)\}$, and $sw(s_j, i - |s_j| + 1, i, b_1, b_2) = 1$, for some $b_1, b_2 \in \{0, 1\}$. Now, we prove the correctness of the dynamic programming recurrence.

**Lemma 1.** $D[i, j, L, b] = 1$ *if and only if there exists a set* $S' \subseteq S$ *of strings uniquely labeled by* $L$ *and covered by a swap ordering* $\mathcal{T}_o[1, i]$ *of* $\mathcal{T}[1, i]$*, such that* $s_j$ *is a maximal substring of* $S'$ *covered by* $\mathcal{T}_o[i - |s_j| + 1, i]$*, for some* $b\{0, 1\}$.

A consequence of Lemma 1 is that there exists a solution $\mathcal{T}_o$ of SWCS over instance $(S, \mathcal{T})$, such that $\mathcal{T}_o$ covers $k$ input strings of $S$ if and only if there exists an entry $D[m, j, \{l_1, \ldots, l_k\}, 1]$, for some $j$ with $1 \leq j \leq n$, such that $D[m, j, \{l_1, \ldots, l_k\}, 1] = 1$. The time complexity of the algorithm is $O(2^{O(k)} m^2 n^2 \log n))$, where $|S| = n$ and $|\mathcal{T}| = m$. Indeed, it is easy to see that the recurrence can be computed in time $O(2^{O(k)} m^2 n^2)$. Since a perfect family $F$ of hash functions from $S$ to $\{1, \ldots, k\}$, having size $O(\log n 2^{O(k)})$ can be constructed in time $O(2^{O(k)} n \log n)$ [1], it follows that the time complexity of the algorithm is $O(2^{O(k)} m^2 n^2 \log n)$.

### 3.2 A Fixed-Parameter Algorithm for $\ell$-RCS

As discussed in the Introduction, the RCS problem is W[1]-hard when parameterized by the size $k$ of the solution [6]. Since $2-$ RCS is NP-hard [6], a natural question is whether $\ell$-RCS parameterized by $k$ and by $\ell$ is fixed-parameter tractable. Here we present a fixed-parameter algorithm for the $\ell$-RCS problem, when both $k$ and $\ell$ are parameters. As in Section 3.1, the algorithm is based on the color-coding technique. However, in this case we combine two families of perfect hash functions to design the algorithm.

Consider a solution $\mathcal{M}_o$ of $\ell$-RCS over instance $(\mathcal{M}, S)$, such that there exists a set $S_k \subseteq S$ of $k$ input strings that are covered by $\mathcal{M}_o$. For each string $s \in S_k$, define the positions $i_{s,l}, i_{s,r}$, $1 \leq i_{s,l} \leq i_{s,r} \leq |\mathcal{M}_o|$, such that $\mathcal{M}_o[i_{s,l}, i_{s,l}]$ is the leftmost occurrence of $s$ in $\mathcal{M}_o$ (notice that there may exist many occurrences of $s$ in $\mathcal{M}_o$). Define the set $AP$ of *applied positions* of $\mathcal{M}_o$ as follows:

$AP = \{j : 1 \leq j \leq |\mathcal{M}_o|, 1 \leq i_{s,l} \leq j \leq i_{s,r} \leq |\mathcal{M}_o|$, for some string $s_i \in S_k\}$.

Then, the following property follows easily.

**Proposition 1.** *Let $S_k$ be a set of $k$ input strings covered by a solution $\mathcal{M}_o$ of $\ell$-RCS. Then, $|AP| \leq \ell \cdot k$.*

The set $AP$ of applied positions of $\mathcal{M}_o$ corresponds to a multiset $AC$ of characters in $\mathcal{M}$ (denoted as the multiset of *applied characters* of $\mathcal{M}$) such that an ordering of $AC$ is sufficient to cover the input strings in $S_k$. A consequence of Prop.1 is that $|AC| \leq \ell k$.

The fixed-parameter algorithm for $\ell$-RCS is obtained by combining two different families of perfect hash functions. As in Section 3.1, the first family of perfect hash functions, denoted by $F_s$, maps the input strings to the set of labels $\{l_1, \dots, l_k\}$. The second family of perfect hash functions, denoted by $F_t$, maps the characters in $\mathcal{M}$ to the set of labels $\{l'_1, \dots, l'_z\}$, for some $1 \leq z \leq \ell k$. Informally, the hash functions in $F_t$ are used to associate a distinct label in $\{l'_1, \dots, l'_z\}$ with each character of the multiset $AC$ of applied characters of $\mathcal{M}$.

The high-level idea of the algorithm is to use dynamic programming to define an ordering $\mathcal{M}_o$ of $\mathcal{M}$ that covers a set $S_k \subseteq S$ of $k$ input strings. Let $f_s \in F_s$ be a function that associates a distinct label in $\{l_1, \dots, l_k\}$ with each covered input string, and let $f_t \in F_t$ be a function that associates a distinct label in $\{l'_1, \dots, l'_z\}$ with each applied character of $\mathcal{M}_o$. The dynamic programming recurrence considers the rightmost string $s_j$, with $1 \leq j \leq n$, covered by $\mathcal{M}_o[1, i]$, with $1 \leq i \leq m$. Since $s_j$ is covered by $\mathcal{M}_o[1, i]$, it follows that: (1) $\mathcal{M}_o[1, i]$ covers $s_j$ (and eventually some input strings which are substrings of $s_j$), and the corresponding set of labels associated by $f_s$; (2) some of the applied characters of $\mathcal{M}$ are sorted in $\mathcal{M}_o[1, i]$ to cover $s_j$, hence the dynamic programming recurrence stores the labels in $\{l'_1, \dots, l'_z\}$ corresponding to those applied characters. Moreover, the dynamic programming recurrence distinguishes two cases: the case that any maximal covered string in $S \setminus \{s_j\}$ does not share any character with $s_j$ (Case 1 of the recurrence), and the case that the rightmost covered string $s_h$

in $S \setminus \{s_j\}$ overlaps with $s_j$ (Case 2 of the recurrence). In the latter case we have to guarantee that $s_j$ and $s_h$ agree on the overlapping part.

Now, we give the formal description of the algorithm. First, let us define some preliminary definitions (see Example 3). Fix a function $f_s \in F_s$, such that each input string covered by $\mathcal{M}_o$ is assigned a unique label in $\{l_1, \ldots, l_k\}$. Furthermore, fix a function $f_t \in F_t$ such that each applied character in $AC$ is assigned a unique label in $\{l'_1, \ldots, l'_z\}$. For a character $\sigma \in \Sigma$, define the set $\Lambda(\sigma)$ of labels as follows:
$\Lambda(\sigma) = \{l'_q \in \{l'_1, \ldots, l'_z\} : \exists$ an occurrence of $\sigma$ in $\mathcal{M}$ associated by $f_t$ with label $l'_q\}$.

Let $s_j$ be an input string and $L'_t \subseteq \{l'_1, \ldots, l'_z\}$. Define $Feas(L'_t, s_j[x,y])$, with $1 \leq x \leq y \leq |s_j|$ and $y - x \leq |L'_t| - 1$, as the collection of subsets $L''_t \subseteq L'_t$, such that there is a bijection $B$ between the characters of $s_j[x,y]$ and the labels in $L''_t$, where $B(s_j[z]) \in \Lambda(s_j[z])$, for each $z$ with $x \leq z \leq y$. Informally, $Feas(L'_t, s_j[x,y])$ contains those subsets of $L'_t$ (of size $y-x+1$) that can be used to uniquely label the characters in $s_j[x,y]$. Next we prove that $Feas(L'_t, s_j[x,y])$ is bounded by $2^{k\ell}$ and that it can be computed in time $O(2^{kl} poly(\ell))$.

**Proposition 2.** *Given a string $s_j$, two positions $x, y$ of $s_j$, with $1 \leq x \leq y \leq |s_j|$, and a set $L'_t \subseteq \{l'_1, \ldots l'_t\}$, $Feas(L'_t, s_j[x,y])$, contains at most $2^{k\ell}$ sets, and it can be computed in time $O(2^{kl} poly(\ell))$.*

Given $L'_s \subseteq \{l_1, \ldots, l_k\}$, two sets $L'_t, L''_t \subseteq \{l'_1, \ldots, l'_z\}$, a string $s_j \in S$ and two positions $x$, $y$ in $s_j$, with $1 \leq x \leq y \leq |s_j|$, define $sm(s[x,y], L''_t, L'_t)$ as follows: $sm(s_j[x,y], L''_t, L'_t) = 1$ if $L''_t$ belongs to $Feas(L'_t, s_j[x,y])$, else $sm(s_j[x,y], L''_t, L'_t) = 0$.

*Example 3.* Let $(\mathcal{M}, S)$ be an instance of $\ell$-RCS (with $\ell = 3$), defined as follows:

$$\mathcal{M} = \{a, a, b, b, c, c, d\} \qquad S = \{s_1 = abc, s_2 = aba, s_3 = dcb\}$$

Let $\mathcal{M}_o = ababccd$ be a solution of $\ell$-RCS that covers the set of string $S_k = \{s_1, s_2\}$. Notice that $s_1$, $s_2$ are covered by $\mathcal{M}_o[1,5]$. It follows that the set $AP$ of applied positions is $AP = \{1, 2, \ldots, 5\}$ and the multiset of applied characters is $AC = \{a, a, b, b, c\}$.

Assume that, given an appropriate function $f_t \in F_t$, $\Lambda(\sigma)$, with $\sigma \in \{a, b, c, d\}$, is defined as follows:

$\Lambda(a) = \{l'_1, l'_2\}$ $\qquad$ $\Lambda(c) = \{l'_2, l'_5\}$
$\Lambda(b) = \{l'_3, l'_4\}$ $\qquad$ $\Lambda(d) = \{l'_3\}$.

Given $s_1[2,3]$ ($s_2[1,3]$ respectively) and the set of labels $\{l'_1, l'_2, l'_3, l'_4, l'_5\}$ ($\{l'_1, l'_2, l'_3, l'_4\}$ respectively), $Feas$ contains the following sets:

$s_1[2,3] = bc \quad Feas(\{l'_1, l'_2, l'_3, l'_4, l'_5\}, s_1[2,3]) = \{\{l'_2, l'_3\}\}, \{l'_2, l'_4\}, \{l'_3, l'_5\}, \{l'_4, l'_5\}\}$
$s_2[1,3] = aba \quad Feas(\{l'_1, l'_2, l'_3, l'_4\}, s_2[1,3]) = \{\{l'_1, l'_2, l'_3\}\}, \{l'_1, l'_2, l'_4\}\}$.

Now, we present the dynamic programming recurrence. Define $D[i, j, L_s, L_t]$, with $1 \leq i \leq l'_t$, $1 \leq j \leq |S|$, $L_s \subseteq \{l_1, \ldots, l_k\}$, $L_t \subseteq \{l'_1, \ldots, l'_z\}$, and $1 \leq i \leq |\mathcal{M}|$, as follows:

- $D[i, j, L_s, L_t] = 1$ if there exists an ordering $\mathcal{M}_o[1, i]$ of a set of $i$ characters of $\mathcal{M}$ uniquely labeled by $L_t$ such that:
  1. $\mathcal{M}_o[1, i]$ covers a set $S^*$ of input strings uniquely labeled by $L_s$,
  2. $s_j$ is a maximal substring of $S^*$ covered by $\mathcal{M}_o[i - |s_j| + 1, i]$;
- else $D[i, j, L_s, L_t] = 0$.

We assume that each entry $D[i, j, L_s, L_t]$ is initialized to 0. We can define $D[i, j, L_s, L_t]$ as the maximum, for each $1 \leq y \leq i$, $1 \leq h \leq |S|$, $L'_t \subseteq L_t$, of the following values:

**Case 1.** $D[y, h, L'_s, L'_t] \wedge sm(s_j[1, |s_j|], L''_t, L_t)$, where $L'_s \subseteq L_s \setminus \{f(s_j)\}$, $L'_s \supseteq L_s \setminus (\{f(s_j)\} \cup \{f(s_p) : s_p \in incl(s_j)\})$, $L'_t = L_t \setminus L''_t$, for some $L''_t$ in $Feas(L_t, s_j[1, |s_j|])$;

**Case 2.** $D[y, h, L'_s, L'_t] \wedge sm(s_j[y - i + |s_j| + 1, |s_j|], L''_t, L_t)$, where the leftmost $y - i + |s_j|$ characters of $s_j$ are identical to the the rightmost $y - i + |s_j|$ characters of $s_h$, $L'_s \subseteq L_s \setminus \{f(s_j)\}$, $L'_s \supseteq L_s \setminus (\{f(s_j)\} \cup \{f(s_p) : s_p \in incl(s_j[y - i + |s_j| + 1, |s_j|])\})$, $L'_t = L_t \setminus L''_t$, for some $L''_t$ in $Feas(L_t, s_j[y - i + |s_j| + 1, |s_j|])$.

For the basic case, given a string $s_j \in S$, it holds $D[i, j, L'_s, L'_t] = 1$, for each $L'_t$ such that $sm(s_j[1, |s_j|], L'_t, L'_t) = 1$, where $i = |s_j|$, and $L'_s = \{f_s(s_j)\} \cup \{f_s(s_p) : s_p \in incl(s_j)\}$. The correctness of the recurrence follows from Lemma 2.

**Lemma 2.** *$D[i, j, L_s, L_t] = 1$ iff there exists an ordering $\mathcal{M}'_o[1, i]$ of a subset of $i$ characters of $\mathcal{M}$ uniquely labeled by $L_t$ such that (1) $\mathcal{M}'_o$ covers a set $S^*$ of input strings uniquely labeled by $L_s$, and (2) $s_j$ is a maximal substring of $S^*$ covered by $\mathcal{M}'_o[i - |s_j| + 1, i]$.*

There exists a solution $\mathcal{M}_o$ of $\ell$-RCS over instance $(S, \mathcal{M})$ that covers $k$ input strings if there exists an entry $D[|\mathcal{M}|, j, \{1, \ldots, k\}, \{1, \ldots, l'_t\}]$, for some $j$ with $1 \leq j \leq n$, and for some $l'_t$ with $1 \leq l'_t \leq lk$, such that $D[|\mathcal{M}|, j, \{1, \ldots, k\}, \{1, \ldots, l'_t\}] = 1$. The time complexity of the algorithm is $O(2^{O(kl)} \text{poly}(np))$, where $p = |\mathcal{M}|$ and $|S| = n$. Indeed, it can be easily proved that the recurrence can be computed in time $O(2^{O(kl)} k \text{poly}(np))$. Since a perfect family $F_s$ of hash functions from $S$ to $\{l_1, \ldots, l_k\}$, having size $O(\log n 2^{O(k)})$ can be constructed in time $O(2^{O(k)} n \log n)$ [1], and since a perfect family $F_t$ of hash functions from $\mathcal{M}$ to $\{l'_1, \ldots, l'_z\}$, with $1 \leq z \leq lk$, having size $O(\log p 2^{O(lk)})$ can be constructed in time $O(2^{O(lk)} p \log p)$ [1], it follows that the time complexity of the algorithm is $O(2^{O(kl)} \text{poly}(np))$.

## 4 Kernelization Complexity

In Section 3 we have given two fixed-parameter algorithms for SWCS and $\ell$-RCS. We complement these results by showing that the two problems are unlikely to admit a polynomial size kernel, by giving two Polynomial Parameter Transformations (PPTs) (see Section 2) from the Longest Path Problem (LONGEST-PATH), which has been proved to not have a polynomial kernel, unless $NP \subseteq coNP/Poly$ [3].

We recall that, given a graph $G = (V, E)$, the LONGEST-PATH problem asks if there exists a simple path, that is a path with no repeated vertices, in $G$ of length at least $k$.

### 4.1 Kernelization Complexity of SWCS

In this section we give a PPT from LONGEST-PATH to SWCS. Let $G = (V, E)$ be an instance of LONGEST-PATH with $V = \{v_1, \ldots, v_q\}$, we define the corresponding instance $(\mathcal{T}, S)$ of SWCS. Define $\Sigma = \{w_{i,1}, w_{i,2} : v_i \in V, 1 \leq i \leq q\} \cup \{w_{0,1}, w_{0,2}, w_{q+1,1}, w_{q+1,2}\} \cup \{a\}$.

Define the string $s^+$ over alphabet $\Sigma$ as follows:

$$s^+ = w_{0,1}w_{0,2}w_{1,1}w_{1,2}\ldots w_{q,1}w_{q,2}w_{q+1,1}w_{q+1,2}a$$

The text $\mathcal{T}$ is a string obtained by concatenating $k$ copies of the string $s^+$, that is $\mathcal{T} = \overbrace{s^+ \cdot s^+ \cdot \ldots \cdot s^+}^{k \text{ times}}$.

Each copy of a string $s^+$ in $\mathcal{T}$ is called a *block* of $\mathcal{T}$. Now, define the set of input strings: $S = \{s_{i,j}, s_{j,i} : \{v_i, v_j\} \in E\} \cup \{s_i : v_i \in V\}$. Given an edge $\{v_i, v_j\} \in E$, the input strings $s_{i,j}$, $s_{j,i}$ are defined as follows:

$$s_{i,j} = w_{i,2}w_{i,1}w_{i+1,1}w_{i+1,2}\cdots w_{q+1,1}w_{q+1,2}aw_{0,1}w_{0,2}\cdots w_{j,2}w_{j,1}$$

$$s_{j,i} = w_{j,2}w_{j,1}w_{j+1,1}w_{j+1,2}\cdots w_{q+1,1}w_{q+2,2}aw_{0,1}w_{0,2}\cdots w_{i,2}w_{i,1}$$

Furthermore, given a vertex $v_i \in V$, define the input string $s_i$ as follows $s_i = w_{0,1}w_{0,2}\ldots\ w_{i,2}w_{i,1}\ldots\ w_{q+1,1}w_{q+1,2}$. The PPT is based on the following result.

**Lemma 3.** *Let $G = (V, E)$ be an instance of* LONGEST-PATH *and let $(S, \mathcal{T})$ be the corresponding instance of* SWCS*. Then: (1) starting from a simple path of length $k$ in $G$, we can compute in polynomial time a solution of* SWCS *over instance $(S, \mathcal{T})$ that covers $2k-1$ input strings; (2) starting from a solution of* SWCS *over instance $(S, \mathcal{T})$ that covers $2k-1$ input strings, we can compute in polynomial time a simple path of length $k$ in $G$.*

*Proof.* Given a path $P$ of length $k$ in $G$, it is easy to compute in polynomial time a solution of SWCS over instance $(S, \mathcal{T})$ that covers $2k-1$ input strings.

Now, consider a solution $\mathcal{T}_o$ of SWCS that covers $2k-1$ strings. Notice that, due to occurrences of character $a$ in $\mathcal{T}$, $\mathcal{T}_o$ must cover exactly $k$ strings $s_i$, where each $s_i$ is associated with a vertex $v_i \in V$. Furthermore, by construction a string $s_{i,j}$ associated with a vertex $\{v_i, v_j\} \in E$ can be covered in two adjacent blocks of $\mathcal{T}_o$. Furthermore, since each block of $\mathcal{T}_o$ covers exactly one string $s_i$, it follows that if blocks $h-1$ and $h$ of $\mathcal{T}_o$ cover a string $s_{i,j}$, with $\{v_i, v_j\} \in E$, and blocks $h$, $h+1$ of $\mathcal{T}_o$ cover a string $s_{x,y}$, with $\{v_x, v_y\} \in E$, then $|\{i, j\} \cap \{x, y\}| = 1$, that is $\{v_i, v_j\}$ and $\{v_x, v_y\}$ are two edges incident on a common vertex. But, then the vertices $v_i \in V$ associated with the covered strings $s_i \in S$ induces a simple path of length $k$. □

As a consequence of Lemma 3 and Theorem 1, we have the following result.

**Theorem 2.** *The* SWCS *problem does not admit a polynomial kernel, unless* $NP \subseteq coNP/Poly$.

### 4.2 Kernelization Complexity of $\ell$-RCS

In this section we give a PPT from LONGEST-PATH to $\ell$-RCS, where $\ell \leq 5$. Let $G = (V, E)$ be an instance of LONGEST-PATH with $V = \{v_1, \ldots, v_q\}$, we define the corresponding instance $(\mathcal{M}, S)$ of $\ell$-RCS.

Define the alphabet $\Sigma$ as follows: $\Sigma = \{w_i : v_i \in V\} \cup \{z\}$. The multiset $\mathcal{M}$ on $\Sigma$ contains one occurrence of $w_i$, for each $v_i \in V$, and $k+1$ occurrences of character $z$.

Now, define the set $\mathcal{S}$ of input strings: $\mathcal{S} = \{s_{i,j}, s_{j,i} : \{v_i, v_j\} \in E\}$, where $s_{i,j} = zw_izw_jz$ and $s_{j,i} = zw_jzw_iz$. Notice that each input string in $\mathcal{S}$ has length bounded by 5. The PPT is based on the following result.

**Lemma 4.** *Let* $G = (V, E)$ *be an instance of* LONGEST-PATH *and let* $(S, \mathcal{M})$ *be the corresponding instance of* $\ell$-RCS. *Then: (1) starting from a simple path of length* $k$ *in* $G$*, we can compute in polynomial time a solution of* $\ell$-RCS *over instance* $(S, \mathcal{M})$ *that covers* $k-1$ *input strings; (2) starting from a solution of* $\ell$-RCS *over instance* $(S, \mathcal{M})$ *that covers* $k-1$ *input strings, we can compute in polynomial time a simple path of length* $k$ *in* $G$.

As a consequence of Lemma 4 and Theorem 1, we have the following result.

**Theorem 3.** *The* $\ell$-RCS *problem does not admit a polynomial kernel, even if* $\ell \leq 5$*, unless* $NP \subseteq coNP/Poly$.

## 5 Conclusion and Open Problems

In this paper we have investigated the parameterized complexity of RCS and SWCS under some natural parameterizations. We have shown that $\ell$-RCS and SWCS, when parameterized by the size of the solution, are in FPT. We have complemented these two results, by showing that $\ell$-RCS and SWCS are unlikely to admit a polynomial kernel.

There are some interesting open problems in the perspective of a multivariate analysis of $\ell$-RCS and SWCS. It would be interesting to investigate the computational complexity of SWCS when the input strings have bounded length or are over a restricted alphabet. We are presently studying these restrictions. Furthermore, it would be interesting to investigate the parameterized complexity of RCS, when parameterized by the size of the solution and by the size of the alphabet.

## References

1. Alon, N., Yuster, R., Zwick, U.: Color-coding. J. ACM 42(4), 844–856 (1995)
2. Blum, A., Jiang, T., Li, M., Tromp, J., Yannakakis, M.: Linear Approximation of Shortest Superstrings. J. ACM 41(4), 630–647 (1994)

3. Bodlaender, H.L., Downey, R.G., Fellows, M.R., Hermelin, D.: On Problems Without Polynomial Kernels. J. Comput. Syst. Sci. 75(8), 423–434 (2009)
4. Bodlaender, H.L., Jansen, B.M.P., Kratsch, S.: Cross-Composition: A New Technique for Kernelization Lower Bounds. In: Proceedings of STACS 2011, pp. 165–176 (2011)
5. Bodlaender, H.L., Thomassé, S., Yeo, A.: Kernel Bounds for Disjoint Cycles and Disjoint Paths. Theor. Comput. Sci. 412(35), 4570–4578 (2011)
6. Clifford, R., Gotthilf, Z., Lewenstein, M., Popa, A.: Restricted Common Superstring and Restricted Common Supersequence. In: Giancarlo, R., Manzini, G. (eds.) CPM 2011. LNCS, vol. 6661, pp. 467–478. Springer, Heidelberg (2011)
7. Fellows, M.R.: Towards Fully Multivariate Algorithmics: Some New Results and Directions in Parameter Ecology. In: Fiala, J., Kratochvíl, J., Miller, M. (eds.) IWOCA 2009. LNCS, vol. 5874, pp. 2–10. Springer, Heidelberg (2009)
8. Fortnow, L., Santhanam, R.: Infeasibility of Instance Compression and Succinct PCPs for NP. J. Comput. Syst. Sci. 77(1), 91–106 (2011)
9. Flum, J., Grohe, M.: Parameterized Complexity Theory. Springer, Heidelberg (2006)
10. Gotthilf, Z., Lewenstein, M., Popa, A.: On Shortest Common Superstring and Swap Permutations. In: Chavez, E., Lonardi, S. (eds.) SPIRE 2010. LNCS, vol. 6393, pp. 270–278. Springer, Heidelberg (2010)
11. Gusfield, D.: Algorithms on Strings, Trees, and Sequences - Computer Science and Computational Biology. Cambridge University Press, New York (1997)
12. Niedermeier, R.: Invitation to Fixed-Parameter Algorithms. Oxford University Press, Oxford (2006)
13. Niedermeier, R.: Reflections on Multivariate algorithmics and Problem Parameterization. In: Proceedings of STACS 2010, pp. 17–32 (2010)
14. Ott, S.: Lower Bounds for Approximating Shortest Superstrings over an Alphabet of Size 2. In: Widmayer, P., Neyer, G., Eidenbenz, S. (eds.) WG 1999. LNCS, vol. 1665, pp. 55–64. Springer, Heidelberg (1999)
15. Storer, J.: Data Compression: Methods and Theory. Computer Science Press, New York (1988)
16. Sweedyk, Z.: A $2\frac{1}{2}$-Approximation Algorithm for Shortest Superstring. SIAM J. Comput. 29(3), 954–986 (1999)
17. Vassilevska, V.: Explicit Inapproximability Bounds for the Shortest Superstring Problem. In: Jedrzejowicz, J., Szepietowski, A. (eds.) MFCS 2005. LNCS, vol. 3618, pp. 793–800. Springer, Heidelberg (2005)

# Nonblocker in $H$-Minor Free Graphs: Kernelization Meets Discharging*

Łukasz Kowalik

Institute of Informatics, University of Warsaw, Poland
kowalik@mimuw.edu.pl

**Abstract.** Perhaps the best known kernelization result is the kernel of size $335k$ for the PLANAR DOMINATING SET problem by Alber et al. [1], later improved to $67k$ by Chen et al. [5]. This result means roughly, that the problem of finding the smallest dominating set in a planar graph is easy when the optimal solution is small. On the other hand, it is known that PLANAR DOMINATING SET parameterized by $k' = |V| - k$ (also known as PLANAR NONBLOCKER) has a kernel of size $2k'$. This means that PLANAR DOMINATING SET is easy when the optimal solution is very large. We improve the kernel for PLANAR NONBLOCKER to $\frac{7}{4}k'$. This also implies that PLANAR DOMINATING SET has no kernel of size at most $(\frac{7}{3} - \epsilon)k$, for any $\epsilon > 0$, unless $\mathbf{P} = \mathbf{NP}$. This improves the previous lower bound of $(2 - \epsilon)k$ of [5]. Both of these results immediately generalize to $H$-minor free graphs (without changing the constants).

In our proof of the bound on the kernel size we use a variant of the *discharging method* (used e.g. in the proof of the four color theorem). We give some arguments that this method is natural in the context of kernelization and we hope it will be applied to get improved kernel size bounds for other problems as well.

As a by-product we show a result which might be of independent interest: every $n$-vertex graph with no isolated vertices and such that every pair of degree 1 vertices is at distance at least 5 and every pair of degree 2 vertices is at distance at least 2 has a dominating set of size at most $\frac{3}{7}n$.

## 1 Introduction

For many NP-complete problems there are kernelization algorithms, i.e. efficient algorithms which replace the input instance with an equivalent, but often much smaller one. More precisely, a *kernelization algorithm* takes an instance $I$ of size $n$ and a parameter $k \in \mathbb{N}$, and after time polynomial in $n$ it outputs an instance $I'$ (called a *kernel*) with a parameter $k'$ such that $I$ is a yes-instance iff $I'$ is a yes instance, $k' \leq k$, and $|I'| \leq f(k)$ for some function $f$ depending only on $k$. The most desired case is when the function $f$ is polynomial, or even linear (then we say that the problem admits a polynomial or linear kernel). In such case, when the parameter $k$ is relatively small, the input instance, possibly very large, is "reduced" to a small one (preferably of size polynomial, or even linear in $k$).

* Work supported by the National Science Centre (grant N206 567140).

D.M. Thilikos and G.J. Woeginger (Eds.): IPEC 2012, LNCS 7535, pp. 61–72, 2012.

**Kernelization and Discharging.** A typical kernelization algorithm processes an instance of an NP-complete graph problem in the following way: roughly, as long as possible it finds a *reducible configuration* in the graph, i.e. a structure which can be replaced by a smaller structure so that the original graph is a yes-instance iff so is the new graph. Then it is shown that the kernel, i.e. a graph which contains no reducible configuration is small.

Many results in graph theory, including the four colour theorem as the best known example, are proven in the following way. Assume we are to show that graphs in some family (e.g. planar graphs) have some property (e.g. are 4-colorable). Then we specify a set of *reducible configurations*, i.e. structures which can be replaced by smaller structures so that the original graph has the desired property iff the new graph also has the property. Now, if a graph in our family contains such a configuration, we can proceed by induction. Otherwise, i.e. if a graph contains no reducible configuration we derive a contradiction. In the known proofs of the four color theorem [2,15] (and many other results, e.g. [4,7]) this second part is realized by so-called *discharging method*.

Since the two situations described above are so similar it is natural to ask whether the discharging method can be used to bound the size of a kernel. In this paper we present a result of that kind. Discharging used in the cited works for planar graphs is based on Euler's formula. Here we do not use the Euler's formula but the common theme is the same: using discharging we show that the graph under consideration cannot be "hard everywhere", i.e. even if it has some parts which are hard to dominate, then it has some parts which are easy, and on the average we get the desired bound. A similar "amortized analysis" has been recently used in the context of kernelization by Kanj and Zhang [10].

**Small Kernels for Planar Graph Problems.** Perhaps the best known kernelization result is the kernel of size $335k$ for the PLANAR DOMINATING SET problem by Alber et al. [1]. This result opened a new research direction, which culminated in general results which show linear/polynomial kernels for large classes of problems in various graph families that contain planar graphs, e.g. bounded genus graphs or even $H$-minor free graphs [8,3]. There are several motivations for restricting the input to planar or $H$-minor free graphs. First, for many problems (including DOMINATING SET) in general graphs no polynomial kernels exist (under appropriate complexity assumptions). Second, even if for some problem there is a polynomial kernel for general graphs, when executed on a planar graph it usually outputs a non-planar kernel, and then we do not want to use it because when we want to *solve* the kernel, we often prefer to use specialized (and faster) algorithms for planar graphs. Finally, it is often the case that for the special case of planar graphs there is a specialized kernelization algorithm which outputs a smaller kernel than that for the general setting. Indeed, as it was shown by Fomin et al. [8] many natural graph problems have a linear kernel for planar graphs. Knowing this, further research is done to reduce the leading constant in the linear function describing the kernel size. For example, the kernel of Alber et al. was later improved to $67k$ by Chen et al. [5]; the first linear kernel for PLANAR CONNECTED VERTEX COVER was that of size

$14k$ due to Guo and Niedermeier [9] and it was then reduced to $4k$ by Wang et al. [16] and even to $\frac{11}{3}k$ by Kowalik et al. [12]. Observe that these constants may be crucial: since we deal with NP-complete problems, in order to find an exact solution in the reduced instance, most likely we need exponential time (or at least superpolynomial, because for planar graphs $2^{O(\sqrt{k})}$-time algorithms are often possible), and these constants appear in the exponents.

**Our Results.** In this paper we study kernelization of the following problem restricted to planar graphs, or more generally to $H$-minor free graphs:

| NONBLOCKER | **Parameter:** $k$ |
|---|---|
| **Input:** an $n$-vertex graph $G = (V, E)$ and an integer $k \in \mathbb{N}$ | |
| **Question:** Is there a dominating set of size $n - k$? | |

This problem can be also defined as DOMINATING SET parameterized by $n-k$, in other words NONBLOCKER is the parametric dual of DOMINATING SET (see [5] for the definition of the parametric dual). NONBLOCKER has a trivial $2k$-kernel for general graphs (and also for any reasonable graph class) since every $n$-vertex graph with no isolated vertices has a dominating set of size at most $n/2$ (consider a spanning forest of $G$, 2-color it and choose the larger color class). This was improved to a $(\frac{5}{3}k + 3)$-kernel by Dehne et al. [6]. Their kernelization algorithm applies the so-called catalytic rule, which identifies the neighbors of two degree 1 vertices, then removes one of the degree 1 vertices and decreases $k$ by 1 (when there is only one degree 1 vertex left, they use a classic result of McCuaig and Shepard [13] which states that any $n$-vertex graph of minimum degree 2 has a dominating set of size at most $\frac{2}{5}n$, for $n$ large enough). As we see the catalytic rule preserves neither planarity nor excluded minors. It follows that the best kernel for PLANAR NONBLOCKER to date is still the trivial $2k$. A natural question arises: can this bound be improved? In this work we answer this question affirmatively: we present a $\frac{7}{4}k$-kernel for PLANAR NONBLOCKER. Since in our reduction rules we only remove edges/vertices or contract edges our result immediately generalizes to $H$-minor-free graphs (with the same constant in the kernel size, which is a rather rare phenomenon in the field).

An important motivation for studying parametric duals, discovered by Chen et al. [5], is that if the dual problem admits a kernel of size at most $\alpha k$, then the original problem has no kernel of size at most $(\alpha/(\alpha-1) - \epsilon)k$, for any $\epsilon > 0$, unless P=NP. Hence, our kernel implies that PLANAR DOMINATING SET has no kernel of size at most $(\frac{7}{3}-\epsilon)k$ for any $\epsilon > 0$ (and the same holds for DOMINATING SET restricted to any graph family closed under taking minors). This is the first improvement over the $(2 - \epsilon)k$ lower bound of Chen et al. [5].

We note here that although using the approach of Dehne et al. [6] one can get a $(\frac{5}{3}k+3)$-kernel for the *annotated version* of PLANAR NONBLOCKER (where the instance is extended by a subset of vertices that do not need to be dominated), it is unclear how to use this result to get an improved lower bound for the kernel size of PLANAR DOMINATING SET.

To bound the size of a kernel means just to give a lower or upper bound for the value of some graph invariant (e.g. the domination number) in a restricted class of graphs. Sometimes it is enough to apply a known combinatorial result, like the lower bound for the domination number of McCuaig and Shepard [13] for graphs of minimum degree 2, used by Dehne et al. [6]. There is also a better bound of $\frac{3}{8}n$ for graphs of minimum degree 3 due to Reed [14] (later improved to $\frac{4}{11}n$ by Kostochka and Stodolsky [11]). However in our kernel there still can be an unbounded number of vertices of degree 1 and 2, though there are some restrictions on them, so a tailor-made bound has to be shown. Applying the approach of Reed we show that every $n$-vertex graph with no isolated vertices and such that every pair of degree 1 vertices is at distance at least 5 and every pair of degree 2 vertices is at distance at least 2 has a dominating set of size at most $\frac{3}{7}n$. We suppose that this result may be of independent interest.

**Terminology and Notation.** By $N_G(v)$ we denote the set of neighbors of vertex $v$, and for a subset of vertices $X \subseteq V(G)$, we denote $N_G(X) = \bigcup_{x \in X} N_G(x) \setminus X$. The subscripts are omitted when it is clear which graph we refer to. $G[S]$ denotes the subgraph of graph $G$ induced by a set of vertices $S$. By a $d$-vertex we mean a vertex of degree $d$. A $d$-neighbor is a neighbor of degree $d$. We also use the Iverson bracket: $[\alpha]$ equals 1 if the condition $\alpha$ holds and 0 otherwise.

## 2 The Kernelization Algorithm

We say that a reduction rule for parameterized graph problem $P$ is *correct* when for every instance $(G, k)$ of $P$ it returns an instance $(G', k')$ such that:

a) $(G', k')$ is an instance of $P$,
b) $(G, k)$ is a yes-instance of $P$ iff $(G', k')$ is a yes-instance of $P$, and
c) $k' \leq k$.

We present six simple reduction rules below. It will be easier for us to formulate and analyze the rules for DOMINATING SET. We will then convert them to rules for NONBLOCKER.

Rule R1. If there is an isolated vertex $v$, then remove $v$ and decrease $k$ by 1.

Rule R2. If there is an isolated edge $vw$, then remove both $v$ and $w$ and decrease $k$ by 1.

Rule R3. If a vertex $v$ has more than one 1-neighbors, then remove all these neighbors except for one.

Rule R4. Assume there is a path $P = abcd$ with $\deg(b) = \deg(c) = 2$. If $a \neq d$, then contract $P$ into a single vertex $v$ and decrease $k$ by one. If $a = d$, then contract the edge $bc$.

Rule R5. If there is a path $abcd$ with $\deg(a) = \deg(d) = 1$, then contract edge $bc$ and decrease $k$ by one.

Rule R6. If there is a path $abcde$ with $\deg(a) = \deg(e) = 1$, then remove edge $bc$.

Now, every Rule R$i$ is converted to Rule R$i'$ as follows. Let $(G, \ell)$ be an instance of NONBLOCKER. Put $k = |V(G)| - \ell$, apply R$i$ to $(G, k)$ and get $(G', k')$. Put $\ell' = |V(G')| - k'$ and return $(G', \ell')$.

**Lemma 1.** *Rules R1′-R6′ are correct for* NONBLOCKER *restricted to any minor-closed graph class.*

Due to the space limitations we skip the proof of the above lemma. We note here that by the Graph Minor Theorem any minor-closed graph class can be characterized by a finite set of forbidden minors, so in particular our rules are correct for $H$-minor-free graphs.

**Observation 1.** *If none of the reduction rules applies to an $n$-vertex graph $G$ then $G$ has no isolated vertices, every pair of 1-vertices is at distance at least 5 and every pair of 2-vertices is at distance at least 2.*

The next section is devoted to the proof of the following theorem, which is the main technical contribution of this work.

**Theorem 1.** *Every graph with no isolated vertices and such that every pair of 1-vertices is at distance at least 5 and every pair of 2-vertices is at distance at least 2 has a dominating set of size at most $\frac{3}{7}n$ and it can be found in polynomial time.*

Let $(G, k)$ be the input instance of NONBLOCKER. Our kernelization algorithm applies rules R1-R6 as long as possible. It is clear that it can be checked in polynomial time whether a particular rule applies, and each rule is applied in linear time. Since in every rule $|V(G)| + |E(G)|$ decreases, it follows that the whole algorithm works in polynomial time (it can be even implemented in $O(n \log n)$ time but we skip the details). Let $(G', k')$ be the resulting instance. Since all the rules are correct from c) it follows that $k' \leq k$. If $k' \leq \frac{4}{7}|V(G')|$ then by Observation 1 and Theorem 1 we know that $G'$ has a dominating set of size at most $\frac{3}{7}|V(G')| \leq |V(G')| - k'$ and the algorithm returns the answer YES. Otherwise $|V(G')| \leq \frac{7}{4}k' \leq \frac{7}{4}k$ so we get a $\frac{7}{4}k$-kernel.

## 3 Proof of Theorem 1, Basic Setup

In our proof we extend the approach of Reed's seminal paper [14]. Let us introduce some basic notation (mostly coming from [14]).

Whenever it does not lead to ambiguity, if $P$ is a path then $P$ refers also to the set of vertices of $P$. The *order* of a path $P$, denoted by $|P|$ is the number of its vertices (as opposed to the *length* of $P$ which is the number of edges, i.e. $|P| - 1$). For $i \in \{0, 1, 2\}$, a path $P$ is an *$i$-path*, if $|P| \equiv i \pmod 3$ (note that we modify the standard definition here but we prefer to be consistent with [14]). A *dangling path* in a graph $G$ is a path of order two with exactly one endpoint of degree 1 in $G$.

If $x$ is a vertex of a path $P$ and $P - x$ consists of an $i$-path and a $j$ -path, then $x$ is called an $(i, j)$-vertex of $P$. An endpoint $x$ of a path $P$ in graph $G$ is an *out-endpoint* if $x$ has a neighbor outside of $P$.

A *vdp-cover* of a graph $G$ is a set $S$ of vertex-disjoint paths that contain all vertices of $G$. By $S_i$ we denote the set of $i$-paths in $S$.

The idea of Reed's paper [14] is to find a carefully selected vdp-cover $S$ and then consider the paths of $S$ one by one and for each such path choose some of its vertices to be in the dominating set. In [14] it is shown that if $G$ is of minimum degree at least 3, then the dominating set is of size at most $3/8n$. Clearly, for any path $P$ of $S$ it is enough to choose $\lceil |P|/3 \rceil$ vertices to dominate the whole $P$. If $P$ is a 0-path, or if $P$ is long enough then this is at most $\frac{3}{8}|P|$. Hence only short 1- and 2-paths remain. A careful analysis in [14] shows that for each short 1-path (resp. 2-path) $P$, if $G[P]$ does not contain a dominating set of size $\lfloor |P|/3 \rfloor$ then one (resp. two) of its endpoints has a neighbor on some path different from $P$ and this neighbor can dominate the endpoint. In our case, when vertices of degree 1 and 2 are allowed this is not always possible: $G[P]$ has fewer edges and it may happen that both endpoints are of degree 1. It turns out that the most troublesome paths are the dangling paths and the paths of order 8. Our strategy is to find a cover that avoids such paths as much as possible. Although we are not able to get rid of them completely, it turns out that it is enough to exclude some configurations that contain these paths.

In the following lemma we describe the properties of the cover we use. It is an extension of the construction in [14]. Our contribution here is the addition of (B4)-(B7) and the explicit statement of the construction algorithm.

**Lemma 2.** *For any graph $G$ one can find in polynomial time a vdp-cover $S$ of $G$ with the following properties. Let $x$ be an out-endpoint of any 1-path or 2-path $P_i$ in $S$. Let $y$ be a neighbor of $x$ on a path $P_j$, $j \neq i$ and let $P_j = P_j' y P_j''$. Then,*

*(B1) $P_j$ is not a 1-path,*
*(B2) if $P_j$ is a 0-path, then both $P_j'$ and $P_j''$ are 1-paths,*
*(B3) if $P_j$ is a 2-path, then both $P_j'$ and $P_j''$ are 2-paths,*
*(B4) if $|P_j| = 8$, then $P_i$ is a 1-path,*
*(B5) if $P_j$ is a 2-path and $P_i$ is a dangling path, then either one of $P_j'$, $P_j''$ is dangling or $|P_j| \in \{11, 17\}$ and $|P_j'| = |P_j''|$,*
*(B6) if $P_j$ is a 2-path and $z$ is the common endpoint of $P_j$ and $P_j'$, then each neighbor of $z$ on $P_j''$ is a $(2,2)$-vertex.*
*(B7) every 0-path in $S$ is of order 3.*

*Proof.* A *potential* of a cover $S$ is a tuple $\Phi(S) = (r_1, r_2, \ldots, r_7)$, where

- $r_1 = 2|S_1| + |S_2|$,
- $r_2 = |S_2|$,
- $r_3 = \sum_{P \in S_0} |P|$,
- $r_4 = \sum_{P \in S_1} |P|$,
- $r_5$ is the number of paths of order 8 in $S$,
- $r_6$ is the number of dangling paths in $S$,
- $r_7 = n - |S_0|$.

For two covers $S$ and $S'$ with potentials $\Phi(S) = (r_1, \ldots, r_7)$ and $\Phi(S') = (r_1', \ldots, r_7')$ we say that $\Phi(S') < \Phi(S)$ if $\Phi(S')$ is smaller than $\Phi(S)$ in lexicographic order, i.e. for some $i = 1, \ldots, 7$ we have $r_j = r_j'$ for $j < i$ and $r_i < r_i'$. We will show that if one of the conditions (B1)-(B7) does not hold then we can modify the vdp-cover $S$ to get a new cover $S'$ with strictly smaller potential. It will be clear from the proof that the modification can be done in linear time. Since for every $i = 1, \ldots, 7$ we have $r_i = O(n)$, it follows that if we start from an arbitrary vdp-cover $S$ then after $O(n^7)$ modifications we get a cover that satisfies all of (B1)-(B7) and the claim of the lemma will follow.

Reed [14] showed that we can decrease the potential if one of (B1)-(B3) does not hold (Observations 1–3 in [14], see also Lemma 1 in [11]).

Assume (B4) does not hold, i.e. $|P_j| = 8$ and $P_i \in S_2$. Then by (B3) both $P'_j$ and $P''_j$ are 2-paths and hence w.l.o.g. $|P'_j| = 2$ and $|P''_j| = 5$. If $|P_i| \neq 5$, we set $S' = S \setminus \{P_i, P_j\} \cup \{P_i y P'_j, P''_j\}$. Note that both $P_i y P'_j$ and $P''_j$ are 2-paths so $r_1, \ldots, r_4$ do not change. Also $|P''_j| = 5$ and $|P_i y P'_j| \neq 8$ so $r_5$ decreases. If $|P_i| = 5$ we set $S' = S \setminus \{P_i, P_j\} \cup \{P'_j, P_i y P''_j\}$. Again, both $P_i y P'_j$ and $P''_j$ are 2-paths so $r_1, \ldots, r_4$ do not change. Also $|P'_j| = 2$ and $|P_i y P'_j| = 11$ so $r_5$ decreases.

Assume (B5) does not hold. Then by (B3) both $P'_j$ and $P''_j$ are 2-paths. By symmetry we can assume that $|P'_j| \neq 5$ since if $|P'_j| = |P''_j| = 5$ then (B5) holds. Also, we can assume that $|P''_j| \neq 8$ since otherwise we know that $|P'_j| \neq 8$ and we can swap the names of $P'_j$ and $P''_j$ and get $|P'_j| = 8 \neq 5$ and $|P''_j| \neq 8$. Then we set $S' = S \setminus \{P_i, P_j\} \cup \{P_i y P'_j, P''_j\}$. Note that both $P_i y P'_j$ and $P''_j$ are 2-paths so $r_1, \ldots, r_4$ do not change. Also $|P''_j| \neq 8$ and $|P_i y P'_j| \neq 8$ so $r_5$ does not increase. Since $|P''_j|$ is not a dangling path (otherwise (B5) holds) and $|P_i y P'_j| \geq 5$, $r_6$ decreases by 1.

Assume (B6) does not hold. Let $P_j = v_1 \ldots v_{3p+2}$ for some $p \geq 1$ where $v_1$ is the common endpoint of $P_j$ and $P'_j$. By (B3) $y = v_{3q}$, $1 \leq q \leq p$. We assumed that for some $r \geq q$ we have $v_1 v_{3r+1} \in E$ or $v_1 v_{3r+2} \in E$. If $v_1 v_{3r+1} \in E$ then we consider the paths $P = v_{3p+2} v_{3p+1} \ldots v_{3r+1} v_1 v_2 \ldots v_{3q} P_i$ and $R = v_{3q+1} \ldots v_{3r}$ (if $q = r$ then $R$ is empty). If $v_1 v_{3r+2} \in E$ then we consider the paths $P = v_{3q+1} v_{3q+2} \ldots v_{3r+2} v_1 v_2 \ldots v_{3q} P_i$ and $R = v_{3r+3} \ldots v_{3p+2}$ (if $r = p$ then $R$ is empty). We set $S' = S \setminus \{P_i, P_j\} \cup \{P, R\}$. Note that $|R| \equiv 0 \pmod 3$ and $|P| \equiv |P_i| + |P_j| \pmod 3$. Hence if $P_i$ is a 1-path then both $P$ and $R$ are 0-paths so $r_1$ decreases and if $P_i$ is a 2-path then $P$ is a 1-path and $R$ is a 0-path, so $r_1$ stays the same and $r_2$ decreases.

If (B7) does not hold, we pick any 0-path $P$ of order at least 6 and replace it by two 0-paths, one of order 3 and one of order $|P| - 3$. Clearly, the potential decreases. □

Let $S$ be the cover from Lemma 2. Similarly as in [14], some of the out-endpoints of the paths in $S$ will be dominated by vertices of other paths which we call accepting. Now we describe our method for finding these paths.

**Accepting Procedure.** First, for every path $P \in S_1$ with at least one out-endpoint we mark exactly one, arbitrarily chosen, out-endpoint. Second, for every path $P$ of order $|P| \in \{2, 5, 8\}$ and with two out-endpoints we mark both of these endpoints.

We say that vertex $v$ is a *neighbor* of path $P$ if $v \notin V(P)$ and $v$ is a neighbor of a vertex of $P$. Path $P \in S$ is *dangerous* if

(i) $|P| = 8$,
(ii) $P$ has exactly one marked neighbor $v$,
(iii) $v$ has exactly one neighbor on $P$,
(iv) $\deg_G(v) > 1$,
(v) the path in $S$ that contains $v$ is of order 1,
(vi) $P$ has at most one out-endpoint.

As long as there is a non-dangerous path $P$ with a marked neighbor, we pick such a path $P$ and for its every marked neighbor $v$ we choose one vertex $w \in N(v) \cap P$ and $w$ *accepts* $v$. Then $v$ becomes unmarked and we call $w$ the *acceptor* of $v$. If $|P| \in \{5, 8\}$ and $P$ has exactly one out-endpoint $x$ we mark $x$, unless $x$ is already accepted. This finishes the description of the accepting procedure.

All vertices that are marked after the above procedure finishes are called *rejected.* A path from $S$ is *rejected* if it contains a rejected vertex. The following observation follows from (*iii*), (*iv*) and (*v*).

**Observation 2.** *Every rejected path is of order 1 and it has at least two neighboring dangerous paths.*

A *weak path* is a path $P \in S$ such that $|P| = 8$, $P$ accepts exactly one neighbor $v$, $v$ has *two* neighbors on $P$, the path in $S$ that contains $v$ is of order 1 and $P$ has no out-endpoints.

Consider a weak path $P = v_1 \ldots v_8$. Then exactly one vertex of $P$ is an acceptor, and by (B3) it is either $v_3$ or $v_6$. By symmetry assume $v_3$ is an acceptor. Then $v_3$ accepts exactly one vertex, say $v$, and $vv_6 \in E$. However, if $\deg_G(v_5) = 2$ then we change the acceptor of $v$ from $v_3$ to $v_6$. Note that then $\deg_G(v_4) \geq 3$. Thus the following invariant holds.

**Invariant 1.** *If we number vertices of a weak path* $P = v_1 \ldots v_8$ *so that* $v_3$ *is an acceptor then* $\deg_G(v_5) \geq 3$.

The intuition behind the notion of dangerous path is that it cannot afford accepting a vertex. As we see, a weak path is very close to being dangerous. A weak path can afford accepting a vertex, but it needs additional help from other paths. This "help" is realized by the following procedure.

**Forcing Procedure.** Now we define a certain set $F \subset V$. The elements of $F$ are called *forced vertices.* The set $F$ is constructed by the following procedure. Begin with empty $F$. Next consider weak paths of $S$, one by one. Let $P$ be such a weak path. If $P \cap F \neq \emptyset$ we skip $P$. Otherwise, let us number vertices of $P = v_1 \ldots v_8$ so that $v_3$ is an acceptor. If $v_5$ has a neighbor outside $P$ then we choose exactly one such neighbor $x$, we add $x$ to $F$ and $x$ becomes *forced by* $P$. This finishes the description of the forcing procedure.

The following observation follows easily from (B3).

**Observation 3.** *If* $w$ *is an endpoint of a path* $P \in S_1 \cup S_2$ *then* $w \notin F$.

In what follows we construct a certain dominating set $D$. As we will see, for some paths $P$ of $S$ the ratio $|P \cap D|/|P|$ is at most $\frac{3}{7}$, and for some of them it is larger than $\frac{3}{7}$. However, we show that the later ones are amortized by the former. To this end we introduce the following discharging procedure (which is *not* a part of the construction of $D$ but it helps to bound $|D|$). We assume that each vertex $v \in V$ and path $P \in S$ is assigned a rational number, called *charge*, which is initially 0. By *sending charge* of value $\alpha$ from $x \in V \cup S$ to $y \in V \cup S$ we mean that the charge of $x$ decreases by $\alpha$ and the charge of $y$ increases by $\alpha$. The charge is sent according to the following rules.

**Rule D1** Let $v$ be an endpoint of a path $P \in S$ such that $v$ is accepted by a vertex $w$. If $P \in S_1$ and $|P| \geq 4$, then $w$ sends $\frac{4}{7}$ to $P$. Otherwise, i.e. if $|P| \in \{1, 2, 5, 8\}$, then $w$ sends $\frac{3}{7}$ to $P$.
**Rule D2** Every rejected path sends $\frac{2}{7}$ to each neighboring dangerous path.
**Rule D3** Every dangling path sends $\frac{1}{7}$ to each neighboring path.
**Rule D4** If a vertex $x$ is forced by a weak path $P$, then $x$ sends $\frac{6}{7}$ to $P$.

After applying all the discharging rules above, each vertex $v$ and each path $P \in S$ ends up with some amount of charge: the total charge it received minus the total charge it sent. For $x \in V \cup S$ let $\mathrm{ch}(x)$ denote the final amount of charge at $x$. For $P \in S$, let $\widehat{\mathrm{ch}}(P) = \mathrm{ch}(P) + \sum_{v \in P} \mathrm{ch}(v)$. Note that the initial total charge in $G$ is equal to 0 and it does not change by applying the discharging rules, so $\sum_{P \in S} \widehat{\mathrm{ch}}(P) = 0$.

Let $A$ be the set of all acceptors. We say that a path $P \in S$ is *safe* when there exists a set $D_P \subseteq P$ such that

a) $D_P \cup A \cup F$ dominates $P$, i.e. $P \subset N[D_P \cup A \cup F]$,
b) $P \cap (A \cup F) \subseteq D_P$,
c) $|D_P| + \widehat{\mathrm{ch}}(P) \leq \frac{3}{7}|P|$.

**Lemma 3.** *If all paths in $S$ are safe, then $G$ has a dominating set of size at most $\frac{3}{7}n$.*

*Proof.* Since all paths in $S$ are safe, for each such path $P$ there is a set $D_P$ that satisfies conditions a)-c). Then we define $D = \bigcup_{P \in S} D_P$. By b), $A \cup F \subseteq D$. This together with a) implies that $D$ is a dominating set of $G$. Since $\sum_{P \in S} \widehat{\mathrm{ch}}(P) = 0$, c) implies that $|D| = \sum_{P \in S} |D_P| \leq \sum_{P \in S} \frac{3}{7}|P| = \frac{3}{7}n$. □

In section 4 we show that $G$ satisfies the assumptions of Theorem 1 then all paths in $S$ are safe. Together with the above lemma that finishes the proof of Theorem 1.

## 4 All Paths Are Safe

From now on we assume that $G$ satisfies the assumptions of Theorem 1. The following lemma follows easily from the discharging rules.

**Lemma 4.** *Let $v$ be a vertex of $G$. If $v$ accepts a path from $S_1$ of order at least 4, then $\mathrm{ch}(v) \leq -\frac{4}{7} - \frac{6}{7}[v \in F]$. If $v$ accepts a path of order 1, 2, 5 or 8, then $\mathrm{ch}(v) \leq -\frac{3}{7} - \frac{6}{7}[v \in F]$. Otherwise $\mathrm{ch}(v) \leq -\frac{6}{7}[v \in F]$.* □

In what follows we will consider various kinds of paths in $S$ and we will show that they are safe. In many cases we will divide these paths into several subpaths, which we call "bricks". Then the safeness of paths from $S$ will be derived from the safeness of bricks, which is defined as follows. We say that a path $P$ in a graph $G$ is *α-safe* when there exists a set $D_P \subseteq P$ such that

a) $D_P \cup A \cup F$ dominates $P$, i.e. $P \subset N[D_P \cup A \cup F]$,
b) $P \cap (A \cup F) \subseteq D_P$, and
c) $|D_P| + \sum_{v \in P} \mathrm{ch}(v) \leq \alpha$.

Note that if a path $P \in S$ is $\frac{3}{7}|P|$-safe it does *not* mean that it is safe, because in the definition of $\alpha$-safeness we ignore the charge of $P$. However the following claim is easy to verify.

**Lemma 5.** *Let $P \in S$ and assume that $P = P_1 \dots P_k$. For $i = 1, \dots, k$ assume that path $P_i$ is $\alpha_i$-safe. If $\sum_{i=1}^{k} \alpha_i + \mathrm{ch}(P) \leq \frac{3}{7}|P|$ then $P$ is safe.* □

**Lemma 6 (3-brick Lemma).** *Any path $P = v_1 v_2 v_3$ in $G$ such that $P \cap A \subseteq \{v_2\}$ is $\frac{8}{7}$-safe. Moreover, if $v_2 \in A$, then $P$ is $\frac{6}{7}$-safe.*

*Proof.* By Lemma 4, for $v \in \{v_1, v_3\}$, $\mathrm{ch}(v) \leq -\frac{6}{7}[v \in F]$.

First assume $v_2$ is an acceptor. We put $D_P = P \cap (A \cup F)$. Note that $|D_P| \leq 1 + |F \cap \{v_1, v_3\}|$ and $\mathrm{ch}(v_2) \leq -\frac{3}{7} - \frac{6}{7}[v \in F]$. Hence, $\sum_{v \in P} \mathrm{ch}(v) \leq -\frac{3}{7} - \frac{6}{7}|F \cap P|$. It follows that $|D_P| + \sum_{v \in P} \mathrm{ch}(v) \leq 1 + |F \cap \{v_1, v_3\}| - \frac{3}{7} - \frac{6}{7}|F \cap P| = \frac{4}{7} + \frac{1}{7}|F \cap \{v_1, v_3\}| \leq \frac{6}{7}$.

Now assume $v_2$ is not an acceptor. By Lemma 4, for any $v \in P$ we have $\mathrm{ch}(v) \leq -\frac{6}{7}[v \in F]$. If $|F \cap P| \geq 2$, we put $D_P = F \cap P$ and then $|D_P| + \sum_{v \in P} \mathrm{ch}(v) \leq |F \cap P| - \frac{6}{7}|F \cap P| \leq \frac{3}{7}$. Otherwise, i.e. when $|F \cap P| \leq 1$, we put $D_P = \{v_2\} \cup (F \cap P)$ and then $|D_P| + \sum_{v \in P} \mathrm{ch}(v) \leq 1 + |F \cap P| - \frac{6}{7}|F \cap P| = 1 + \frac{1}{7}|F \cap P| \leq \frac{8}{7}$. □

**Lemma 7 (4-brick Lemma).** *Any path $P = v_1 v_2 v_3 v_4$ in $G$ such that $P \cap A \subseteq \{v_3\}$ is 2-safe. Moreover, if $v_3 \in A$, then $P$ is $(\frac{11}{7} + \frac{1}{7}|F \cap \{v_1, v_4\}|)$-safe (and hence $\frac{13}{7}$-safe).*

*Proof.* By Lemma 4, for $v \in \{v_1, v_2, v_4\}$, $\mathrm{ch}(v) \leq -\frac{6}{7}[v \in F]$.

First assume $v_3$ is an acceptor. We put $D_P = \{v_2\} \cup P \cap (A \cup F)$. Note that $|D_P| \leq 2 + |F \cap \{v_1, v_4\}|$ and $\mathrm{ch}(v_3) \leq -\frac{3}{7} - \frac{6}{7}[v \in F]$ by Lemma 4. Hence, $\sum_{v \in P} \mathrm{ch}(v) \leq -\frac{3}{7} - \frac{6}{7}|F \cap P|$. It follows that $|D_P| + \sum_{v \in P} \mathrm{ch}(v) \leq 2 + |F \cap \{v_1, v_4\}| - \frac{3}{7} - \frac{6}{7}|F \cap P| \leq \frac{11}{7} + \frac{1}{7}|F \cap \{v_1, v_4\}| \leq \frac{13}{7}$.

Now assume $v_3$ is not an acceptor. By Lemma 4, for $v \in P$ we have $\mathrm{ch}(v) \leq -\frac{6}{7}[v \in F]$ and hence $\sum_{v \in P} \mathrm{ch}(v) = -\frac{6}{7}|F \cap P|$. If $F \cap P = \emptyset$, we put $D_P = \{v_2, v_3\}$ and then $|D_P| + \sum_{v \in P} \mathrm{ch}(v) = 2$. Finally assume $F \cap P \neq \emptyset$. If $\{v_1, v_2\} \cap F \neq \emptyset$ we put $D_P = \{v_3\} \cup (F \cap P)$ and otherwise we put $D_P = \{v_2\} \cup (F \cap P)$. Then, $|D_P| + \sum_{v \in P} \mathrm{ch}(v) \leq 1 + |F \cap P| - \frac{6}{7}|F \cap P| = 1 + \frac{1}{7}|F \cap P| \leq \frac{11}{7}$. □

**Lemma 8.** *Every 0-path $P$ is safe.*

*Proof.* By (B8) $|P| = 3$. Clearly, $P$ may get charge only by Rule D3. Moreover, Rule D3 applies at most once to $P$ because otherwise by (B2) there are two dangling paths neighboring with the only $(1,1)$-vertex of $P$ and then there are 1-vertices at distance 4, a contradiction. It follows that $\mathrm{ch}(P) \leq \frac{1}{7}$. By Lemma 6 path $P$ is $\frac{8}{7}$-safe. Since $\frac{8}{7} + \mathrm{ch}(P) \leq \frac{9}{7} = \frac{3}{7}|P|$ so by Lemma 5 path $P$ is safe. □

**Lemma 9.** *Every path $P$ of order 1 is safe.*

*Proof.* Let $V(P) = \{v\}$. Since there are no isolated vertices, $v$ is an out-endpoint. By (B1) $P$ does not receive charge by Rule D3.

Note that $P \cap A = \emptyset$ by (B1) and $F \cap P = \emptyset$ by Observation 3. Hence, by Lemma 4, $\mathrm{ch}(v) = 0$.

First assume $v$ is accepted. Then $P$ gets exactly $\frac{3}{7}$ by Rule D1, and Rule D2 does not apply, so $\mathrm{ch}(P) = \frac{3}{7}$. We put $D_P = \emptyset$. It follows that $|D_P| + \widehat{\mathrm{ch}}(P) = \frac{3}{7} = \frac{3}{7}|P|$.

If $v$ is not accepted, Rule D1 does not apply. Moreover, then $P$ is rejected, so by Observation 2 it sends $2 \cdot \frac{2}{7} = \frac{4}{7}$ by Rule D2. Hence, $\mathrm{ch}(P) \leq -\frac{4}{7}$. We put $D_P = \{v\}$. It follows that $|D_P| + \widehat{\mathrm{ch}}(P) \leq 1 - \frac{4}{7} = \frac{3}{7}|P|$. □

**Lemma 10.** *Every 1-path $P$, $|P| \geq 4$, with an out-endpoint is safe.*

*Proof.* Assume $P = v_0v_1 \ldots v_{3k}$ for some $k \geq 1$. By the accepting procedure, since $P$ has an out-endpoint, $P$ has an accepted out-endpoint and $P$ gets $\frac{4}{7}$ by D1. Assume w.l.o.g. $v_0$ is the accepted out-endpoint of $P$. By (B1), D3 does not apply to $P$. It follows that $\mathrm{ch}(P) = \frac{4}{7}$. Note that $P \cap A = \emptyset$ by (B1). We partition $P$ into $k+1$ paths: $P = P_0P_1 \ldots P_k$, where $P_0 = v_0$ and $P_i = v_{3i-2}v_{3i-1}v_{3i}$ for any $i = 1, \ldots, k$. By Observation 3 and Lemma 4 we have $\mathrm{ch}(v_0) = 0$, so we see that $P_0$ is 0-safe (by choosing $D_{P_0} = \emptyset$). By Lemma 6 for every $i = 1, \ldots, k$ the path $P_i$ is $\frac{8}{7}$-safe. Since $0 + k \cdot \frac{8}{7} + \mathrm{ch}(P) = \frac{8k+4}{7} \leq \frac{9k+3}{7} = \frac{3}{7}|P|$, by Lemma 5 path $P$ is safe. □

**Lemma 11.** *If a path $P \in S$ of order 4 has no out-endpoint then $P = N[x]$ for some $x \in P$.*

*Proof.* Let $P = v_1v_2v_3v_4$. Since $P$ has no out-endpoint, $N(v_1) \subseteq \{v_2, v_3, v_4\}$ and $N(v_4) \subseteq \{v_1, v_2, v_3\}$. If $v_1v_3 \in E(G)$ then $N(v_3) = \{v_1, v_2, v_4\}$, so we can take $v_3$ as $x$. Hence $v_1v_3 \notin E(G)$ and by symmetry also $v_4v_2 \notin E(G)$. If $v_1v_4 \in E(G)$ then $\deg_G(v_1) = \deg_G(v_4) = 2$ and we have 2-vertices at distance 1, a contradiction. It follows that $\deg_G(v_1) = \deg_G(v_4) = 1$, and we have 1-vertices at distance 3, a contradiction. □

**Lemma 12.** *Every path $P$ of order 4 is safe.*

*Proof.* Let $P = v_1v_2v_3v_4$. By Lemma 10 we can assume that $P$ has no out-endpoint. Then $\mathrm{ch}(P) = 0$, since D3 does not apply by (B1). Let $x \in P$ be a vertex such that $P = N[x]$, as guaranteed by Lemma 11. Note that $P \cap A = \emptyset$ by (B1).

We put $D_P = \{x\} \cup (F \cap P)$. By Lemma 4 and Observation 3, $\mathrm{ch}(v_1) = \mathrm{ch}(v_4) = 0$ and $\mathrm{ch}(v_2), \mathrm{ch}(v_3) \leq -\frac{6}{7}[v \in F]$. By Observation 3, $|F \cap P| \leq 2$. Hence, $|D_P| + \widehat{\mathrm{ch}}(P) \leq 1 + |F \cap P| - \frac{6}{7}|F \cap P| = 1 + \frac{1}{7}|F \cap P| \leq \frac{9}{7} < \frac{3}{7}|P|$. □

Because of the space limitations the safeness of the remaining paths is deferred to the journal version.

**Acknowledgement.** I thank Michał Dębski and Marcin Mucha for helpful discussions. I am also very grateful to anonymous reviewers for careful reading and many helpful remarks.

## References

1. Alber, J., Fellows, M.R., Niedermeier, R.: Polynomial-time data reduction for dominating set. J. ACM 51(3), 363–384 (2004)
2. Appel, K., Haken, W.: Every planar map is four colorable part I. Discharging. Illinois J. of Math. 21, 429–490 (1977)
3. Bodlaender, H.L., Fomin, F.V., Lokshtanov, D., Penninkx, E., Saurabh, S., Thilikos, D.M.: (meta) kernelization. In: Proc. FOCS 2009, pp. 629–638 (2009)
4. Borodin, O.V., Kostochka, A.V., Woodall, D.R.: List edge and list total colourings of multigraphs. J. Comb. Theory, Ser. B 71(2), 184–204 (1997)
5. Chen, J., Fernau, H., Kanj, I.A., Xia, G.: Parametric duality and kernelization: Lower bounds and upper bounds on kernel size. SIAM J. Comput. 37(4), 1077–1106 (2007)
6. Dehne, F., Fellows, M., Fernau, H., Prieto, E., Rosamond, F.: NONBLOCKER: Parameterized Algorithmics for MINIMUM DOMINATING SET. In: Wiedermann, J., Tel, G., Pokorný, J., Bieliková, M., Štuller, J. (eds.) SOFSEM 2006. LNCS, vol. 3831, pp. 237–245. Springer, Heidelberg (2006)
7. Dvorak, Z., Skrekovski, R., Valla, T.: Planar graphs of odd-girth at least 9 are homomorphic to the petersen graph. SIAM J. Disc. Math. 22(2), 568–591 (2008)
8. Fomin, F.V., Lokshtanov, D., Saurabh, S., Thilikos, D.M.: Bidimensionality and kernels. In: Charikar, M. (ed.) SODA, pp. 503–510. SIAM (2010)
9. Guo, J., Niedermeier, R.: Linear Problem Kernels for NP-Hard Problems on Planar Graphs. In: Arge, L., Cachin, C., Jurdziński, T., Tarlecki, A. (eds.) ICALP 2007. LNCS, vol. 4596, pp. 375–386. Springer, Heidelberg (2007)
10. Kanj, I.A., Zhang, F.: On the Independence Number of Graphs with Maximum Degree 3. In: Kolman, P., Kratochvíl, J. (eds.) WG 2011. LNCS, vol. 6986, pp. 238–249. Springer, Heidelberg (2011)
11. Kostochka, A.V., Stodolsky, B.Y.: An upper bound on the domination number of n-vertex connected cubic graphs. Discrete Mathematics 309(5), 1142–1162 (2009)
12. Kowalik, L., Pilipczuk, M., Suchan, K.: Towards optimal kernel for connected vertex cover in planar graphs. CoRR abs/1110.1964 (2011)
13. McCuaig, W., Shepherd, B.: Domination in graphs with minimum degree two. J. Graph Th. 13(6), 749–762 (1989)
14. Reed, B.A.: Paths, stars and the number three. Combinatorics, Probability & Computing 5, 277–295 (1996)
15. Robertson, N., Sanders, D.P., Seymour, P.D., Thomas, R.: The four-colour theorem. J. Comb. Theory, Ser. B 70(1), 2–44 (1997)
16. Wang, J., Yang, Y., Guo, J., Chen, J.: Linear Problem Kernels for Planar Graph Problems with Small Distance Property. In: Murlak, F., Sankowski, P. (eds.) MFCS 2011. LNCS, vol. 6907, pp. 592–603. Springer, Heidelberg (2011)

# Some Definitorial Suggestions for Parameterized Proof Complexity

Jörg Flum[1] and Moritz Müller[2]

[1] Universität Freiburg, Germany
joerg.flum@math.uni-freiburg.de
[2] Kurt Gödel Research Center, Universität Wien, Austria
moritz.mueller@univie.ac.at

**Abstract.** We introduce a (new) notion of parameterized proof system. For parameterized versions of standard proof systems such as Extended Frege and Substitution Frege we compare their complexity with respect to parameterized simulations.

## 1 Introduction

Consider the following problems for graphs: the vertex cover problem VC, the clique problem CLIQUE, and the dominating set problem DS; they ask, given a graph $G$ and a natural number $k$, whether $G$ contains a cardinality $k$ vertex cover, clique, and dominating set, respectively. All three problems are NP-complete and hence, from the point of view of polynomial reductions any two of them have the same computational complexity.

Taking in each case the natural number $k$ as the parameter of an instance we get the parameterized problems $p$-VC, $p$-CLIQUE, and $p$-DS. In parameterized complexity there is not only a new notion of tractability, namely fixed-parameter tractability, but also the notion of reducibility has been adapted so that it preserves fixed-parameter tractability; the new notion being that of fpt-reduction. One knows that $p$-VC $\le_{\text{fpt}}$ $p$-CLIQUE (that is, $p$-VC is fpt-reducible to $p$-CLIQUE) and $p$-CLIQUE $\le_{\text{fpt}}$ $p$-DS. However, accepting the hypotheses FPT $\neq$ W[1] and W[1] $\neq$ W[2] (which are fundamental hypotheses of parameterized complexity and each of them implies P $\neq$ NP) neither $p$-CLIQUE $\le_{\text{fpt}}$ $p$-VC nor $p$-DS $\le_{\text{fpt}}$ $p$-CLIQUE. As Downey and Fellows write in [7]:

> *Parameterized reductions tend to be much more* structure preserving *than classical reductions, and certainly most classical reductions ... are definitely not parameterized reductions. ... Parameterized reductions are sufficientlly refined that instead of one large class of naturally intractable problems all of the same complexity, there seem to be many sets of natural combinatorial problems, all intractable in the parameterized sense, and yet of differing parameterized complexity*

In proof theory among the proof systems best studied there are Frege systems, Extended Frege systems, and Substitution Frege systems. Classically, they are compared

D.M. Thilikos and G.J. Woeginger (Eds.): IPEC 2012, LNCS 7535, pp. 73–84, 2012.

via polynomial simulations. It is known that there are polynomial simulations between any Extended Frege system and any Substitution Frege system, while it is not known whether Extended Frege systems and Substitution Frege systems may be simulated by Frege systems. The question arises whether also in this context parameterized complexity yields new insights or even allows a more fine-grained analysis. In this note we want to lay down the conceptual framework for such an analysis. Furthermore, we give some positive and some negative answers and state some open problems.

What are natural parameterizations of proof systems? Recall that the definitions of parameterized complexity are tailored to address complexity issues in situations where we know that the parameter is relatively small. We believe that for Extended Frege systems the number of extension axioms used in a proof could be a natural parameter. At least, if we start with an arbitrary, say, random tautology, it does not seem plausible that many extension axioms can be used in a proof with advantage. We should emphasize the word "random" here. For example, in a standard example often mentioned to motivate the use of extension axioms, namely the formalization of the pigeon-principle in propositional logic, the number of extension axioms used to derive the $n$ pigeonhole principle by a straightforward induction on $n$ is $\Omega(n^3)$ and hence, certainly not small.[1] Similarly the number of applications of the substitution rule seems to be a natural parameter for Substitution Frege Systems.

As proof systems are functions, simulations between them should be value-preserving functions (as are the standard polynomial simulations). We believe that this fact has not been taken into account appropriately in the approaches to proof theory using parameterized complexity. Taking this fact seriously, we define the notion of fpt-simulation. When we realized that our notion coincides with the notion of parsimonious reduction between parameterized counting functions, we were confirmed in our belief that this is the appropriate definition.

We show that under fpt-simulations the parameterized versions of Extended Frege and Substitution Frege are both equivalent to Frege. In this sense, the notion of fpt-simulation does not offer a more fine-grained complexity analysis of these proof systems; or, expressing it in positive terms, we gain the insight that there is a simulation, say, of an Extended Frege system in a Frege system whose superpolynomial running time is confined to a factor depending only on the number of extension axioms used in the original proof. Similarly, we see that there is a simulation of Substitution Frege in Extended Frege where the number of extension axioms is bounded in terms of the number of applications of the substitution rule.

Having in mind the goal of a more refined analysis, we propose to study the relationship between these proof systems under parameterized polynomial simulations, a notion that in some sense refines both polynomial simulations and fpt-simulations: such a simulation is a polynomial simulation with the additional property that it increases the parameter at most polynomially. We do not see any way to simulate Substitution Frege in Extended Frege in this sense (while the converse is easy). However, we construct a parameterized polynomial simulation of treelike Substitution Frege in treelike Extended Frege.

[1] It is well-known that Buss [3] gave polynomial proofs of the pigeon-principle in Frege systems.

*Related work.* A different approach to introduce parameterizations into proof complexity has been initiated by Dantchev et al. [6]. They introduced parameterized proof systems for *parameterized* problems. They considered the following parameterized problem: given a pair $(\alpha, k)$ of a CNF $\alpha$ and $k \in \mathbb{N}$, where $k$ is the parameter, decide whether $\alpha$ has no satisfying assignment of Hamming weight at most $k$. The proof systems they had in mind are classical refutation systems such as Resolution that may freely use additional clauses expressing the constraint on the Hamming weight. The goal of this approach is to strengthen lower bounds of classical refutation systems by showing that their parameterized counterparts are not *fpt bounded*[2]. It can be understood as a parameterized analogue of Cook's program, here trying to prove coW[2] $\nsubseteq$ paraNP. For this approach Beyersdorff et al. [1] lack an interpretation of the parameterization of the proof system and argue that it can be dispensed with.

## 2 Preliminaries

In this section we fix some notations and recall some definitions and results, in the first part of parameterized complexity theory and in the second part of proof theory.

### 2.1 Parameterized Complexity

Formally, a *parameterized problem* is a pair $(Q, \kappa)$ consisting of a (classical) problem $Q \subseteq \{0,1\}^*$ and a polynomial time computable *parameterization* $\kappa : \{0,1\}^* \to \mathbb{N}$ that maps any input $x \in \{0,1\}^*$ to its *parameter* $\kappa(x) \in \mathbb{N}$. A parameterized problem $(Q, \kappa)$ is *fixed-parameter tractable*, that is, tractable from the point of view of parameterized complexity, if there is an algorithm solving $x \in Q$ in $\leq f(\kappa(x)) \cdot |x|^{O(1)}$ steps for some computable $f : \mathbb{N} \to \mathbb{N}$.

A function $R : \{0,1\}^* \to \{0,1\}^*$ is *fpt-computable* with respect to a parameterization $\kappa$ if $R(x)$ can be computed in time $f(\kappa(x)) \cdot |x|^{O(1)}$, where again $f : \mathbb{N} \to \mathbb{N}$ is computable.

Also the notion of polynomial reduction, that is, the natural notion of reduction preserving classical tractability, has to be adapted so that it preserves fixed-parameter tractability. An *fpt-reduction* $R$ from a parameterized problem $(Q, \kappa)$ to another $(Q', \kappa')$ is an fpt-computable (with respect to $\kappa$) reduction from $Q$ to $Q'$ such that $\kappa'(R(x)) \leq g(\kappa(x))$ for some computable $g : \mathbb{N} \to \mathbb{N}$ and all $x \in \{0,1\}^*$. We write $(Q, \kappa) \leq_{\text{fpt}} (Q', \kappa')$ if there is an fpt-reduction from $(Q, \kappa)$ to $(Q', \kappa')$.

### 2.2 Proof Theory

A *proof system* for a problem $Q \subseteq \{0,1\}^*$ is a polynomial time computable surjection $P$ from $\{0,1\}^*$ onto $Q$. If $P(w) = x$, then $w$ is a *P-proof* of $x$. In case $Q$ = TAUT, we call $P$ *propositional*. A proof system $P$ is *p-bounded* if any $x \in Q$ has a $P$-proof of size $|x|^{O(1)}$. Cook and Reckhow [5] observed that a p-bounded propositional proof system exists if and only if NP = coNP. Cook's program aims to prove that natural propositional proof systems are not p-bounded.

[2] As pointed out in [1] one should restrict attention to instances $(\alpha, k)$ with contradictory $\alpha$.

Proof systems for a problem $Q$ are compared in strength via p-simulations: a *p-simulation* of a proof system $P'$ in a proof system $P$ is a polynomial time computable function $R$ such that $P(R(w')) = P'(w')$ for all $w' \in \{0,1\}^*$; in case such an $R$ exists, we say $P$ *p-simulates* $P'$ and write $P' \leq_{\mathrm{pol}} P$; if additionally, $P'$ p-simulates $P$, we call $P$ and $P'$ *p-equivalent*.

A *Frege system* $F$ is a propositional proof system given by finitely many axiom schemes (in the de Morgan language) and finitely many rules including, for simplicity, modus ponens. An $F$*-proof* of a (propositional) formula $\alpha$ from a set of formulas $\Gamma$ is a sequence of formulas such that each of them is either a member of $\Gamma$ or a substitution instance of an axiom scheme or follows from earlier formulas in the sequence by one of the rules of $F$; furthermore, the last formula of the sequence is $\alpha$. An $F$-proof of $\alpha$ is an $F$-proof of $\alpha$ from the empty set of formulas. Frege systems are assumed to be *implicationally complete*, that is, whenever a set of formulas $\Gamma$ logically implies $\alpha$, then there exists an $F$-proof of $\alpha$ from $\Gamma$.

For a Frege system $F$ we denote by $F^*$ the proof system *treelike* $F$: an $F$-proof $\pi$ is *treelike* if every occurrence of a formula in $\pi$ is used as an hypothesis in an application of a rule at most once; equivalently, $\pi$ is treelike if it can be written as a tree labeled by the formulas in $\pi$ such that the leaves are labeled by the substitution instances of the axiom schemes and the labels of inner nodes are obtained by one of the rules from their immediate predecessors.

The following are well-known [10,5].

**Theorem 1.** *(1) (Cook, Reckhoff) Any two Frege systems are p-equivalent.*
*(2) (Krajíček) $F$ and $F^*$ are p-equivalent for every Frege system $F$.*

By part (1) of this theorem we get that, instead of (2), we could claim

$$F_1 \text{ and } F_2^* \text{ are } p\text{-equivalent for Frege systems } F_1 \text{ and } F_2.$$

The same observation applies to all equivalences mentioned in this paper (not only to p-equivalences but also to fpt-equivalences and pp-equivalences introduced later).

There are two well-studied extensions of a Frege system $F$:

*Extension Frege.* Let $F$ be a Frege system. The *Extension Frege system EF* adds to $F$ the *extension rule*: It allows to add in a proof of $\alpha$ (without any hypotheses) an *extension axiom* $(r \leftrightarrow \sigma)$ where $\sigma$ is a propositional formula and the *extension variable* $r$ neither occurs in $\sigma$ nor in $\alpha$ nor in any earlier line of the proof.

Equivalently, an *EF*-proof of $\alpha$ is an $F$-proof of $\alpha$ from an extension sequence whose extension variables do not occur in $\alpha$. Here, an *extension sequence* (for $\alpha$) of length $k$ is a sequence of the form

$$(r_1 \leftrightarrow \sigma_1), \ldots, (r_k \leftrightarrow \sigma_k)$$

with pairwise distinct *extension variables* $r_1, \ldots, r_k$ such that $r_i$ does not occur in $\sigma_j$ for $1 \leq j \leq i$.

By *EF*$^*$ we denote the treelike version of *EF*.

*Substitution Frege.* Let $F$ be a Frege system. The *Substitution Frege system SF* adds to $F$ the *substitution rule* that allows to derive from the formula $\alpha$ the formula $\alpha[x/\sigma]$

where $\alpha[x/\sigma]$ is obtained from $\alpha$ by substituting the variable $x$ by the formula $\sigma$. By *SF** we denote the treelike version of *SF*.

In [2] Buss introduces two restrictions of *SF*:

- *Boolean Substitution Frege BSF* requires that in any application of the substitution rule the formula $\sigma$ is the Boolean constant $\top$ (TRUE) or $\bot$ (FALSE);
- *Renaming Frege RF* requires $\sigma$ to be a variable.

Again, *BSF** and *RF** denote the treelike versions of these systems.

Natural simulations of *EF* and *SF* in *F* roughly proceed as follows:

- Let $\pi$ be an *EF*-proof. To delete the first extension axiom $(r \leftrightarrow \sigma)$ substitute everywhere in $\pi$ the formula $\sigma$ for $r$; this transforms the extension axiom into the tautology $(\sigma \leftrightarrow \sigma)$ for which we add a linear size $F$-proof. Proceed like this with the second extension axiom and so on. If $\pi$ contains $k$ extension axioms, the resulting $F$-proof has size $|\pi|^{O(k)}$.
- Let $\pi$ be an *SF*-proof. Let the first application in $\pi$ of the substitution rule yield $\alpha[x/\sigma]$ from $\alpha$. Replace it by a proof of $\alpha[x/\sigma]$ obtained by applying the substitution $x/\sigma$ to the initial segment of $\pi$ up to $\alpha$. If $\pi$ contains $k$ substitution inferences, the resulting $F$-proof has size $|\pi|^{O(k)}$.

Hence, both simulations are not polynomial ones. In fact, it is open whether $EF \leq_{\text{pol}} F$ and whether $SF \leq_{\text{pol}} F$. However, the following is known [12,2]:

**Theorem 2.** *(1) EF, EF*, SF, SF*, RF, BSF are p-equivalent for every Frege system F.* *(2) RF*, BSF* and F are p-equivalent for every Frege system F.*

Comparing their status with that of *RF** and of *BSF** we see that perhaps *RF* and *BSF* are proof systems where the ability to reuse already derived lines adds power. We shall see a similar phenomenon for *SF* in the parameterized setting.

## 3 Parameterized Proof Systems and fpt-Simulations

In this section we introduce the main new concepts of this paper, parameterized proof systems and simulations between them.

**Definition 3.** A *parameterized proof system for Q* is a pair $(P, \kappa)$ such that $P$ is a proof system for $Q$ and $\kappa$ a parameterization.

Having in mind, as we do, to compare Frege systems, Extended Frege systems, and Substitution Frege systems, it does not seem natural to consider a more general notion of parameterized proof systems where $P$ is only required to be an fpt-computable (with respect to $\kappa$) function from $\{0,1\}^*$ onto $Q$ instead of a polynomial time computable one.

We identify a (classical) proof system $P$ for $Q$ with the parameterized proof system $(P, 0)$, i.e., $P$ with the parameterization that is constantly 0.

For an Extended Frege system *EF* we denote by $\kappa_{EF}$ the parameterization

$$\kappa_{EF}(w) := \text{number of extension axioms in } w.$$

Similarly, for a Substitution Frege system *SF* we denote by $\kappa_{SF}$ the parameterization

$$\kappa_{SF}(w) := \text{number of applications of the substitution rule in } w.$$

We consider the restriction $EF^*$ of *EF* with the parameterization $\kappa_{EF}$ and the restrictions $SF^*$, $BSF^{(*)}$, and $RF^{(*)}$ of *SF* with the parameterization $\kappa_{SF}$. We denote the resulting parameterized proof systems by *p-EF*, *p-EF*$^*$, *p-SF*, *p-RF*, *p-BSF*, *p-SF*$^*$, *p-RF*$^*$ and *p-BSF*$^*$.

In order to compare parameterized proof systems in strength we use the following notion of simulation. We already mentioned that for parameterized counting problems the notion coincides with that of fpt parsimonious reduction introduced in [8, Definition 14.10].

**Definition 4.** Let $(P,\kappa)$ and $(P',\kappa')$ be parameterized proof systems for $Q \subseteq \{0,1\}^*$. An *fpt-simulation* of $(P',\kappa')$ in $(P,\kappa)$ is a function $R : \{0,1\}^* \to \{0,1\}^*$ such that
(a) $R$ is *fpt*-computable with repect to $\kappa'$;
(b) $P'(w') = P(R(w'))$ for all $w' \in \{0,1\}^*$;
(c) $\kappa(R(w')) \leq g(\kappa'(w'))$ for some computable $g : \mathbb{N} \to \mathbb{N}$ and all $w' \in \{0,1\}^*$.
In case such an $R$ exists, we say that $(P,\kappa)$ *fpt-simulates* $(P',\kappa')$ and write $(P',\kappa') \leq_{\text{fpt}} (P,\kappa)$. The problems $(P,\kappa)$ and $(P',\kappa')$ are *fpt-equivalent*, written $(P,\kappa) \equiv_{\text{fpt}} (P,\kappa)$, if $(P,\kappa) \leq_{\text{fpt}} (P',\kappa')$ and $(P',\kappa') \leq_{\text{fpt}} (P,\kappa)$.

Note that if $P$ and $P'$ are classical proof systems for a problem $Q$, then $P$ fpt-simulates $P'$ if and only if $P$ p-simulates $P'$. However, in general, neither $(P,\kappa) \leq_{\text{fpt}} (P',\kappa')$ implies $P \leq_{\text{pol}} P'$ nor $P \leq_{\text{pol}} P'$ implies $(P,\kappa) \leq_{\text{fpt}} (P',\kappa')$.

**Lemma 5.** *If* $(P,\kappa) \leq_{\text{fpt}} (P',\kappa')$ *and* $(P',\kappa') \leq_{\text{fpt}} (P'',\kappa'')$, *then* $(P,\kappa) \leq_{\text{fpt}} (P'',\kappa'')$.

## 4 Comparing Proof Systems via fpt-Simulations

By the following result all parameterized proof systems introduced so far are fpt-equivalent.

**Theorem 6.** *p-EF*, *p-SF*, *and F are pairwise fpt-equivalent.* [3]

As $F \leq_{\text{fpt}}$ *p-EF*, the theorem follows from the following three propositions showing (among others):

$$p\text{-}EF \leq_{\text{fpt}} p\text{-}SF \leq_{\text{fpt}} p\text{-}BSF \leq_{\text{fpt}} F.$$

In Proposition 7 and Proposition 8 we obtain the first two 'inequalities' by merely observing that known p-simulations already are fpt-simulations.

[3] The second author gave a talk at the workshop *Proof complexity* (11w5103, Banff International Research Station) on this subject mentioning that at that time we didn't know whether *p-EF* $\leq_{\text{fpt}}$ $F$. Kaveh Ghasemloo pointed out that he was convinced that such a simulation could be constructed via the system $G_1^*$ (cf. [4, p.179]).

**Proposition 7.** *p-EF* $\leq_{\text{fpt}}$ *p-SF and p-EF** $\leq_{\text{fpt}}$ *p-SF**.

*Proof.* Cook and Reckhow's original p-simulation [5] of *EF* in *SF* is an fpt-simulation of *p-EF* in *p-SF*; this yields the first assertion.

We turn to the second claim. An *EF**-proof $\pi$ of $\alpha$ is an $F^*$-proof of $\alpha$ from an extension sequence $(r_1 \leftrightarrow \sigma_1), \ldots, (r_k \leftrightarrow \sigma_k)$ (recall that the $r_i$ have to be paiwise distinct and that $r_i$ neither occurs in $\sigma_j$ for $1 \leq j \leq i$ nor in $\alpha$). By the Deduction Theorem for $F$ (see [11, Lemma 4.4.10]) there is an $F$-proof $\pi'$ of

$$(r_k \leftrightarrow \sigma_k) \to (r_{k-1} \leftrightarrow \sigma_{k-1}) \to \cdots \to (r_1 \leftrightarrow \sigma_1) \to \alpha \tag{1}$$

(where the iterated implications are associated to the right) of size $|\pi|^{O(1)}$. By part (2) of Theorem 1 we can assume that $\pi'$ is treelike.

By our assumption on the extension variables, the variable $r_k$ occurs exactly once in (1). We apply the substitution rule and substitute $\sigma_k$ for $r_k$ in (1); hence we get the formula obtained from (1) by replacing the equivalence $(r_k \leftrightarrow \sigma_k)$ by $(\sigma_k \leftrightarrow \sigma_k)$. We add a short $F^*$-proof of $(\sigma_k \leftrightarrow \sigma_k)$ and apply modus ponens to arrive at formula (1) with $k-1$ instead of $k$. Repeating this process gives an *SF**-proof of $\alpha$ of size $O(k \cdot |\pi'|)$. We observe that in this simulation $k$ extension axioms are simulated in *SF** by $k$ applications of the substitution rule. Therefore, this is an fpt-simulation. □

**Proposition 8.** *p-SF* $\leq_{\text{fpt}}$ *p-BSF.*

*Proof.* Buss [2] simulates an application of the substitution rule $\frac{\alpha}{\alpha[x/\sigma]}$ as follows: first, he applies twice the *BSF*-substitution rule to get

$$\alpha[x/\top] \quad \text{and} \quad \alpha[x/\bot]$$

from $\alpha$; then he adds short proofs of

$$((\sigma \wedge \alpha[x/\top]) \to \alpha[x/\sigma]) \quad \text{and} \quad ((\neg\sigma \wedge \alpha[x/\bot]) \to \alpha[x/\sigma]).$$

Finally, he derives $\alpha[x/\sigma]$ from these four formulas.

In this way, an *SF*-proof with $k$ applications of the substitution rule is transformed in polynomial time into an *BSF*-proof with $2k$ applications of the *BSF*-substitution rule. Hence, this is an fpt-simulation. □

**Proposition 9.** *p-BSF* $\leq_{\text{fpt}}$ *F.*

*Proof.* Let $\pi$ be an *BSF*-proof of $\beta$ with $k$ applications of the *BSF*-substitution rule. Let $\pi_1$ be the initial segment of $\pi$ that ends in the premise $\alpha$ of the first application $\frac{\alpha}{\alpha[x/\sigma]}$ with $\sigma \in \{\top, \bot\}$ of this rule. We obtain the $F$-proof $\pi_1'$ of $\alpha[x/\sigma]$ by applying the substitution $x/\sigma$ to every line of $\pi_1$. Furthermore, delete all occurrences of $\alpha[x/\sigma]$in $\pi$, thus getting $\pi'$. Then $\pi_1', \pi'$ is a *BSF*-proof of $\beta$ with $(k-1)$ applications of the *BSF*-substitution rule and of size at most $2|\pi|$. Repeating this process we finally obtain an $F$-proof of $\beta$ of size $2^k \cdot |\pi|$. □

**Remark 10.** As an analysis of the previous proofs shows, for every *EF*-proof of size $n$ with $k$ extension axioms there exists an $F$-proof $\pi$ of the same formula with $|\pi| \leq 2^{2k} \cdot n^{O(1)}$.

Standard p-simulations of *SF* in *EF* (e.g., see [12]) map an *SF*-proof $\pi$ of a formula $\alpha(\bar{x})$ (where $\bar{x}$ are the propositional variables in $\alpha$) with $k$ applications of the substitution rule and $\ell$ lines to an *EF*-proof with $\ell \cdot |\bar{x}|$ extension axioms. They are not fpt-simulations. By the previous theorem there is an fpt-simulation of *p-SF* in *p-EF*. We encourage the reader to give a 'direct' one.

## 5 Comparing Proof Systems via Parameterized Polynomial Simulations

In the previous section we have seen that fpt-simulations are too coarse in the sense that they do not distinguish any two of the parameterized proof system considered so far. In this section therefore we analyze these proof systems under a notion of simulation which strengthens both the notion of p-simulation and that of fpt-simulation. For parameterized decision problems this concept was introduced in [9].

**Definition 11.** Let $(P, \kappa)$ and $(P', \kappa')$ be parameterized proof systems for $Q \subseteq \{0,1\}^*$. A *pp-simulation* (or, *parameterized polynomial simulation*) of $(P', \kappa')$ in $(P, \kappa)$ is a p-simulation $R$ of $P'$ in $P$ such that

$$\kappa(R(w')) \leq q(\kappa'(w')) \text{ for some polynomial } q \text{ and all } w' \in \{0,1\}^*.$$

In case such an $R$ exists, we say that $(P, \kappa)$ *pp-simulates* $(P', \kappa')$ and write $(P', \kappa') \leq_{pp} (P, \kappa)$. The problems $(P, \kappa)$ and $(P', \kappa')$ are *pp-equivalent*, written $(P, \kappa) \equiv_{pp} (P, \kappa)$, if $(P, \kappa) \leq_{pp} (P', \kappa')$ and $(P', \kappa') \leq_{pp} (P, \kappa)$.

Clearly, if $(P', \kappa') \leq_{pp} (P, \kappa)$, then $P' \leq_{pol} P$ and $(P', \kappa') \leq_{fpt} (P, \kappa)$.

As the proofs of Proposition 7 and of Proposition 8 show, we get:

**Proposition 12.** *p-EF* $\leq_{pp}$ *p-SF, p-EF** $\leq_{pp}$ *p-SF**, *and p-SF* $\leq_{pp}$ *p-BSF.*

**Example 13.** The $p$-simulation of *BSF* in *RF* from [2] maps a *BSF*-proof with $k$ substitution inferences of a formula with $m$ variables to an *RF*-proof with $k \cdot (m-1)$ substitution inferences. This is not a pp-simulation (not even an fpt-simulation).

By the results of the previous section there is an fpt-simulation of *p-SF* in *p-EF* even though (as mentioned at the end of that section) standard p-simulations of *SF* in *EF* are not fpt-simulations. We do not know whether *p-SF* $\leq_{pp}$ *p-EF*. However, this holds for the tree-like versions of these proof systems:

**Theorem 14.** *p-SF** $\leq_{pp}$ *p-EF**.

*Proof.* We say that an *SF**-proof of $\beta$ from an extension sequence (for $\beta$) is an *ESF**-proof of $\beta$ if every application of the substitution rule has the form

$$\frac{\alpha}{\alpha[x/\sigma]}$$

where the formula $x \wedge \sigma$ does not contain any extension variable.

Clearly, an *EF**-proof of $\beta$ is an *ESF**-proof of $\beta$ without applications of the substitution rules.

We now describe how to stepwise eliminate applications of the substitution rule in *ESF**-proofs. So, let $\pi$ be an *ESF**-proof of $\beta$ with $k$ applications of the substitution rule. We depict $\pi$ as a labeled tree $T$ with $\beta$ at the root; for any node $t$ of $T$ labeled by $\gamma$ the subtree $T_t$ rooted at this node (and consisting of the predecessors of this node) constitutes an *ESF**-proof of $\gamma$. Consider a node $t$ such that

- $t$ is labeled by a formula $\alpha[x/\sigma]$ obtained from its predecessor $t^-$ labeled by $\alpha$ by an application of the substitution rule (via the substitution $x/\sigma$);
- no further applications of the substitution rule occur in $T_t$.

Let $r$ be a variable not occuring in $\pi$ and obtain $T_{t^-}(x/r)$ by substituting $x$ by $r$ in all formulas of $T_{t^-}$. By the proviso on the applications of the substitution rule in an *ESF**-proof, the variable $x$ is not a substitution variable and hence extension axioms of $T$ are transformed into extension axioms in $T_{t^-}(x/r)$. Hence, $T_{t^-}(x/r)$ is an $F^*$-proof of $\alpha[x/r]$ from a set of extension axioms.

Let $\pi'$ be a short $F^*$-proof of

$$(\alpha[x/r] \to ((r \leftrightarrow \sigma) \to \underbrace{\alpha[x/r][r/\sigma]}_{=\alpha[x/\sigma]}))$$

Using the new extension axiom $(r \leftrightarrow \sigma)$ (and adding some applications of modus ponens) we merge this $F^*$-proof with $T_{t^-}(x/r)$ to get a $F^*$-proof of $\alpha[x/\sigma]$ from an extension sequence.

$$\dfrac{\dfrac{\begin{array}{c}\vdots\, T_{t^-}(x/r)\\ \alpha[x/r]\end{array} \qquad \begin{array}{c}\vdots\, \pi'\\ (\alpha[x/r] \to ((r \leftrightarrow \sigma) \to \alpha[x/\sigma])\end{array}}{((r \leftrightarrow \sigma) \to \alpha[x/\sigma])} \qquad (r \leftrightarrow \sigma)}{\alpha[x/\sigma]}$$

Replace in the original proof $\pi$ the subtree $T_t(x/r)$ by this new proof, thus obtaining a proof $\pi''$. It should be clear that $\pi''$ is an *ESF**-proof of $\beta$ with $k-1$ applications of the substitution rule.

Iterating this process $k$ times we finally get an $F^*$-proof $\pi^*$ of $\beta$ from an extension sequence (for $\beta$) consisting of $k$ extension axioms. As $\pi^*$ is obtained from $\pi$ in polynomial time the mapping $\pi \mapsto \pi^*$ is the desired pp-simulation of *p-SF** in *p-EF**. □

Note that in the previous proof we have used that the *SF*-proof we start with is treelike: the simulation replaces all predecessors of a formula obtained by a substitution rule. In an arbitrary *SF*-proof some later inferences may be based on some formulas not further available.

We prove the following result by standard means:

**Proposition 15.** *p-EF* $\leq_{\text{pp}}$ *p-EF**.

*Proof.* Let $\pi = \alpha_1, \ldots, \alpha_s$ be an *EF*-proof with $k$ extension axioms. For $1 \leq i \leq s$ we set $\gamma_i := \bigwedge_{j=1}^{i} \alpha_j$. We construct for $i = 1, \ldots, s$ successively *EF**-proofs $\pi_i$ of $\gamma_i$ such that the variables in $\pi_i$ are precisely those in $\alpha_1, \ldots, \alpha_i$ and the extension axioms in $\pi_i$ are the same as in $\alpha_1, \ldots, \alpha_i$.

The tree $\pi_1$ just consists of the root labeled by $\alpha_1$. Assume that we have already constructed the *EF**-proof $\pi_i$ of $\gamma_i$. To construct $\pi_{i+1}$ we first consider the case where $\alpha_{i+1}$ is an extension axiom or a substitution instance of an axiom of $F$. Let $\pi^1$ be a short $F^*$-proof of $(u \rightarrow (v \rightarrow (u \wedge v)))$. Then $\pi^1[u/\gamma_i, v/\alpha_{i+1}]$ is an $F^*$-proof of $(\gamma_i \rightarrow (\alpha_{i+1} \rightarrow \gamma_{i+1}))$ of size $O(|\gamma_{i+1}|)$. As an intermediate step we get an $F^*$-proof $\pi^2$ of $(\alpha_{i+1} \rightarrow \gamma_{i+1})$ from the $F^*$-proofs $\pi_i$ and $\pi^1[u/\gamma_i, v/\alpha_{i+1}]$ by an application of modus ponens. A further modus ponens inference yields from $\pi^2$ and the 'leaf' $\alpha_{i+1}$ the desired $F^*$-proof $\pi_{i+1}$ of $\gamma_{i+1}$.

$$\dfrac{\dfrac{\dfrac{\vdots\ \pi^1[u/\gamma_i, v/\alpha_{i+1}]}{(\gamma_i \rightarrow (\alpha_{i+1} \rightarrow \gamma_{i+1}))} \qquad \dfrac{\vdots\ \pi_i}{\gamma_i}}{(\alpha_{i+1} \rightarrow \gamma_{i+1})} \qquad \alpha_{i+1}}{\gamma_{i+1}}$$

Now assume that $\alpha_{i+1}$ is obtained by one of the rules of $F$. The general case being analogous, we treat the case where this rule is modus ponens. So assume $\alpha_{i+1}$ is obtained from $\alpha_k$ and $\alpha_\ell$ (where $1 \leq k, \ell \leq i$) by modus ponens. Let $\pi^1$ be an $F^*$-proof of $(\bigwedge_{j=1}^{i} u_j \rightarrow (u_k \wedge u_\ell))$ of size polynomial in $i$. Substituting in $\pi^1$ the $u_j$s by the $\alpha_j$s yields an $F^*$-proof $\pi^2$ of $(\gamma_i \rightarrow (\alpha_k \wedge \alpha_\ell))$ of size polynomial in $|\gamma_i|$.

To a short $F^*$-proof of $((u \rightarrow v) \rightarrow ((v \rightarrow w) \rightarrow (u \rightarrow (u \wedge w))))$ we apply the substitution $[u/\gamma_i, v/(\alpha_k \wedge \alpha_\ell), w/\alpha_{i+1}]$ obtaining an $F^*$-proof $\pi^3$ of size $O(|\gamma_{i+1}|)$ of

$$((\gamma_i \rightarrow (\alpha_k \wedge \alpha_\ell)) \rightarrow ((\alpha_k \wedge \alpha_\ell) \rightarrow \alpha_{i+1}) \rightarrow (\gamma_i \rightarrow \gamma_{i+1}))).$$

Finally, let $\pi^4$ be an $F^*$-proof of $((\alpha_k \wedge \alpha_\ell) \rightarrow \alpha_{i+1})$ of size $O(|\alpha_k| + |\alpha_\ell| + |\alpha_{i+1}|)$ (recall that $\alpha_{i+1}$ was obtained from $\alpha_k$ and $\alpha_\ell$ by modus ponens). Now it is easy to merge $\pi_i$, $\pi^1$, $\pi^2$, $\pi^3$, and $\pi^4$ to an $F^*$-proof $\pi_{i+1}$ of $\gamma_{i+1}$.

It is easy to construct a treelike proof $\pi^*$ of $\alpha_s$ from $\pi_s$. It is clear that $\pi^*$ can be computed from $\pi$ in polynomial time. □

**Theorem 16.** $F \equiv_{\text{pp}}$ *p-BSF** $\equiv_{\text{pp}}$ *p-RF** $\leq_{\text{pp}}$ *p-EF* $\equiv_{\text{pp}}$ *p-EF** $\equiv_{\text{pp}}$ *SF** $\leq_{\text{pp}}$ *p-SF* $\equiv_{\text{pp}}$ *p-BSF*.

*Proof.* The first two equivalences are easy to see. The third equivalence follows from the preceding proposition. The equivalence *p-EF** $\equiv_{\text{pp}}$ *p-SF** follows from Proposition 12 and Theorem 14. The last equivalence also follows from Proposition 12. □

Hence, the proof systems mentioned in the previous theorem belong to at most three distinct pp-degrees. Are these degrees distinct? Note that this theorem does not mention *p-RF*. Does it belong to any of these degrees? Of course, $F \leq_{pp}$ *p-RF* $\leq_{pp}$ *p-SF*. Furthermore, we can show the following:

**Proposition 17.** *If p-RF $\leq_{pp}$ p-EF, then p-SF $\leq_{pp}$ p-EF.*

*Proof.* Assume *p-RF* $\leq_{pp}$ *p-EF*. By Proposition 12 it suffices to show *p-BSF* $\leq_{pp}$ *p-EF*. So let $\pi = \alpha_1, \dots, \alpha_s$ be a *BSF*-proof with $k$ substitution inferences (substituting a variable by $\bot$ or by $\top$). Let $y_1, \dots, y_k$ and $z_1, \dots, z_k$ be new variables (not occurring in $\pi$) and let

$$\delta := \bigwedge_{i=1}^{k} \neg y_i \wedge \bigwedge_{i=1}^{k} z_i.$$

Consider the sequence

$$(\delta \rightarrow \alpha_1), \dots, (\delta \rightarrow \alpha_s).$$

This sequence can be "filled up" to an *RF*-proof with $k$ substitution inferences (substituting a variable by another variable): if $\alpha_i$ in $\pi$ is a substitution instance of an axiom, replace $(\delta \rightarrow \alpha_i)$ by a short $F$-proof of $(\delta \rightarrow \alpha_i)$. If $\alpha_i$ is obtained by modus ponens from $\alpha_j, \alpha_{j'}$ with $j, j' < i$, then replace $(\delta \rightarrow \alpha_i)$ by a short $F$-proof of $(\delta \rightarrow \alpha_i)$ from $(\delta \rightarrow \alpha_j)$ and $(\delta \rightarrow \alpha_{j'})$. Finally, if $\alpha_i$ is obtained by a substitution inference, then there is $j < i$ such that $\alpha_i = \alpha_j[x/\bot]$ or $\alpha_i = \alpha_j[x/\top]$ for some variable $x$. Assume this is the $\ell$th substitution inference ($1 \leq \ell \leq k$) in $\pi$ and that $\alpha_i = \alpha_j[x/\bot]$ (the other case $\alpha_i = \alpha_j[x/\top]$ is similar). Replace $(\delta \rightarrow \alpha_i)$ by the following *RF*-proof: give a short $F$-proof of $(\delta \wedge \alpha_j[x/y_\ell] \rightarrow \alpha_i)$ (note that $\neg y_\ell$ is a conjunct of $\delta$) and derive $\alpha_j[x/y_\ell]$ from $\alpha_j$ by an *RF* substitution inference; from these two formulas it is easy to derive $(\delta \rightarrow \alpha_i)$.

Clearly, this *RF*-proof can be computed from $\pi$ in polynomial time. By assumption we can in polynomial time compute from this *RF*-proof an *EF*-proof $\pi'$ of $(\delta \rightarrow \alpha_s)$ with $k^{O(1)}$ extension axioms. Since the $y_i$'s and the $z_i$'s occur in $\delta$, they are not used as extension variables in $\pi'$. Let $\pi''$ result from $\pi'$ by substituting $\bot$ for all occurrences of the $y_i$'s and $\top$ for all occurrences of the $z_i$'s. Then (note the $y_i$'s and the $z_i$'s do not occur in $\alpha_s$) $\pi''$ is an *EF*-proof of $(\delta' \rightarrow \alpha_s)$ where $\delta'$ is a true Boolean sentence (a true formula without variables). Adding a short proof of $\delta'$ and an application of modus ponens gives an *EF*-proof of $\alpha_s$. □

**Acknowledgements.** The authors thank the John Templeton Foundation for its support through Grant #13152, *The Myriad Aspects of Infinity*. The second author thanks the FWF (Austrian Research Fund) for its support through Grant P 23989 - N13.

## References

1. Beyersdorff, O., Galesi, N., Lauria, M., Razborov, A.: Parameterized Bounded-Depth Frege Is Not Optimal. In: Aceto, L., Henzinger, M., Sgall, J. (eds.) ICALP 2011. LNCS, vol. 6755, pp. 630–641. Springer, Heidelberg (2011)
2. Buss, S.: Some remarks on the lengths of propositional proofs. Archive for Mathematical Logic 34, 377–394 (1995)

3. Buss, S.: Polynomial size proofs of the propositional pigeon principle. Journal of Symbolic Logic 52, 916–927 (1987)
4. Cook, S., Nguyen, P.: Logical Foundations of Proof Complexity. Cambridge University Press (2010)
5. Cook, S., Reckhow, R.: The relative efficiency of propositional proof systems. The Journal of Symbolic Logic 44, 36–50 (1979)
6. Dantchev, S.S., Martin, B., Szeider, S.: Parameterized proof complexity. Computational Complexity 20(1), 51–85 (2011)
7. Downey, R.G., Fellows, M.R.: Parameterized Complexity. Springer (1999)
8. Flum, J., Grohe, M.: Parameterized Complexity Theory. Springer (2006)
9. Fortnow, L., Santhanam, R.: Infeasibility of instance compression and succinct PCPs for NP. Journal of Computer and System Sciences 77(1), 91–106 (2011)
10. Krajíček, J.: On the number of steps in proofs. Annals of Pure and Applied Logic 41, 153–178 (1989)
11. Krajíček, J.: Bounded arithmetic, propositional logic, and complexity theory. Cambridge University Press (1995)
12. Krajíček, J., Pudlák, P.: Propositional proof systems, the consistency of first order theories and the complexity of computations. The Journal of Symbolic Logic 54, 1063–1088 (1989)

# An Exact Algorithm for Subset Feedback Vertex Set on Chordal Graphs*

Petr A. Golovach[1], Pinar Heggernes[1], Dieter Kratsch[2], and Reza Saei[1]

[1] Department of Informatics, University of Bergen, Norway
{petr.golovach,pinar.heggernes,reza.saeidinvar}@ii.uib.no
[2] LITA, Université de Lorraine - Metz, France
kratsch@univ-metz.fr

**Abstract.** Given a graph $G = (V, E)$ and a set $S \subseteq V$, a set $U \subseteq V$ is a subset feedback vertex set of $(G, S)$ if no cycle in $G[V \setminus U]$ contains a vertex of $S$. The SUBSET FEEDBACK VERTEX SET problem takes as input $G$, $S$, and an integer $k$, and the question is whether $(G, S)$ has a subset feedback vertex set of cardinality or weight at most $k$. Both the weighted and the unweighted versions of this problem are NP-complete on chordal graphs, even on their subclass split graphs. We give an algorithm with running time $O(1.6708^n)$ that enumerates all minimal subset feedback vertex sets on chordal graphs with $n$ vertices. As a consequence, SUBSET FEEDBACK VERTEX SET can be solved in time $O(1.6708^n)$ on chordal graphs, both in the weighted and in the unweighted case. On arbitrary graphs, the fastest known algorithm for the problems has $O(1.8638^n)$ running time.

## 1 Introduction

Given a graph $G = (V, E)$ and a set $S \subseteq V$, a set $U \subseteq V$ is a subset feedback vertex set of $(G, S)$ if no cycle in $G[V \setminus U]$ contains a vertex of $S$. A subset feedback vertex set $U$ is minimal if no subset feedback vertex set of $(G, S)$ is a proper subset of $U$. The SUBSET FEEDBACK VERTEX SET problem takes as input $G$, $S$, and an integer $k$, and the question is whether $(G, S)$ has a subset feedback vertex set of cardinality at most $k$. In the weighted version of the problem, every vertex of $G$ has a weight, and the question is whether there is a subset feedback vertex set of total weight at most $k$.

SUBSET FEEDBACK VERTEX SET was introduced by Even et al. [4], and it generalizes several well-studied problems. When $S = V$, it is equivalent to the classical FEEDBACK VERTEX SET problem [11], and when $|S| = 1$, it generalizes the MULTIWAY CUT problem [7]. Weighted SUBSET FEEDBACK VERTEX SET admits a polynomial-time constant-factor approximation algorithm [4]. The unweighted version of the problem is fixed parameter tractable [3]. The only exact algorithm known for its weighted version is by Fomin et al. [7] and it runs in $O(1.8638^n)$ time and solves the problem by enumerating all minimal subset feedback vertex sets.

As a comparison, the unweighted version of FEEBACK VERTEX SET can be solved in time $O(1.7347^n)$ [9], whereas the best algorithm for its weighted version

* This work has been supported by the European Research Council, the Research Council of Norway, and the French National Research Agency.

D.M. Thilikos and G.J. Woeginger (Eds.): IPEC 2012, LNCS 7535, pp. 85–96, 2012.

runs in time $O(1.8638^n)$ and enumerates all minimal feedback vertex sets [5]. FEEBACK VERTEX SET has also been studied on many graph classes, like chordal graphs and AT-free graphs [1,15], and several positive results exist. This is not yet the case for SUBSET FEEDBACK VERTEX SET, and no algorithm with a running time of $O(c^n)$ such that $c < 1.8637$ is known for any significant graph class. Interestingly, whereas both the weighted and the unweighted versions of FEEBACK VERTEX SET are solvable in polynomial time on chordal graphs [1,19], even the unweighted version of SUBSET FEEDBACK VERTEX SET is NP-complete on chordal graphs; in fact on their restricted subclass split graphs, by a standard reduction from VERTEX COVER [7].

In this paper we give an algorithm with running time $O(1.6708^n)$ that enumerates all minimal subset feedback vertex sets when the input graph is chordal. As a consequence, SUBSET FEEDBACK VERTEX SET can be solved in time $O(1.6708^n)$ on chordal graphs, both in the weighted and in the unweighted case. Our algorithm differs completely from the $O(1.8638^n)$ time algorithm of [7] for the general case, and it heavily uses the structure of chordal graphs. Chordal graphs form one of the most studied graph classes; they have extensive practical applications in several fields [10,12,18], and they are crucial in characterizing and understanding fundamental algorithmic tools, like treewidth.

Enumeration algorithms are central in the field of Exact Exponential Algorithms, as the running times of many exact exponential time algorithms rely on the maximum number of various objects in graphs [8]. A classical example is the widely used result of Moon and Moser [16], showing that the maximum number of maximal cliques or maximal independent sets in an $n$-vertex graph is $3^{n/3}$. More recently, the maximum numbers and enumeration of objects like minimal dominating sets, minimal feedback vertex sets, minimal subset feedback vertex sets, minimal separators, and potential maximal cliques, have been studied; see e.g., [5,6,7,9,13,14,17]. The maximum number of such objects in graphs have traditionally found independent interest also in graph theory and combinatorics.

The results we present in this paper give an upper bound of $O(1.6708^n)$ on the maximum number of minimal subset feedback vertex sets a chordal graph can have. A tight bound on the maximum number of minimal feedback vertex sets on chordal graphs is known to be $1.5848^n$ [2], and this thus gives a lower bound on the maximum number of minimal subset feedback vertex sets on chordal graphs. Consequently, our results tighten the gap between the upper and lower bounds on the maximum number of subset feedback vertex sets on chordal graphs. The corresponding gap is much larger on general graphs. There, the maximum numbers of minimal feedback and subset feedback vertex sets are both $O(1.8638^n)$ [5,7], but no examples of graphs having $1.5927^n$ or more minimal feedback or subset feedback vertex sets are known [5]. Note that the maximum number of minimal subset feedback vertex sets can be dramatically different from the maximum number of minimal feedback vertex sets. Split graphs, which form a subclass of chordal graphs, have at most $n^2$ minimal feedback vertex sets, whereas they can have $3^{n/3}$ minimal subset feedback vertex sets [7].

## 2 Preliminaries

We work with simple undirected graphs. We denote such a graph by $G = (V, E)$, where $V$ is the set of vertices and $E$ is the set of edges of $G$. We adhere to the convention that $n = |V|$. The set of neighbors of a vertex $v \in V$ is denoted by $N_G(v)$. The degree of $v$, $|N_G(v)|$, is denoted by $d_G(v)$. The *closed neighborhood* of $v$ is $N_G[v] = N(v) \cup \{v\}$. For a vertex subset $X \subseteq V$, the subgraph of $G$ induced by $X$ is denoted by $G[X]$. For ease of notation, we use $G - v$ to denote the graph $G[V \setminus \{v\}]$, and $G - X$ to denote the graph $G[V \setminus X]$.

A *path* in $G$ is a sequence of distinct vertices such that the next vertex in the sequence is adjacent to the previous vertex. A *cycle* is a path with at least three vertices such that the last vertex is in addition adjacent to the first. Given a subset $S \subseteq V$, we call a cycle an *S-cycle* if it contains a vertex of $S$. For a cycle or an $S$-cycle $C$, we use $V(C)$ to denote the set of vertices in $C$. A subset $F \subseteq V$ will be called a *forest* if $G[F]$ contains no cycle. Similarly, $F$ is an $S$-forest if no cycle in $G[F]$ contains a vertex of $S$. A graph is *connected* if there is a path between every pair of its vertices. A maximal connected subgraph of $G$ is called a *connected component* of $G$. A set $X \subseteq V$ is a *clique* if $uv \in E$ for every pair of vertices $u, v \in X$; and $X$ is an *independent set* if $uv \notin E$ for every pair of vertices $u, v \in X$.

A *chord* of a cycle is an edge between two non consecutive vertices of the cycle. A graph is *chordal* if every cycle of length at least 4 contains a chord. Induced subgraphs of chordal graphs are also chordal [12]. A vertex $v$ is called *simplicial* if $N(v)$ is a clique. Every chordal has a simplicial vertex [12]. A graph is a *split graph* if its vertex set can be partitioned into a clique and an independent set. Split graphs are chordal.

Given a set $S \subseteq V$, a set $U \subseteq V$ is a *subset feedback vertex set (sfvs)* of $(G, S)$ if no cycle in $G - U$ contains a vertex of $S$. Observe that $U$ is a sfvs of $(G, S)$ if and only if $V \setminus U$ is an $S$-forest. If $S = V$ then $U$ is a *feedback vertex set (fvs)* of $G$, and $V \setminus U$ is a forest. A sfvs $U$ is *minimal* if no proper subset of $U$ is a sfvs of $(G, S)$, and an $S$-forest is *maximal* if it cannot be extended to a larger $S$-forest by including more vertices of $G$. Clearly, $U$ is a minimal sfvs of $(G, S)$ if and only if $V \setminus U$ is a maximal $S$-forest of $G$. Consequently, the number of minimal sfvs of $(G, S)$ is equal to the number of maximal $S$-forests of $G$.

Let $\mu(G, S)$ denote the number of minimal sfvs of $(G, S)$, equivalently the number of maximal $S$-forests of $G$. Observe that $\mu(G, S) = \prod_{i=1}^{t} \mu(G_i, S)$, where $G_1, G_2, \ldots, G_t$ are the connected components of $G$. This is because every maximal $S$-forest of $G$ is the union of maximal $S$-forests of the connected components of $G$.

Let $\mu(G) = \max\{\mu(G, S) \mid S \subseteq V\}$. Note that $\mu(G)$ is lower bounded by the number of minimal fvs of $G$. Let $H$ be the complete graph on 5 vertices. This graph has 10 minimal fvs [2]. Let $H_\ell$ be the graph obtained by taking $\ell$ disjoint copies of $H$, for $\ell \geq 1$. The number of minimal feedback vertex sets of $H_\ell$ is thus $10^\ell = 10^{n/5} \approx 1.5848^n$. Any graph $H_\ell$ is chordal and hence $10^{n/5}$ is a lower bound on the number of minimal sfvs of chordal graphs, i.e., there is a chordal graph $G = (V, E)$ and a set $S \subseteq V$ such that $(G, S)$ has $10^{n/5}$ minimal sfvs. When it comes to the maximum number of minimal fvs in chordal graphs, Couturier et al. showed that the above lower bound is also the upper bound [2]. An upper bound on the number of minimal sfvs of chordal graphs better than the one for general graphs has not been known until the result we present below.

## 3 Enumerating Minimal Subset Feedback Vertex Sets in Chordal Graphs

This section is devoted to proving the following theorem.

**Theorem 1.** *All minimal subset feedback vertex sets of a chordal graph on $n$ vertices can be listed in $O(1.6708^n)$ time.*

Two corollaries follow from the above result. Corollary 1 follows immediately, whereas Corollary 2 follows by noting that any sfvs of minimum cardinality or minimum weight is a minimal sfvs. Hence we can check the cardinality or weight of each generated minimal sfvs, and compare the smallest one with the given bound $k$ of the input.

**Corollary 1.** *A chordal graph on $n$ vertices has at most $O(1.6708^n)$ minimal subset feedback vertex sets.*

**Corollary 2.** *Both weighted and unweighted versions of* Subset Feedback Vertex Set *can be solved in $O(1.6708^n)$ time on chordal graphs.*

To prove Theorem 1, we will describe an algorithm that takes as input a chordal graph $G = (V, E)$ and a vertex subset $S \subseteq V$, and lists all maximal $S$-forests of $G$. Our algorithm is a recursive branching algorithm; every maximal $S$-forest of $G$ will be present at some leaf of the corresponding branching tree, whereas some of the leaves might not correspond to maximal $S$-forests. Every recursive call has input $(G', F, U, R)$, where $F$ is the set of vertices of $G$ placed so far in an $S$-forest of $G$, $U$ is the set of vertices so far deleted from $G$ and hence placed in the corresponding sfvs, $R \subseteq F$ is the set of vertices that are placed in $F$ and that are no longer relevant for making further decisions, and $G' = G - (U \cup R)$. We call the vertices in $R$ *hidden*. The vertices in $V \setminus (U \cup F)$ are called *undecided* vertices. As $G$ and $S$ do not change throughout the algorithm, they are not parts of the input to the recursive calls. Given $G$ and $S$, the main program runs the recursive branching algorithm on $(G, \emptyset, \emptyset, \emptyset)$.

If at some call $(G', F, U, R)$, the graph $G'$ has no undecided vertices, then we are at a leaf of the branching tree, and the algorithm stops after checking whether $F$ is a maximal $S$-forest of $G$. If $F$ is a maximal $S$-forest, it is added to the list of $S$-forests that will be output. If $G'$ has undecided vertices, the algorithm continues, but first it checks whether $F$ is an $S$-forest. If not, then the algorithm stops, discards $F$ since it can never lead to a maximal $S$-forest, and no new subproblems are generated from this instance. If the algorithm continues, then since $G'$ is chordal, we know that it has a simplicial vertex. The algorithm chooses an arbitrary simplicial vertex $v$ of $G'$ and makes choices depending on $v$. Vertex $v$ might already be placed in $F$ or not; these two cases will be handled separately in the first two subsections below. The following operations will be used in our algorithm:

- *Deleting* a vertex $x$: deletes $x$ from $G'$ and adds it to $U$. Vertex $x$ will be permanently deleted from $G'$ and it will be a part of the suggested sfvs $U$ in all subsequent subproblems.
- *Adding* a vertex $x$ to $F$: adds $x$ to $F$. Vertex $x$ will be a part of $F$ in all subsequent subproblems, and will never be considered for deletion.
- *Hiding* a vertex $x$ of $F$: this operation is only applicable on some simplicial vertices of $G'$ that are already placed in $F$. We apply it when $x$ is no longer relevant for making further decisions on the remaining vertices of $G' - F$. When $x$ is hidden, it is added to $R$ and removed from $G'$ but it remains a part of $F$ in all subsequent subproblems, and in particular it remains in $G - U$.

Throughout the algorithm we will keep the following invariant.

**Invariant 1.** *Let $(G', F, U, R)$ be an instance. For any $S$-cycle $C$ in $G - U$ that contains a vertex of $R$, there is an $S$-cycle $C'$ in $G'$ such $V(C') = V(C) \setminus R$.*

Invariant 1 is clearly true when $R$ is empty. Whenever we hide a vertex $v$, we will argue that the invariant is still true after $v$ is hidden. The next lemma shows that we can safely ignore the vertices in $R$ when we make further decisions on $G - U$, and hence it is safe to work on $G' = G - (U \cup R)$ instead of $G - U$.

**Lemma 1.** *Let $(G', F, U, R)$ be an instance. Under Invariant 1, $F'$ is a maximal $S$-forest of $G - U$ such that $F \subseteq F'$ if and only if $F' \setminus R$ is a maximal $S$-forest of $G'$.*

*Proof.* Let $F'$ be a maximal $S$-forest of $G - U$ such that $F \subseteq F'$. Then clearly $F' \setminus R$ is an $S$-forest in $G'$. Let us argue for maximality. Since $F'$ is maximal, for any vertex $x$ of $G - (U \cup F')$, $x$ is involved in an $S$-cycle $C$ in $G - U$ such that $V(C) \subseteq F' \cup \{x\}$. Observe that since $R \subseteq F \subseteq F'$, any such vertex $x$ is also a vertex in $G'$. By Invariant 1, $x$ is involved in an $S$-cycle $C'$ in $G'$ such that $V(C') = V(C) \setminus R$. Since $G' = G - (U \cup R)$, it follows that $V(C') \subseteq (F' \setminus R) \cup \{x\}$. Hence $x$ cannot be added to $F' \setminus R$, which is thus a maximal $S$-forest of $G'$.

For the other direction, assume that $F' \setminus R$ is a maximal $S$-forest of $G'$. Hence every vertex $x$ in $G'$ outside of $F' \setminus R$ is involved in an $S$-cycle $C$ in $G'$ such that $V(C) \subseteq (F' \setminus R) \cup \{x\}$. Since $G - U$ is a supergraph of $G'$, $C$ is also an $S$-cycle in $G - U$. Hence no more vertices can be added to $F'$, which is thus maximal. Let us argue that $F'$ is an $S$-forest. Assume for contradiction that it is not. Then a vertex $y$ of $R$ is involved in an $S$-cycle $C$ in $G - U$ such that $V(C) \subseteq F'$. Then by Invariant 1, there is an $S$-cycle $C'$ in $G'$ such that $V(C') \subseteq F' \setminus R$, which contradicts the assumption that $F' \setminus R$ is an $S$-forest of $G'$. □

The *measure* of an instance $(G', F, U, R)$ is the number of undecided vertices, i.e., the vertices in $G' - F$. In the beginning of the algorithm all vertices are undecided and hence the measure of $(G, \emptyset, \emptyset, \emptyset)$ is $n$. The measure drops by the number of vertices deleted from $G'$ plus the number of vertices added to $F$. Hiding a vertex does not affect the measure of an instance. In the call with input $(G', F, U, R)$, the algorithm will further branch into subproblems in which some vertices will be deleted from $G'$ and some vertices will be placed in $F$, and the measure will drop accordingly. If at a step, we branch into $t$ new subproblems, where the measure decreases by $c_1, c_2, \ldots, c_t$ in each subproblem, respectively, we get the *branching vector* $(c_1, c_2, \ldots, c_t)$. At each branching point, we will give the corresponding branching vector to prepare for the running time analysis, which will be given in the last subsection of this section.

We now describe the reduction and the branching rules of the algorithm when $G'$ has undecided vertices and $F$ is an $S$-forest. Let $(G', F, U, R)$ be a call of the algorithm satisfying this. In the below, we let $N(v) = N_{G'}(v)$, $N[v] = N_{G'}[v]$, and $d(v) = d_{G'}(v)$. First, we state three reduction rules. These rules are applied recursively on the considered instance as long as it is possible to apply at least one of them. It is easy to see that the first reduction rule is safe:

**Rule A.** *If in $G'$ an undecided vertex $v$ is adjacent to vertices $u, w \in F$ such that $uw \in E$ and $\{u, v, w\} \cap S \neq \emptyset$, then delete $v$, i.e., reduce to the subproblem $(G' - v, F, U \cup \{v\}, R)$.*

The following observation immediately results in the next reduction rule: Rule B.

**Observation 1** *Let $v$ be a vertex of $G'$ such that no $S$-cycle of $G'$ contains $v$. Then $v$ must be added to $F$ if it is not in $F$, and it is then safe to hide $v$.*

**Rule B.** *If $G'$ has a vertex $v$ with $d(v) \leq 1$, then add $v$ to $F$ if $v$ is undecided, and when $v \in F$ then hide $v$, i.e., reduce to the subproblem $(G' - v, F \cup \{v\}, U, R \cup \{v\})$.*

Since $G'$ is not empty and it is chordal, it has a simplicial vertex. With the following observation we obtain the next reduction rule: Rule C.

**Observation 2** *Let $v$ be a simplicial vertex of $G'$. If $N[v] \cap S = \emptyset$, then $v$ must be added to $F$ if it is not already in $F$, and it is then safe to hide $v$.*

**Rule C.** *If there is a simplicial vertex $v$ such that $N[v] \cap S = \emptyset$, then add $v$ to $F$ if $v$ is undecided, and when $v \in F$ then hide $v$, i.e., reduce to the subproblem $(G' - v, F \cup \{v\}, U, R \cup \{v\})$.*

If we cannot apply Rules A–C, then we start branching. To do it, we pick a simplicial vertex $v$, hence $N(v)$ is a clique. If vertex $v$ is undecided then we proceed as described in the first subsection below. If $v \in F$ then we proceed as described in the second subsection below. Notice that by Rule B, $d(v) \geq 2$.

### 3.1 The Chosen Simplicial Vertex $v$ Is Undecided

**Case 3.1.1:** $v \notin F$, $v \in S$, and $N(v) \cap F = \emptyset$.

If $d(v) = 2$ then let $u_1$ and $u_2$ be the two neighbors of $v$. Since $v \in S$, at most two vertices from $\{v, u_1, u_2\}$ can be added to $F$. Note however that, if exactly one of $u_1, u_2$ is added to $F$ and the other one is deleted, then $v$ must also be added to $F$ by Observation 1. This implies that if $v$ is deleted then both $u_1$ and $u_2$ must be added to $F$. Consequently, we branch into the following subproblems, which cover all possibilities, and we obtain $(3, 3, 3, 3)$ as the branching vector:

- Vertex $v$ is deleted from $G'$ and added to $U$; vertices $u_1$ and $u_2$ are added to $F$: the decrease in the measure is 3.
- Vertex $u_1$ is deleted from $G'$ and added to $U$; vertices $v$ and $u_2$ are added to $F$: the decrease is 3.
- Vertex $u_2$ is deleted from $G'$ and added to $U$; vertices $v$ and $u_1$ are added to $F$: the decrease is 3.
- Vertices $u_1$ and $u_2$ are deleted from $G'$ and added to $U$; vertex $v$ is added to $F$: the decrease is 3.

If $d(v) = 3$ then let $u_1, u_2, u_3$ be the three neighbors of $v$. Again, at most two vertices from $\{v, u_1, u_2, u_3\}$ can be added to $F$. As above, we will branch on the possibilities of adding $v$ and at most one of its neighbors into $F$ and deleting the other neighbors, or deleting $v$. For the choice of deleting $v$, we observe the following: either $u_1$ is added to $F$ or $u_1$ is also deleted. If both $v$ and $u_1$ are deleted, then both $u_2$ and $u_3$ must be added to $F$, by Observation 1. Consequently, we branch into the following subproblems, which cover all possibilities, and we obtain $(4, 4, 4, 4, 2, 4)$ as the branching vector:

- Vertices $u_2$ and $u_3$ are deleted from $G'$ and added to $U$; vertices $v$ and $u_1$ are added to $F$: the decrease is 4.
- Vertices $u_1$ and $u_3$ are deleted from $G'$ and added to $U$; vertices $v$ and $u_2$ are added to $F$: the decrease is 4.
- Vertices $u_1$ and $u_2$ are deleted from $G'$ and added to $U$; vertices $v$ and $u_3$ are added to $F$: the decrease is 4.
- Vertices $u_1$, $u_2$, and $u_3$ are deleted from $G'$ and added to $U$; vertex $v$ is added to $F$: the decrease is 4.
- Vertex $v$ is deleted from $G'$ and added to $U$; vertex $u_1$ is added to $F$: the decrease is 2.
- Vertices $v$ and $u_1$ are deleted from $G'$ and added to $U$; vertices $u_2$ and $u_3$ are added to $F$: the decrease is 4.

In the rest we assume that $t = d(v) \geq 4$. By the same arguments as above, either $v$ is deleted or it is added to $F$ with at most one of its neighbors. Consequently, we branch into the following subproblems, where $u_1, u_2, \ldots, u_t$ are the neighbors of $v$ in $G'$:

- Vertex $v$ is deleted from $G'$ and added to $U$; nothing else changes: the decrease in the measure is 1.
- Vertex $v$ is added to $F$; all of its neighbors are deleted from $G'$ and added to $U$: the decrease in the measure is $t + 1$.

– Vertices $v$ and $u_1$ are added to $F$; all other neighbors of $v$ are deleted from $G'$ and added to $U$: the decrease is $t+1$.
– The last step above is repeated with each of the other neighbors of $v$ instead of $u_1$: the decrease is $t+1$ in each of these $t-1$ additional cases.

The branching vector is $(1, t+1, t+1, \ldots, t+1)$, where the term $t+1$ appears $t+1$ times, and $t \geq 4$.

**Case 3.1.2:** $v \notin F$, $v \in S$, and $N(v) \cap F \neq \emptyset$.

As we cannot apply Rule A for the considered instance, $|N(v) \cap F| = 1$. Since $t = d(v) \geq 2$, we know that $v$ has exactly one neighbor in $F$, say $u_1 \in F$, whereas the rest of its neighbors $u_2, \ldots, u_t$ are undecided. We branch into the two possibilities of adding $v$ to $F$ or deleting $v$. If we add $v$ to $F$, since one neighbor is already in $F$ then none of the $t-1$ undecided neighbors can be added, and therefore we delete them from $G'$ and add them to $U$. We get the following two subproblems: $(G' - v, U \cup \{v\}, F, R)$ and $(G' - \{u_2, \ldots, u_t\}, U \cup \{u_2, \ldots, u_t\}, F \cup \{v\}, R)$. In the first subproblem the measure decreases by 1, and in the second it decreases by $t$. We get the branching vector $(1, t)$ with $t \geq 2$.

**Case 3.1.3:** $v \notin F$, $v \notin S$, and $N(v) \cap F = \emptyset$.

Since we cannot apply Rule C, $v$ has at least one neighbor belonging to $S$.

If $d(v) = 2$, let $u_1$ and $u_2$ be the neighbors of $v$. Since $u_1$ or $u_2$ belongs to $S$, we know that at most two vertices from $\{v, u_1, u_2\}$ can be added to $F$. Consequently, this case is identical to the subcase of Case 3.1.1 handling $d(v) = 2$. We branch into the same subproblems and we obtain $(3, 3, 3, 3)$ as the branching vector.

If $d(v) = 3$, let $u_1, u_2, u_3$ be the neighbors of $v$. Assume without loss of generality that $u_1 \in S$. This case is very similar to the subcase of Case 1 handling $d(v) = 3$, but now we branch on $u_1$ instead of $v$. If $u_1$ is added to $F$ then at most one of $v, u_2, u_3$ can be added to $F$. If $u_1$ is deleted then either $v$ is added to $F$ or $v$ is also deleted. If $v$ is also deleted then both $u_2$ and $u_3$ must be added to $F$, by Observation 1. Consequently, we branch into the following subproblems, which cover all possibilities, and we obtain $(4, 4, 4, 4, 2, 4)$ as the branching vector:

– Vertices $u_2$ and $u_3$ are deleted from $G'$ and added to $U$; vertices $u_1$ and $v$ are added to $F$: the decrease is 4.
– Vertices $v$ and $u_3$ are deleted from $G'$ and added to $U$; vertices $u_1$ and $u_2$ are added to $F$: the decrease is 4.
– Vertices $v$ and $u_2$ are deleted from $G'$ and added to $U$; vertices $u_1$ and $u_3$ are added to $F$: the decrease is 4.
– Vertices $v$, $u_2$, and $u_3$ are deleted from $G'$ and added to $U$; vertex $u_1$ is added to $F$: the decrease is 4.
– Vertex $u_1$ is deleted from $G'$ and added to $U$; vertex $v$ is added to $F$: the decrease is 2.
– Vertices $u_1$ and $v$ are deleted from $G'$ and added to $U$; vertices $u_2$ and $u_3$ are added to $F$: the decrease is 4.

If $t = d(v) \geq 4$, then let $u_1, u_2, \ldots, u_t$ be the neighbors of $v$ in $G'$, and assume without loss of generality that $u_1 \in S$. We will branch on the two possibilities of adding $u_1$ to $F$ and deleting $u_1$. If we add $u_1$ to $F$ then we can add at most one other vertex of $N[v]$ to $F$ and all others must be deleted. Consequently, we branch into the following subproblems:

- Vertex $u_1$ is deleted from $G'$ and added to $U$; nothing else changes: the decrease in the measure is 1.
- Vertex $u_1$ is added to $F$; vertices $v, u_2, \ldots, u_t$ are deleted from $G'$ and added to $U$: the decrease in the measure is $t+1$.
- Vertices $u_1$ and $v$ are added to $F$; all other neighbors of $v$ are deleted from $G'$ and added to $U$: the decrease is $t+1$.
- Vertices $u_1$ and $u_2$ are added to $F$; $v$ and all other neighbors of $v$ are deleted from $G'$ and added to $U$: the decrease is $t+1$.
- The last step above is repeated with each of the neighbors $u_3, \ldots, u_t$ of $v$ instead of $u_2$: the decrease is $t+1$ in each of these $t-2$ additional cases.

The branching vector is $(1, t+1, t+1, \ldots, t+1)$, where the term $t+1$ appears $t+1$ times, with $t \geq 4$.

**Case 3.1.4:** $v \notin F$, $v \notin S$, and $N(v) \cap F \neq \emptyset$.

As we cannot apply Rule C, $N(v) \cap S \neq \emptyset$. Suppose that $|N(v) \cap F| \geq 2$. If there is a vertex $u \in (N(v) \setminus F) \cap S$, then Rule A can be applied for $u$. Consequently, there is a vertex $u \in N(v) \cap F \cap S$, but then Rule A can be applied for $v$. It means that $v$ has exactly one neighbor $u$ in $F$. We take action depending on whether or not $u$ belongs to $S$:

If $u \in S$, then at most one more vertex from $N[v]$ can be added to $F$, and all others must be deleted from $G'$ and added to $U$. We get $t = d(v)$ subproblems in each of which a vertex of $N[v] \setminus \{u\}$ is added to $F$ and all others are deleted from $G'$ and added to $U$. Observe that we do not get a subproblem where all vertices of $N[v] \setminus \{u\}$ are deleted from $G'$, due to Observation 1. Thus we get $(t, \ldots, t)$ as the branching vector, where the term $t$ is repeated $t$ times, and $t \geq 2$.

If $u \notin S$, then we know that $v$ has another neighbor $w \in S$. We branch into two subproblems resulting from adding $w$ to $F$ or deleting $w$ from $G'$. If we add $w$ to $F$, then since $u$ is also in $F$, no other vertex from $N[v]$ can be added to $F$ and hence they must all be deleted from $G'$ and added to $U$. We get a subproblem in which the measure decreases by $t = d(v)$. In the other subproblem we simply delete $w$ from $G'$ and add it to $U$; the decrease is 1. Hence we get $(1, t)$ as the branching vector for this case, where $t \geq 2$.

### 3.2 The Chosen Simplicial Vertex $v$ Belongs to $F$

**Case 3.2.1:** $v \in F$ and $v \in S$.

Because $G[F]$ has no $S$-cycles, $|N(v) \cap F| \leq 1$. If $N(v) \cap F \neq \emptyset$, then Rule A can be applied for the vertices $N(v) \setminus F$. It follows that $N(v) \cap F = \emptyset$. Since $v \in S$ and $v \in F$, at most one vertex of $N(v)$ can be added to $F$, regardless of how many of these are in $S$.

If $d(v) = 2$ then let $u$ and $w$ be the two neighbors of $v$. We branch on the two possibilities of either adding $u$ to the $S$-forest $F$ or adding $u$ to the subset feedback vertex set $U$. In the latter subproblem we delete $u$ from $G'$ and add it to $U$; the decrease is 1. In the first subproblem, we add $u$ to $F$, and consequently we must delete $w$ from $G'$ and add it to $U$; the decrease is 2. We get $(1, 2)$ as the branching vector.

If $t = d(v) \geq 3$ then we branch into the possibilities of adding exactly one vertex of $N(v)$ to $F$ and deleting all others from $G'$, or deleting all vertices of $N(v)$ from $G'$. We get $t$ subproblems in which one vertex is added to $F$ and all other vertices of $N(v)$ are deleted from $G'$ and added to $U$, and one subproblem in which all vertices of $N(v)$ are deleted from $G'$ and added to $U$. In each of these $t+1$ subproblems the decrease is $t$. Hence we get $(t, t, t, \ldots, t)$ as the branching vector, where the term $t$ is repeated $t+1$ times, and $t \geq 3$.

**Case 3.2.2:** $v \in F$ and $v \notin S$.

Suppose that $N(v) \cap F \neq \emptyset$. If a neighbor $u$ of $v$ is both in $F$ and in $S$, then all other neighbors of $v$ are undecided, since $G[F]$ has no $S$-cycles. Then we can apply Rule A for these neighbors of $v$. If there is $u \in (N(v) \cap F) \setminus S$, then Rule A can be applied for all $w \in N(v) \cap S$. It means that $N(v) \cap S = \emptyset$, but in this case we can apply Rule C. Therefore, $N(v) \cap F = \emptyset$. Because we cannot apply Rule C, $v$ has at least one neighbor that is undecided and belongs to $S$.

Recall that $t = d(v) \geq 2$, and let $u_1, u_2, \ldots, u_t$ be the neighbors of $v$, and assume without loss of generality that $u_1 \in S$. We branch into the two possibilities of either deleting $u_1$ from $G'$ and adding it to $U$, or adding $u_1$ to $F$. In the latter case, no other neighbor of $N(v)$ can be added to $F$, since they all form $S$-cycles with $v$ and $u_1$, and hence they must all be deleted from $G'$ and added to $U$. We get one subproblem where the decrease is 1, and one subproblem where the decrease is $t$. This gives us the branching vector $(1, t)$ with $t \geq 2$.

The description of the algorithm is now complete. The correctness of the algorithm follows from Invariant 1, Lemma 1, Observations 1, 2, and the arguments given for each case, observing that we have taken care of all possible cases. In the next section, we analyze the running time.

### 3.3 Running Time Analysis

In each of the branching rules, the measure decreases as described, and in each of the reduction rules, either the measure decreases or at least one vertex of $F$ is deleted from $G'$. When all vertices of $G'$ are either in $U$ or in $F$, then the recurrence stops. At this point we need to check whether $F$ is a maximal $S$-forest of $G$. This can easily be done in polynomial time; $F$ is an $S$-forest if and only if every vertex of $S \cap F$ is incident in $G$ to edges that are bridges. Maximality is also easy to check since if a subset $X$ of $V \setminus F$ can be added to $F$ to obtain a larger $S$-forest, then also a single vertex of $X$ can be added, so we can repeatedly check possible extensions by single vertices. Consequently, the running time will be upper bounded by the number of leaves in the search tree.

For the analysis of the number of leaves $T(n)$ in the search tree, we use standard terminology [8]. In particular, a branching vector $(c_1, c_2, \ldots, c_t)$ results in the recurrence $T(n) \leq T(n-c_1)+T(n-c_2)+\ldots+T(n-c_t)$. In this case $T(n) =$

$O^*(\alpha^n)$, where $\alpha$ is the unique positive real root of $x^n - x^{n-c_1} - \ldots - x^{n-c_t} = 0$ [8], and the $O^*$-notation suppresses polynomial factors. The number $\alpha$ is called the *branching number* of this branching vector. It is common to round $\alpha$ to the fourth digit after the decimal point. By rounding the last digit up, we can use $O$-notation instead of $O^*$-notation [8]. As different branching vectors are involved at different steps of our algorithm, the branching vector with the highest branching number gives an upper bound on $T(n)$.

We now list the branching vectors that have appeared during the description of the algorithm, in the order of first appearence. We give the branching number for each of them; however we do not include here the explicit calculations.

- $(3, 3, 3, 3)$: the branching number is $\approx 1.5875$.
- $(4, 4, 4, 4, 2, 4)$: the branching number is $\approx 1.6708$.
- $(1, t, t, t, t, \ldots, t)$, where the term $t$ appears $t$ times, and $t \geq 5$: $(1, 5, 5, 5, 5, 5)$ gives the maximum branching number for this vector, which is $\approx 1.6595$.
- $(1, t)$, $t \geq 2$: $(1, 2)$ gives the maximum branching number for this branching vector, which is $\approx 1.6181$.
- $(t, \ldots, t)$, where the term $t$ is repeated $t$ times, and $t \geq 2$: $(3, 3, 3)$ gives the maximum branching number for this vector, which is $\approx 1.4423$.
- $(t, t, \ldots, t)$, where the term $t$ is repeated $t + 1$ times, and $t \geq 3$: $(3, 3, 3, 3)$ gives the maximum branching number for this vector, which is $\approx 1.5875$.
- $(1, 2)$: the branching number is $\approx 1.6181$.

The largest branching number is 1.6708, and it is obtained for $(4, 4, 4, 4, 2, 4)$. Thus the running time of our algorithm is $O(1.6708^n)$.

## 4 Concluding Remarks

As mentioned earlier, there are chordal graphs with $10^{n/5} \approx 1.5848$ minimal sfvs. We have shown that the maximum number of minimal sfvs in chordal graphs is $O(1.6708^n)$. Could it be that the lower bound is also an upper bound or are there chordal graphs with more than $10^{n/5}$ minimal sfvs? Is there an algorithm for Subset Feeback Vertex Set on chordal graphs with running time $O(c^n)$ such that $c < 1.6707^n$?

The lower bound on the maximum number of minimal sfvs of a split graph is $3^{n/3}$ [7], and it is obtained when $S$ is equal to the independent set. Is there a better upper bound for split graphs than for chordal graphs? Does Subset Feeback Vertex Set admit a faster solution on split graphs than on chordal graphs?

We conclude by asking whether all minimal sfvs can be enumerated in time that is polynomial in the number of minimal sfvs. Such an algorithm is known for enumerating minimal fvs in general graphs [17]. It would be very interesting to have such an algorithm for sfvs, even on chordal graphs or split graphs.

## References


1. Corneil, D.G., Fonlupt, J.: The complexity of generalized clique covering. Disc. Appl. Math. 22, 109–118 (1988/1989)
2. Couturier, J.-F., Heggernes, P., van't Hof, P., Villanger, Y.: Maximum number of minimal feedback vertex sets in chordal graphs and cographs. In: Proceedings. of COCOON 2012. LNCS (to appear, 2012)
3. Cygan, M., Pilipczuk, M., Pilipczuk, M., Wojtaszczyk, J.O.: Subset Feedback Vertex Set Is Fixed-Parameter Tractable. In: Aceto, L., Henzinger, M., Sgall, J. (eds.) ICALP 2011, Part I. LNCS, vol. 6755, pp. 449–461. Springer, Heidelberg (2011)
4. Even, G., Naor, J., Zosin, L.: An 8-approximation algorithm for the subset feedback vertex set problem. SIAM J. Comput. 30(4), 1231–1252 (2000)
5. Fomin, F.V., Gaspers, S., Pyatkin, A.V., Razgon, I.: On the minimum feedback vertex set problem: Exact and enumeration algorithms. Algorithmica 52(2), 293–307 (2008)
6. Fomin, F.V., Grandoni, F., Pyatkin, A.V., Stepanov, A.A.: Combinatorial bounds via measure and conquer: Bounding minimal dominating sets and applications. ACM Trans. Algorithms 5(1) (2008)
7. Fomin, F.V., Heggernes, P., Kratsch, D., Papadopoulos, C., Villanger, Y.: Enumerating Minimal Subset Feedback Vertex Sets. In: Dehne, F., Iacono, J., Sack, J.-R. (eds.) WADS 2011. LNCS, vol. 6844, pp. 399–410. Springer, Heidelberg (2011)
8. Fomin, F.V., Kratsch, D.: Exact Exponential Algorithms. Texts in Theoretical Computer Science. Springer (2010)
9. Fomin, F.V., Villanger, Y.: Finding induced subgraphs via minimal triangulations. In: Proceedings of STACS 2010, pp. 383–394 (2010)
10. George, J.A., Liu, J.W.H.: Computer Solution of Large Sparse Positive Definite Systems. Prentice-Hall Inc. (1981)
11. Garey, M.R., Johnson, D.S.: Computers and Intractability. Freeman and Co. (1978)
12. Golumbic, M.C.: Algorithmic Graph Theory and Perfect Graphs. Annals of Disc. Math. 57 (2004)
13. Gaspers, S., Mnich, M.: Feedback Vertex Sets in Tournaments. In: de Berg, M., Meyer, U. (eds.) ESA 2010, Part I. LNCS, vol. 6346, pp. 267–277. Springer, Heidelberg (2010)
14. Kanté, M.M., Limouzy, V., Mary, A., Nourine, L.: Enumeration of Minimal Dominating Sets and Variants. In: Owe, O., Steffen, M., Telle, J.A. (eds.) FCT 2011. LNCS, vol. 6914, pp. 298–309. Springer, Heidelberg (2011)
15. Kratsch, D., Müller, H., Todinca, I.: Feedback vertex set on AT-free graphs. Disc. Appl. Math. 156, 1936–1947 (2008)
16. Moon, J.W., Moser, L.: On cliques in graphs. Israel J. Math. 3, 23–28 (1965)
17. Schwikowski, B., Speckenmeyer, E.: On enumerating all minimal solutions of feedback problems. Disc. Appl. Math. 117, 253–265 (2002)
18. Semple, C., Steel, M.: Phylogenetics. Oxford lecture series in mathematics and its applications (2003)
19. Spinrad, J.P.: Efficient graph representations. Fields Institute Monograph Series, vol. 19. AMS (2003)

# Preprocessing Subgraph and Minor Problems: When Does a Small Vertex Cover Help?*

Fedor V. Fomin[1], Bart M.P. Jansen[2], and Michał Pilipczuk[1]

[1] Department of Informatics, University of Bergen, N-5020 Bergen, Norway
{fomin,michal.pilipczuk}@ii.uib.no

[2] Utrecht University, P.O. Box 80.089, 3508 TB Utrecht, The Netherlands
b.m.p.jansen@uu.nl

**Abstract.** We prove a number of results around kernelization of problems parameterized by a vertex cover of the input graph. We provide two simple general conditions characterizing problems admitting kernels of polynomial size. Our characterizations not only give generic explanations for the existence of many known polynomial kernels for problems like Odd Cycle Transversal, Chordal Deletion, $\eta$-Transversal, Long Path, Long Cycle, or $H$-packing, parameterized by the size of a vertex cover, they also imply new polynomial kernels for problems like $\mathcal{F}$-Minor-Free Deletion, which is to delete at most $k$ vertices to obtain a graph with no minor from a fixed finite set $\mathcal{F}$.

While our characterization captures many interesting problems, the kernelization complexity landscape of problems parameterized by vertex cover is much more involved. We demonstrate this by several results about induced subgraph and minor containment, which we find surprising. While it was known that testing for an induced complete subgraph has no polynomial kernel unless NP $\subseteq$ coNP/poly, we show that the problem of testing if a graph contains a complete graph on $t$ vertices as a minor admits a polynomial kernel. On the other hand, it was known that testing for a path on $t$ vertices as a minor admits a polynomial kernel, but we show that testing for containment of an induced path on $t$ vertices is unlikely to admit a polynomial kernel.

## 1 Introduction

Kernelization is an attempt at providing rigorous mathematical analysis of preprocessing algorithms. While the initial interest in kernelization was driven mainly by practical applications, it turns out that kernelization provides a deep insight into the nature of fixed-parameter tractability. In the last few years, kernelization has transformed into a major research domain of Parameterized Complexity and many important advances in the area are on kernelization. These

* This work was supported by the Netherlands Organization for Scientific Research (NWO), project "KERNELS: Combinatorial Analysis of Data Reduction", and by the European Research Council (ERC) grant "Rigorous Theory of Preprocessing", reference 267959.

D.M. Thilikos and G.J. Woeginger (Eds.): IPEC 2012, LNCS 7535, pp. 97–108, 2012.

advances include general algorithmic findings on problems admitting kernels of polynomial size [1,2,3] and frameworks for ruling out polynomial kernels under certain complexity-theoretic assumptions [4,5,6,7].

A recent trend in the development of Parameterized Complexity, and more generally, Multivariate Analysis [8], is the study of the contribution of various secondary measurements (i.e., different than just the total input size or solution size) to problem complexity. Not surprisingly, the development of kernelization followed this trend resulting in various kernelization algorithms and complexity lower bounds for different kinds of parameterizations. In parameterized graph algorithms, one of the most important and relevant *complexity measures* of the graph is its treewidth. The algorithmic properties of problems parameterized by treewidth are, by now, well-understood. However, from the perspective of kernelization, this complexity measure is too general to obtain positive results: it is known for a multitude of graph problems such as VERTEX COVER, DOMINATING SET, and 3-COLORING, that there are no polynomial kernels parameterized by the treewidth of the input graphs unless NP $\subseteq$ coNP/poly [4]. This is why parameterization by more restrictive complexity measures, like the minimum size of a feedback vertex set or a vertex cover, is much more fruitful for kernelization.

In particular, kernelization of graph problems parameterized by the *vertex cover number*, which is the size of the smallest vertex set meeting all edges, was studied intensively [5,9,10,11,12]. For example, it has been shown that several graph problems such as VERTEX COVER, TREEWIDTH, and 3-COLORING, admit polynomial kernels parameterized by the size of a given vertex cover. On the other hand, under certain complexity-theoretic assumptions it is possible to show that a number of problems including DOMINATING SET [11], CLIQUE [5], and CHROMATIC NUMBER [5], do not admit polynomial kernels for this parameter. While different kernelization algorithms for various problems parameterized by vertex cover are known, we lack general a characterization of such problems. The main motivation of our work on this paper is the quest for meta-theorems on kernelization algorithms for problems parameterized by vertex cover.

According to Grohe [13], meta-theorems expose the deep relations between logic and combinatorial structures, which is a fundamental issue of computational complexity. Such theorems also yield a better understanding of the scope of general algorithmic techniques and the limits of tractability. The canonical example here is Courcelle's Theorem which states that all problems expressible in Monadic Second-Order Logic are linear-time solvable on graphs of bounded treewidth (see also [14,15]). In kernelization there are meta-theorems showing polynomial kernels for restricted graph families [1,2]. A systematic way to understand the kernelization complexity of parameterizations by vertex cover would therefore be to obtain a meta-theorem capturing a large class of problems admitting polynomial kernels. But is there a logic capturing the known positive results we are interested in? If such a logic exist, it would have to be able to express VERTEX COVER, which admits polynomial kernel, but not CLIQUE, which does not [16]; it should capture ODD CYCLE TRANSVERSAL and LONG CYCLE [16] but not DOMINATING SET [11]; and TREEWIDTH [9] but not CUTWIDTH [10]. As a consequence, if a logic capturing the phenomenon of polynomial kernelizability for problems parameterized by vertex cover exists, it should be a very strange logic and we therefore take a different approach.

In this paper, we provide two theorems with general conditions capturing a wide variety of known kernelization results about vertex cover parameterization. It has been observed before that reduction rules which identify irrelevant vertices by marking a polynomial number of vertices for each constant-sized subset of the vertex cover, lead to a polynomial kernel for several problems [17,12]. Our first contribution here is to uncover a characteristic of graph problems which explains their amenability to such reduction strategies, and to provide theorems using this characteristic. Roughly speaking, the problem of finding a minimum-size set of vertices which hits all induced subgraphs belonging to some family $\Pi$ has a polynomial kernel parameterized by vertex cover, if membership in $\Pi$ is invariant under changing the presence of all but a constant number of (non)edges incident to each vertex (and some technical conditions are met). The problem of finding the largest induced subgraph belonging to $\Pi$ has a polynomial kernel parameterized by vertex cover under similar conditions. Our general theorems not only capture a wide variety of known results, they also imply results which were not known before. For example, as a corollary to our theorems we establish that the $\mathcal{F}$-Minor-Free Deletion deletion problem (i.e., for a fixed, finite list $\mathcal{F}$ of graphs, can we delete $k$ vertices from $G$ to ensure that the remaining graph does not contain a graph from $\mathcal{F}$ as a minor?) has a polynomial kernel for every fixed $\mathcal{F}$, when parameterized by the size of a vertex cover.

After studying the kernelization complexity of vertex-deletion and largest induced subgraph problems, we turn to two basic cases of property $\Pi$: containing some graph as an induced subgraph or minor. It is known that testing for a clique as an induced subgraph (when the desired size of the subgraph is part of the input) does not admit a polynomial kernel parameterized by vertex cover unless NP $\subseteq$ coNP/poly [5]. This is why we find the following result surprising: testing for a clique as a minor admits a polynomial kernel under the chosen parameterization. Driven by our desire to obtain a better understanding of the kernelization complexity of problems parameterized by vertex cover, we investigate induced subgraph testing and minor testing for other classes of graphs such as cycles, paths, matchings and stars. The kernelization complexity of induced subgraph testing and minor testing turns out to be exactly opposite for all these classes. For example, testing for a star minor does not have a polynomial kernel due to its equivalence to Connected Dominating Set [11], but we provide a polynomial kernel for testing the existence of an induced star subgraph by using a guessing step to reduce it to cases which are covered by our general theorems.

The paper is organized as follows. In Section 3 we describe a general reduction scheme and use it to derive sufficient conditions for the existence of polynomial kernels. In Section 4 we investigate the kernelization complexity of induced subgraph versus minor testing for various graph families. A succinct overview of our results is given in Tables 1, 2, and 3. Several proofs had to be deferred to the full version of this paper [18] due to space restrictions.

## 2 Preliminaries

**Parameterized Complexity and Kernels.** A parameterized problem $Q$ is a subset of $\Sigma^* \times \mathbb{N}$, the second component being the *parameter* which expresses some structural measure of the input. A parameterized problem is (strongly uniformly) *fixed-parameter tractable* if there exists an algorithm to decide whether $(x, k) \in Q$ in time $f(k)|x|^{\mathcal{O}(1)}$ where $f$ is a computable function. We refer to the textbooks by Downey and Fellows, Flum and Grohe, and Niedermeier, for more background on parameterized complexity.

A *kernelization algorithm* (or *kernel*) for a parameterized problem $Q$ is a polynomial-time algorithm which transforms an instance $(x, k)$ into an equivalent instance $(x', k')$ such that $|x'|, k' \leq f(k)$ for some computable function $f$, which is the *size* of the kernel. If $f \in k^{\mathcal{O}(1)}$ then this is a *polynomial kernel* (cf. [19]).

**Graphs.** All graphs we consider are finite, simple, and undirected. An undirected graph $G$ consists of a vertex set $V(G)$ and a set of edges $E(G)$. A graph property $\Pi$ is a (possibly infinite) set of graphs. The maximum degree of a vertex in $G$ is denoted by $\Delta(G)$. A graph $G$ is *empty* if $E(G) = \emptyset$. A vertex $v$ is *simplicial* in graph $G$ if $N_G(v)$ is a clique. A *minor model* of a graph $H$ in a graph $G$ is a mapping $\phi$ from $V(H)$ to subsets of $V(G)$ (called *branch sets*) which satisfies the following conditions: (a) $\phi(u) \cap \phi(v) = \emptyset$ for distinct $u, v \in V(H)$, (b) $G[\phi(v)]$ is connected for $v \in V(H)$, and (c) there is an edge between a vertex in $\phi(u)$ and a vertex in $\phi(v)$ for all $uv \in E(H)$. An *$H$-packing* in $G$ is a set of vertex-disjoint subgraphs of $G$, each of which is isomorphic to $H$. An $H$-packing is *perfect* if the subgraphs cover the entire vertex set. The minimum size of a vertex cover in a graph $G$ is denoted by $\textsc{vc}(G)$. For a set of vertices $X$ in a graph $G$ we use $G - X$ to denote the graph which results after deleting all vertices of $X$ and their incident edges. We use the terms $K_t$ and $P_t$ to denote a clique or path on $t$ vertices, respectively, whereas $K_{s,t}$ is a *biclique* (complete bipartite graph) whose partite sets have sizes $s$ and $t$. The disjoint union of $t$ copies of a graph $G$ is represented by $t \cdot G$. The set $\{1, 2, \ldots, n\}$ is abbreviated as $[n]$. If $X$ is a finite set then $\binom{X}{n}$ denotes the collection of all subsets of $X$ which have size exactly $n$. Similarly we use $\binom{X}{\leq n}$ for the subsets of size at most $n$ (including $\emptyset$). The following proposition will be useful in showing that $\mathcal{F}$-Minor-Free Deletion can be captured by our general theorems.

**Proposition 1.** *If $G$ contains $H$ as a minor, then there is a subgraph $G^* \subseteq G$ containing an $H$-minor such that $\Delta(G^*) \leq \Delta(H)$ and $|V(G^*)| \leq |V(H)| + \textsc{vc}(G^*) \cdot (\Delta(H) + 1)$.*

## 3 General Kernelization Theorems

### 3.1 Characterization by Few Adjacencies

In this section we introduce a general reduction rule for problems parameterized by vertex cover, and show that the rule preserves the existence of certain kinds of induced subgraphs. The central concept is the following.

**Algorithm 1.** REDUCE(Graph $G$, Vertex cover $X \subseteq V(G), \ell \in \mathbb{N}, c_\Pi \in \mathbb{N}$)

**for each** $Y \in \binom{X}{\leq c_\Pi}$ and partition of $Y$ into $Y^+ \dot\cup Y^-$ **do**
  let $Z$ be the vertices in $V(G) \setminus X$ adjacent to all of $Y^+$ and to none of $Y^-$
  mark $\ell$ arbitrary vertices from $Z$ (if $|Z| < \ell$ then mark all of them)
delete from $G$ all unmarked vertices which are not contained in $X$

**Definition 1.** *A graph property $\Pi$ is* characterized by $c_\Pi \in \mathbb{N}$ adjacencies *if for all graphs $G \in \Pi$, for every vertex $v \in V(G)$, there is a set $D \subseteq V(G) \setminus \{v\}$ of size at most $c_\Pi$ such that all graphs $G'$ which are obtained from $G$ by adding or removing edges between $v$ and vertices in $V(G) \setminus D$, are also contained in $\Pi$.*

As an example of a property characterized by few adjacencies, consider the Hamiltonian graphs, i.e., the graphs which have a Hamiltonian cycle. This property is characterized by two adjacencies: given a graph $G$ with a Hamiltonian cycle $C$ and a vertex $v$, it is easy to see that as long as we preserve the edges between $v$ and its predecessor and successor on $C$, changing the adjacency between $v$ and other vertices preserves the Hamiltonicity of $G$. There are numerous other graph properties which are characterized by few adjacencies.

**Proposition 2.** *The following properties are characterized by constantly many adjacencies: (for any fixed finite set $\mathcal{F}$, graph $H$, or $\ell \geq 4$, respectively)*

1. *Containing $H \in \mathcal{F}$ as a minor ($c_\Pi = \max_{H \in \mathcal{F}} \Delta(H)$).*
2. *Having a perfect $H$-packing ($c_\Pi = \Delta(H)$).*
3. *Having a chordless cycle of length at least $\ell$ ($c_\Pi = \ell - 1$).*
4. *Having a Hamiltonian path (resp. cycle), or having an odd cycle ($c_\Pi = 2$).*

As an illustrative non-example, note that the properties of being a cycle, of having chromatic number at least four, or of *not* being a perfect graph, cannot be characterized by a constant number of adjacencies.

The single reduction rule that we will use to derive our general kernelization theorems, is the REDUCE procedure presented as Algorithm 1. Its utility for kernelization stems from the fact that it efficiently shrinks a graph to a size bounded polynomially in the cardinality of the given vertex cover $X$.

**Observation 1.** *For every fixed constant $c_\Pi$, REDUCE$(G, X, \ell, c_\Pi)$ runs in polynomial time and results in a graph on $\mathcal{O}(|X| + \ell \cdot 2^{c_\Pi} \cdot |\binom{X}{\leq c_\Pi}|) = \mathcal{O}(|X| + \ell \cdot |X|^{c_\Pi})$ vertices.*

The soundness of the REDUCE procedure for many types of kernelization comes from the following lemma. It shows that for graph properties $\Pi$ which are characterized by few adjacencies, an application of REDUCE with parameter $\ell = s + p$ preserves the existence of induced $\Pi$ subgraphs of size up to $p$ that avoid any set of size at most $s$.

**Lemma 1.** *Let $\Pi$ be characterized by $c_\Pi$ adjacencies, and let $G$ be a graph with vertex cover $X$. If $G[P] \in \Pi$ for some $P \subseteq V(G) \setminus S$ and $S \subseteq V(G)$, then for any $\ell \geq |S| + |P|$ the graph $G'$ resulting from* REDUCE$(G, X, \ell, c_\Pi)$ *contains $P' \subseteq V(G') \setminus S$ such that $G'[P'] \in \Pi$ and $|P'| = |P|$.*

**Table 1.** Problems which admit polynomial kernels when parameterized by the size of a given vertex cover, by applying Theorem 1

| Problem | $\Pi$ | $c_\Pi$ |
|---|---|---|
| VERTEX COVER | $\{K_2\}$ | 1 |
| ODD CYCLE TRANSVERSAL | Graphs containing an odd cycle | 2 |
| CHORDAL DELETION | Graphs with a chordless cycle | 3 |
| PLANARIZATION | Graphs with a $K_5$ or $K_{3,3}$ minor | 4 |
| $\eta$-TRANSVERSAL (cf. [20]) | Graphs of treewidth $> \eta$ | $f(\eta)$ |
| $\mathcal{F}$-MINOR-FREE DELETION | Graphs with an $H \in \mathcal{F}$-minor | $\max_{H \in \mathcal{F}} \Delta(H)$ |

### 3.2 Kernelization for Vertex-Deletion Problems

We will present a general theorem which gives polynomial kernels for vertex-deletion problems of the following form.

DELETION DISTANCE TO $\Pi$-FREE (VC)
**Input:** A graph $G$ with vertex cover $X \subseteq V(G)$, and an integer $k \geq 1$.
**Parameter:** The size $|X|$ of the vertex cover.
**Question:** Is there a set $S \subseteq V(G)$ of size at most $k$ such that $G - S$ does not contain a graph in $\Pi$ as an induced subgraph?

Observe that $\Pi$ need not be finite or decidable. The condition that a vertex cover is given along with the input is present for technical reasons; to apply the data reduction schemes presented in this paper, one may simply compute a 2-approximate vertex cover and use that as $X$.

**Theorem 1.** *If $\Pi$ is a graph property such that:*

*(i) $\Pi$ is characterized by $c_\Pi$ adjacencies,*
*(ii) every graph in $\Pi$ contains at least one edge, and*
*(iii) there is a non-negative polynomial $p\colon \mathbb{N} \to \mathbb{N}$ such that all graphs $G \in \Pi$ contain an induced subgraph $G' \in \Pi$ such that $V(G') \leq p(\text{VC}(G'))$,*

*then* DELETION DISTANCE TO $\Pi$-FREE (VC) *has a kernel with $\mathcal{O}((x+p(x))x^{c_\Pi})$ vertices, where $x := |X|$.*

Before proving the theorem, let us briefly discuss its preconditions. We cannot drop Property (ii), as otherwise the theorem would capture the CLIQUE problem (taking $\Pi := \{2 \cdot K_1\}$), for which a lower bound exists [5]. If Property (i) is dropped, then the theorem would capture problems such as PERFECT DELETION for which the kernelization complexity is still open. We require the third condition to make the proof go through.

*Proof (of Theorem 1).* Consider some input instance $(G, X, k)$. Firstly, observe that if $k \geq |X|$, then we clearly have a YES-instance: removal of $X$ results in an edgeless graph, which is guaranteed not to contain induced subgraphs from $\Pi$ due to Property (ii). Therefore, we may assume that $k < |X|$ as otherwise we output a trivial YES-instance.

We let $G'$ be the result of $\textsc{Reduce}(G, X, k + p(|X|), c_\Pi)$ and return the instance $(G', X, k)$, which gives the right running time and size bound by Observation 1. We need to prove that the output instance $(G', X, k)$ is equivalent to the input instance $(G, X, k)$. As $G'$ is an induced subgraph of $G$, it follows that if $G - S$ does not contain any graph in $\Pi$, then neither does $G' - (S \cap V(G'))$. Therefore, if $(G, X, k)$ is a YES-instance, then so is $(G', X, k)$. Assume then, that $(G', X, k)$ is a YES-instance and let $S$ be a subset of vertices with $|S| \leq k$ such that $G' - S$ does not contain any induced subgraph from $\Pi$. We claim that $G - S$ does not contain such induced subgraphs either, i.e., that $S$ is also a feasible solution for the instance $(G, X, k)$.

Assume for the sake of contradiction that there is a set $P \subseteq V(G) \setminus S$ such that $G[P] \in \Pi$. Consider a *minimal* such set $P$, which ensures by Property (iii) that $|P| \leq p(\mathrm{vc}(G[P]))$. As $P \cap X$ is a vertex cover of $G[P]$, it follows that $|P| \leq p(|P \cap X|) \leq p(|X|)$. As we executed the reduction with parameter $\ell = k + p(|X|)$, Lemma 1 guarantees the existence of a set $P' \subseteq V(G') \setminus S$ such that $G'[P'] \in \Pi$. But this shows that the graph $G' - S$ contains an induced $\Pi$ subgraph, contradicting the assumption that $S$ is a solution for $G'$ and thereby concluding the proof. □

**Corollary 1.** *All problems in Table 1 fit into the framework of Theorem 1 and admit polynomial kernels parameterized by the size of a given vertex cover.*

*Proof.* We give the proof for $\mathcal{F}$ MINOR-FREE DELETION; the proofs for the other items can be found in the full version. If we let $\Pi$ contain all graphs that contain a member of $\mathcal{F}$ as a minor, then a graph is $\Pi$-induced-subgraph-free if and only if it is $\mathcal{F}$-minor-free. By Proposition 2 this class $\Pi$ is characterized by $c_\Pi := \max_{H \in \mathcal{F}} \Delta(H)$ adjacencies, so we satisfy Property (i). If $\mathcal{F}$ contains an empty graph, then $\mathcal{F}$-minor-free graphs have constant size and the problem is polynomial-time solvable; hence in interesting cases the graphs containing a minor from $\mathcal{F}$ have at least one edge (Property (ii)). Finally, consider a vertex-minimal graph $G^*$ which contains a graph $H \in \mathcal{F}$ as a minor. By Proposition 1 we have $|V(G^*)| \leq |V(H)| + \mathrm{vc}(G^*) \cdot (\Delta(H) + 1)$. As $\mathcal{F}$ is fixed, the maximum degree and size of graphs in $\mathcal{F}$ are constants which shows that Property (iii) is satisfied, resulting in a kernel with $\mathcal{O}(|X|^{\Delta+1})$ vertices for $\Delta := \max_{H \in \mathcal{F}} \Delta(H)$. □

### 3.3 Kernelization for Largest Induced Subgraph Problems

In this section we study the following class of problems, which is in some sense dual to the class considered previously.

LARGEST INDUCED $\Pi$-SUBGRAPH (VC)
**Input:** A graph $G$ with vertex cover $X \subseteq V(G)$, and an integer $k \geq 1$.
**Parameter:** The size $|X|$ of the vertex cover.
**Question:** Is there a set $P \subseteq V(G)$ of size at least $k$ such that $G[P] \in \Pi$?

The following theorem gives sufficient conditions for the existence of polynomial kernels for such problems.

**Table 2.** Problems which admit polynomial kernels when parameterized by the size of a given vertex cover, by applying Theorem 2

| Problem | $\Pi$ | $c_\Pi$ |
|---|---|---|
| LONG CYCLE | Graphs with a Hamiltonian cycle | 2 |
| LONG PATH | Graphs with a Hamiltonian path | 2 |
| $H$-PACKING for nonempty $H$ | Graphs with a perfect $H$-packing | $\Delta(H)$ |

**Theorem 2.** *If $\Pi$ is a graph property such that:*

*(i) $\Pi$ is characterized by $c_\Pi$ adjacencies, and*
*(ii) there is a non-negative polynomial $p\colon \mathbb{N} \to \mathbb{N}$ such that for all graphs $G \in \Pi$, $|V(G)| \leq p(\text{VC}(G))$,*

*then* LARGEST INDUCED $\Pi$-SUBGRAPH (VC) *has a kernel with $\mathcal{O}(p(|X|)|X|^{c_\Pi})$ vertices.*

The proof is in the full version [18], and is similar to the proof of Theorem 1.

## 4 Subgraph Testing versus Minor Testing

Several important graph problems such as CLIQUE, LONG PATH, and LONG INDUCED PATH, can be stated in terms of testing for the existence of a certain graph $H$ as an induced subgraph, or as a minor. Note that for these problems, the size of the graph whose containment in $G$ is tested is part of the input as the problem is polynomial-time solvable for each constant size. We compared the kernelization complexity of induced subgraph- versus minor testing for various types of graphs, parameterized by vertex cover, and found the surprising outcome that the kernelization complexity is often opposite: one variant admits a polynomial kernel while the other does not, assuming NP $\not\subseteq$ coNP/poly. We discuss our findings separately for each type of graph whose containment is tested.

### 4.1 Testing for Cliques

The CLIQUE problem (i.e., testing for $K_t$ as an induced subgraph) was one of the first problems known not to admit a polynomial kernel parameterized by the size of a given vertex cover [5, Theorem 11]. Our main result of this section is a polynomial kernel for the related minor testing problem.

CLIQUE MINOR TEST (VC)
**Input:** A graph $G$ with vertex cover $X \subseteq V(G)$, and an integer $t \geq 1$.
**Parameter:** The size $|X|$ of the vertex cover.
**Question:** Does $G$ contain $K_t$ as a minor?

All problems we study in Section 4 are defined similarly in the obvious way, and we will not define them explicitly. Our polynomial kernel uses reduction rules based on simplicial vertices, inspired by the recent work on kernels for TREEWIDTH [9].

**Table 3.** Kernelization complexity of testing for induced $H$ subgraphs versus testing for $H$ as a minor, when the graph $H$ is given as part of the input by specifying $t$. The problems are parameterized by the size of a given vertex cover. Kernel lower bounds are under the assumption that NP $\not\subseteq$ coNP/poly.

| Graph $H$ | Testing for induced $H$ | | Testing for $H$-minor | |
|---|---|---|---|---|
| $K_t$ | $\neg\ \exists \lvert X\rvert^{\mathcal{O}(1)}$ kernel | [5] | $\exists \lvert X\rvert^{\mathcal{O}(1)}$ kernel | (Thm. 3) |
| $K_{1,t}$ | $\exists \lvert X\rvert^{\mathcal{O}(1)}$ kernel | (Sect. 4.2) | $\neg\ \exists \lvert X\rvert^{\mathcal{O}(1)}$ kernel | [11] |
| $K_{s,t}$ | $\neg\ \exists \lvert X\rvert^{\mathcal{O}(1)}$ kernel | (Sect. 4.2) | $\neg\ \exists \lvert X\rvert^{\mathcal{O}(1)}$ kernel | [11] |
| $P_t$ | $\neg\ \exists \lvert X\rvert^{\mathcal{O}(1)}$ kernel | (Sect. 4.3) | $\exists \lvert X\rvert^{\mathcal{O}(1)}$ kernel | ([17] or Thm. 2) |
| $t \cdot K_2$ | $\neg\ \exists \lvert X\rvert^{\mathcal{O}(1)}$ kernel | (Sect. 4.4) | P-time solvable | |

**Theorem 3.** CLIQUE MINOR TEST (VC) *admits a kernel with* $\mathcal{O}(|X|^4)$ *vertices.*

Firstly, observe that if a graph has a clique $K_t$ as a minor, then its vertex cover number is at least $t-1$: taking a minor does not increase the vertex cover number, and $\text{VC}(K_t) = t-1$. Therefore, we assume that $t \leq |X|+1$, as otherwise we may output a trivial NO-instance. Our algorithm is based on three reduction rules. In the following, we assume the reduction rules are exhaustively applied in their given order.

**Reduction Rule 1.** *For every distinct pair* $v, w \in X$ *such that* $vw \notin E(G)$, *if there are more than* $(|X|+1)^2$ *vertices in* $V(G) \setminus X$ *adjacent both to* $v$ *and* $w$, *then add the edge* $vw$. *Output the resulting instance* $(G', X, t)$.

**Lemma 2.** *Rule 1 is safe.*

*Proof.* As $G$ is a subgraph of $G'$, any clique minor in $G$ is also contained in $G'$. Therefore we need to argue that if $G'$ admits a $K_t$ minor, then so does $G$.

Assume that $G'$ has a $K_t$ minor, and let $G^*$ be a subgraph of $G'$ containing a $K_t$ minor model $\phi$ such that $|V(G^*)| \leq |V(K_t)| + \text{VC}(G') \cdot (\Delta(K_t)+1) = t + \text{VC}(G') \cdot t$, whose existence is guaranteed by Proposition 1. As $\text{VC}(G') \leq |X|$ it follows that $|\bigcup_{v \in K_t} \phi(v)| \leq t + |X| \cdot t$, and since $t \leq |X|+1$ the number of vertices involved in the minor model is at most $(|X|+1)^2$. Hence by the precondition to the reduction rule, there is a vertex $y$ adjacent to both $v$ and $w$ which is not used in the minor model.

Observe that if $\phi$ avoids one of $v$ and $w$, it is also a clique model in $G$. Assume then that $v \in \phi(u_1)$ and $w \in \phi(u_2)$; it may happen that $u_1 = u_2$. Now we can transform $\phi$ into a clique minor model $\phi'$ in $G$, by adding $y$ to $\phi(u_1)$: contraction of the edge $vy$ in this branch set creates the edge $vw$ that was missing in $G$. □

**Reduction Rule 2.** *If there exists a simplicial vertex* $s \in V(G) \setminus X$ *such that* $\deg(s) \geq t-1$, *output a trivial* YES-*instance.*

**Reduction Rule 3.** *If there exists a simplicial vertex* $s \in V(G) \setminus X$ *such that* $\deg(s) < t-1$, *delete it. Output the resulting instance* $(G', X, t)$.

Correctness of Rule 2 is obvious, as $s$ together with its neighborhood already forms a $K_t$. The correctness proof for Rule 3 can be found in the full version. The running time of the kernelization algorithm is polynomial, as the presented reduction rules can only add edges inside $X$ and remove vertices from $V(G) \setminus X$. Exhaustive application of the reduction rules results in an instance with at most $(|X| + 1)^4$ vertices, which proves Theorem 3.

### 4.2 Testing for Bicliques

We now consider testing for a biclique as an induced subgraph or minor. Observe first that if $G$ is a connected graph on at least three vertices, then the following conditions are equivalent: graph $G$ has a (a) spanning tree with $t$ or more leaves, (b) $K_{1,t}$ minor, (c) connected dominating set of size at most $|V(G)| - t$. Hence there is a trivial polynomial-parameter transformation [19] from CONNECTED DOMINATING SET (VC) to $K_{1,t}$ MINOR TEST (VC). Dom et al. [11, Theorem 5] showed[1] that the former problem does not admit polynomial kernels unless NP $\subseteq$ coNP/poly, and hence the same lower bound holds for the latter.

The situation is more diverse when testing for a biclique as an induced subgraph. If we fix a constant $c$ and wish to test for a biclique $K_{c,t}$ as induced subgraph, where $t$ is part of the input, then this problem admits a polynomial kernel parameterized by vertex cover. Our main insight is a polynomial-size compression which is obtained by guessing the model of the constant-size partite set within the vertex cover, reducing the problem to the OR of $\binom{|X|}{c}$ instances of INDEPENDENT SET parameterized by vertex cover. As INDEPENDENT SET parameterized by vertex cover is equivalent to VERTEX COVER parameterized by the size of a given (suboptimal) vertex cover, each of these can be compressed to a size polynomial in $|X|$ using Theorem 1, and the NP-completeness transformation then results in an instance of the original problem of size $\mathcal{O}(|X|^{\mathcal{O}(1)})$ which forms the kernel.

If the sizes of both partite sets are part of the input, then we can no longer obtain a polynomial kernel. We employ a cross-composition [5] from BALANCED BICLIQUE IN BIPARTITE GRAPHS to show that testing for an induced $K_{s,t}$ subgraph, parameterized by vertex cover, does not admit a polynomial kernel unless NP $\subseteq$ coNP/poly.

### 4.3 Testing for Paths

Since a graph contains $P_t$ as a minor if and only if it contains $P_t$ as a subgraph, testing for a $P_t$ minor is equivalent to the LONG PATH problem and hence has a polynomial kernel parameterized by vertex cover, through Theorem 2. The related induced subgraph testing problem, however, is unlikely to admit a polynomial kernel. We cross-compose $t$ instances of HAMILTONIAN $s - t$ PATH into a single instance of LONG INDUCED PATH (VC). The main idea behind the construction is to create an instance containing three paths $P_A, P_B, P_C$ of consecutive degree-two vertices, such that any sufficiently long induced path

[1] The lower bound they give is for DOMINATING SET, but a trivial transformation extends it to CONNECTED DOMINATING SET.

traverses all these paths. The only connections between $P_A$ and $P_B$ can be made by visiting a vertex $v_i$ outside the vertex cover; there is one such vertex $v_i$ for each input instance. To make a suitably long path, a solution must traverse $P_A$, then visit some $v_i$, and then continue traversing $P_B$. The inducedness requirement ensures the path cannot visit neighbors of $v_i$ except for its predecessor on $P_A$ and successor on $P_B$. This allows us to encode the adjacency matrix of the input graph corresponding to $v_i$ into the set of edges incident on $v_i$. The proof in the full version shows that LONG INDUCED PATH (VC) does not admit a polynomial kernel unless NP $\subseteq$ coNP/poly.

### 4.4 Testing for Matchings

Matchings (i.e., disjoint unions of $K_2$'s) are the last type of graphs whose containment testing we consider. It is not difficult to see that $G$ has a $t \cdot K_2$ minor if and only if $G$ has a matching of size $t$, and hence we can solve the minor-testing variant of this containment problem in polynomial time by simply computing a maximum matching. On the other hand, finding an induced matching is a classic NP-complete problem and we give evidence that it does not admit a polynomial kernel parameterized by vertex cover. We use a bit-selector strategy to cross-compose MAXIMUM INDUCED MATCHING IN BIPARTITE GRAPHS into our target problem, exploiting the inducedness requirement to allow the bitselector to isolate a solution corresponding to a single input instance. Hence we prove that MAXIMUM INDUCED MATCHING (VC) does not admit a polynomial kernel unless NP $\subseteq$ coNP/poly.

## 5 Conclusion

We have studied the existence of polynomial kernels for graph problems parameterized by vertex cover. The general theorems we presented unify known positive results for many problems, and the characterization in terms of forbidden or desired induced subgraphs from a class characterized by few adjacencies gives a common explanation for the results obtained earlier. Our comparison of induced subgraph and minor testing problems shows that the kernelization complexity landscape of problems parameterized by vertex cover is rich and difficult to capture with a single meta-theorem. The kernel lower bounds for induced subgraph testing show that besides connectivity and domination requirements, an inducedness requirement can form an obstacle to kernelizability for parameterizations by vertex cover.

An obvious direction for further work is to find even more general kernelization theorems which can also encompass the known positive results for problems like TREEWIDTH (VC) [9] and CLIQUE MINOR TEST (VC). There are also various problems for which the kernelization complexity parameterized by vertex cover is still open; among these are PERFECT DELETION, INTERVAL DELETION, BANDWIDTH and GENUS. Finally, one may investigate whether Theorem 1 has an analogue for edge-deletion problems, and whether our positive results can be transferred to the smaller parameter twin cover [15].

## References

1. Bodlaender, H.L., Fomin, F.V., Lokshtanov, D., Penninkx, E., Saurabh, S., Thilikos, D.M.: (Meta) Kernelization. In: Proc. 50th FOCS, pp. 629–638 (2009)
2. Fomin, F., Lokshtanov, D., Saurabh, S., Thilikos, D.M.: Bidimensionality and kernels. In: Proc. 21st SODA, pp. 503–510 (2010)
3. Kratsch, S., Wahlström, M.: Compression via matroids: a randomized polynomial kernel for odd cycle transversal. In: Proc. 23rd SODA, pp. 94–103 (2012)
4. Bodlaender, H.L., Downey, R.G., Fellows, M.R., Hermelin, D.: On problems without polynomial kernels. J. Comput. Syst. Sci. 75, 423–434 (2009)
5. Bodlaender, H.L., Jansen, B.M.P., Kratsch, S.: Cross-composition: A new technique for kernelization lower bounds. In: Proc. 28th STACS, pp. 165–176 (2011)
6. Dell, H., van Melkebeek, D.: Satisfiability allows no nontrivial sparsification unless the polynomial-time hierarchy collapses. In: Proc. 42nd STOC, pp. 251–260 (2010)
7. Fortnow, L., Santhanam, R.: Infeasibility of instance compression and succinct PCPs for NP. J. Comput. Syst. Sci. 77, 91–106 (2011)
8. Niedermeier, R.: Reflections on multivariate algorithmics and problem parameterization. In: Proc. 27th STACS, pp. 17–32 (2010)
9. Bodlaender, H.L., Jansen, B.M.P., Kratsch, S.: Preprocessing for Treewidth: A Combinatorial Analysis through Kernelization. In: Aceto, L., Henzinger, M., Sgall, J. (eds.) ICALP 2011, Part I. LNCS, vol. 6755, pp. 437–448. Springer, Heidelberg (2011)
10. Cygan, M., Lokshtanov, D., Pilipczuk, M., Pilipczuk, M., Saurabh, S.: On Cutwidth Parameterized by Vertex Cover. In: Marx, D., Rossmanith, P. (eds.) IPEC 2011. LNCS, vol. 7112, pp. 246–258. Springer, Heidelberg (2012)
11. Dom, M., Lokshtanov, D., Saurabh, S.: Incompressibility through Colors and IDs. In: Albers, S., Marchetti-Spaccamela, A., Matias, Y., Nikoletseas, S., Thomas, W. (eds.) ICALP 2009, Part I. LNCS, vol. 5555, pp. 378–389. Springer, Heidelberg (2009)
12. Jansen, B.M.P., Kratsch, S.: Data Reduction for Graph Coloring Problems. In: Owe, O., Steffen, M., Telle, J.A. (eds.) FCT 2011. LNCS, vol. 6914, pp. 90–101. Springer, Heidelberg (2011)
13. Grohe, M.: Logic and Automata: History and Perspectives. In: Logic, Graphs, and Algorithms, pp. 357–422. Amsterdam University Press (2007)
14. Lampis, M.: Algorithmic Meta-theorems for Restrictions of Treewidth. In: de Berg, M., Meyer, U. (eds.) ESA 2010, Part I. LNCS, vol. 6346, pp. 549–560. Springer, Heidelberg (2010)
15. Ganian, R.: Twin-Cover: Beyond Vertex Cover in Parameterized Algorithmics. In: Marx, D., Rossmanith, P. (eds.) IPEC 2011. LNCS, vol. 7112, pp. 259–271. Springer, Heidelberg (2012)
16. Jansen, B.M.P., Kratsch, S.: On Polynomial Kernels for Structural Parameterizations of Odd Cycle Transversal. In: Marx, D., Rossmanith, P. (eds.) IPEC 2011. LNCS, vol. 7112, pp. 132–144. Springer, Heidelberg (2012)
17. Bodlaender, H.L., Jansen, B.M.P., Kratsch, S.: Kernel Bounds for Path and Cycle Problems. In: Marx, D., Rossmanith, P. (eds.) IPEC 2011. LNCS, vol. 7112, pp. 145–158. Springer, Heidelberg (2012)
18. Fomin, F.V., Jansen, B.M.P., Pilipczuk, M.: Preprocessing subgraph and minor problems: When does a small vertex cover help? CoRR abs/1206.4912 (2012)
19. Bodlaender, H.L.: Kernelization: New Upper and Lower Bound Techniques. In: Chen, J., Fomin, F.V. (eds.) IWPEC 2009. LNCS, vol. 5917, pp. 17–37. Springer, Heidelberg (2009)
20. Cygan, M., Lokshtanov, D., Pilipczuk, M., Pilipczuk, M., Saurabh, S.: On the Hardness of Losing Width. In: Marx, D., Rossmanith, P. (eds.) IPEC 2011. LNCS, vol. 7112, pp. 159–168. Springer, Heidelberg (2012)

# A Polynomial-Time Algorithm for Planar Multicuts with Few Source-Sink Pairs*

Cédric Bentz

LRI, Univ. Paris-Sud & CNRS, 91405 Orsay Cedex, France
cedric.bentz@lri.fr

**Abstract.** Given an edge-weighted undirected graph and a list of $k$ source-sink pairs of vertices, the well-known *minimum multicut problem* consists in selecting a minimum-weight set of edges whose removal leaves no path between every source and its corresponding sink. We give the first polynomial-time algorithm to solve this problem in planar graphs, when $k$ is fixed. Previously, this problem was known to remain **NP**-hard in general graphs with fixed $k$, and in trees with arbitrary $k$; the most noticeable tractable case known so far was in planar graphs with fixed $k$ and sources and sinks lying on the outer face.

## 1 Introduction

In this paper, we are interested in the study of the minimum multicut problem in undirected graphs (no directed version is considered). This fundamental problem has been extensively studied, and is well-known to be **NP**-hard even in very restricted classes of graphs.

Assume we are given a $n$-vertex $m$-edge undirected graph $G = (V, E)$, a *weight function* $w : E \to \mathbb{Z}^+$ and a list $\mathcal{L}$ of pairs (source $s_i$, sink $s'_i$) of *terminal* vertices. Each pair $(s_i, s'_i)$ defines a *commodity*. The *minimum multicut problem* (MINMC) consists in selecting a minimum weight set of edges whose removal separates $s_i$ from $s'_i$ for each $i$. The *minimum multiterminal cut problem* (MINMTC) is a special case of MINMC in which, given a set of vertices $\mathcal{T} = \{t_1, \ldots, t_{|\mathcal{T}|}\}$, the source-sink pairs are $(t_i, t_j)$ for $i \neq j$.

For $|\mathcal{L}| = 1$, the problem is the classical minimum cut problem. For $|\mathcal{L}| = 2$, the problem can be solved in polynomial time by solving two minimum cut problems [18]. However, Dahlhaus et al. showed that, for any fixed $|\mathcal{L}| \geq 3$, MINMTC (and hence MINMC) becomes **NP**-hard (and even **APX**-hard) in general graphs [9]. When $|\mathcal{L}|$ is not fixed, MINMC is **APX**-hard even in unweighted stars [11] and **NP**-hard even in unweighted binary trees [6], while MINMTC is **NP**-hard in planar graphs [9]. We also mention that, in bounded tree-width graphs, MINMTC (resp. MINMC) is polynomial-time solvable when $|\mathcal{L}|$ is arbitrary [12] (resp. when $|\mathcal{L}|$ is fixed [3]). There have been recent results concerning FPT algorithms for MINM(T)C: however, the parameter considered in these papers is the size of the solution, and hence we shall not mention them here.

* This research work was supported by the French ANR project *DOPAGE* (ANR-09-JCJC-0068).

D.M. Thilikos and G.J. Woeginger (Eds.): IPEC 2012, LNCS 7535, pp. 109–119, 2012.

In their seminal paper, Dahlhaus et al. also showed that MinMTC can be solved in polynomial time in planar graphs if $|\mathcal{L}|$ is fixed, but they left as open three important questions: first, does MinMTC admit a polynomial-time approximation scheme (PTAS)? Second, is MinMTC FPT in planar graphs, if $|\mathcal{L}|$ is viewed as the parameter [10]? Third, is MinMC also polynomial-time solvable in planar graphs if $|\mathcal{L}|$ is fixed? The first open question was recently addressed by Bateni et al. [1]. The second one was even more recently addressed by Marx [15], and we answer the third question in this paper (while the case where all the sources and sinks lie on the outer face was already solved in [4]).

It should be noticed that Hartvigsen [13] and Yeh [19] later provided other algorithms to solve MinMTC in planar graphs when $|\mathcal{L}|$ is fixed (none of them being FPT with respect to $|\mathcal{L}|$). Moreover, it was observed in [2] and [5] that unfortunately the proof of Yeh's algorithm is not correct, and later it was proved in [7] that the algorithm itself is not correct. The main mistake in the proof of this algorithm was to assume that, when replacing the boundary of any connected component by a minimum cut between some well-chosen vertices, we still obtain a single connected component. More recently, Marx and Klein gave an even faster algorithm to solve MinMTC in planar graphs when $|\mathcal{L}|$ is fixed [14], but Marx also managed to prove that, assuming the *Exponential Time Hypothesis*, this problem is *not* FPT with respect to $|\mathcal{L}|$ [15]. This latter result immediately implies that MinMC in planar graphs is not FPT with respect to $|\mathcal{L}|$.

In this paper, we give an algorithm based, on the one hand, on a revised and generalized Yeh-like approach, and, on the other hand, on shortest homotopic paths methods, and show that this algorithm can be used to solve MinMC in polynomial time when the graph is planar and $|\mathcal{L}|$ is fixed. (Obviously, this also provides an alternative polynomial-time algorithm to solve MinMTC in planar graphs when $|\mathcal{L}|$ is fixed.) It is worth noticing that our major tool is a new characterization of optimal solutions for this problem. Moreover, although homotopic routing methods have already been used to solve planar disjoint paths problems (see [16] and [17] for instance), to the best of our knowledge they have never been used to solve (multi)cut problems so far. (Our algorithm is not FPT, but the recent result of Marx [15] implies that unfortunately this is essentially the best one can hope for.)

The paper is organized as follows. In Section 2, we describe the starting point of our algorithm. Then, in Section 3, we give some preliminary definitions and results, that will be useful in Section 4. Finally, in Section 4, we describe our algorithm, and prove its correctness.

## 2 The Starting Point

The first step of our algorithm is a simple idea presented in [4]. Given a MinMC instance $I = (G = (V, E), w, \mathcal{L})$ and any of its optimal multicuts $C$, one can define the clustering of the terminals associated with the connected components of $G' = (V, E \setminus C)$ (we also say that this particular clustering *induces* these connected components). The $i$th cluster of this clustering, denoted by $\mathcal{T}_i$, contains all the terminals lying in the $i$th connected component of $G'$. Once this clustering has been defined (although, so far, we need to know $C$ in order to do it), finding an optimal solution to $I$ is equivalent to removing a minimum-weight set of edges $C$ whose removal separates all the terminals in $\mathcal{T}_i$ from all the terminals in $\mathcal{T}_j$ for each $i \neq j$.

In this paper, we will refer to this problem as the *minimum multi-cluster cut problem* (MinMCC). This problem has been defined as the *Colored Multiterminal Cut problem* in [9], where it is shown to be **NP**-hard in planar graphs, even with only four clusters (and it is claimed that this is also true for three clusters). Note that, in general graphs, MinMCC and MinMTC are equivalent, since from a MinMCC instance we can obtain an equivalent MinMTC instance by adding one new terminal vertex for each cluster, and linking all the terminals in this cluster (which will no longer be terminals in the MinMTC instance) to this new vertex by sufficiently heavy edges. However, this reduction does *not* necessarily preserve planarity. Given a MinMC instance, we can build an equivalent MinMCC instance by enumerating all the possible clusterings of the terminals (such a clustering can contain up to $2|\mathcal{L}|$ clusters): when $|\mathcal{L}|$ is fixed, this can be done in constant time, and so this yields the following lemma.

**Lemma 1.** *When $|\mathcal{L}|$ is fixed,* MinMC *can be polynomially reduced to* MinMCC, *and this reduction preserves planarity.*

Since we enumerate all the possible clusterings in order to guess the right one, we can also assume without loss of generality that the one we chose has the property that no clustering associated with an optimal solution induces more connected components than this one does. In other words, in the (planar) MinMCC instance we obtain, every cluster induces exactly one connected component in any optimal solution. In the remainder of the paper, we design an efficient algorithm to solve MinMCC in planar graphs when the sum of the sizes of the clusters is fixed (otherwise, from [9], the problem is **NP**-hard); from the above enumeration argument, we can assume that every cluster induces only one connected component (note that this problem generalizes planar MinMTC with a fixed number of terminals). To do this, we will make use of some notions and results related to planarity, planar curves and planar duality, which we introduce in the next section.

## 3 Preliminary Definitions and Results

Throughout the paper, each time we consider a MinMCC instance in a planar graph $G$, we assume without loss of generality that $G$ is simple, loopless, connected (otherwise, we can consider each connected component independently), and even 2-vertex-connected (from [4]), but also that some planar embedding of $G$ is given. Recall that to any planar graph $G$ (embedded in the plane) we can associate a dual (planar) graph $G^*$: each face (including the outer face) of the initial (or primal) graph $G$ is associated with one vertex in the dual graph $G^*$, and there is an edge between two vertices in the dual graph iff the associated faces are adjacent (i.e., share an edge) in the primal graph. (If an edge belongs to only one face, then it corresponds to a loop in the dual graph.) As a consequence, there is a one-to-one correspondence between primal faces (resp. vertices) and dual vertices (resp. faces).

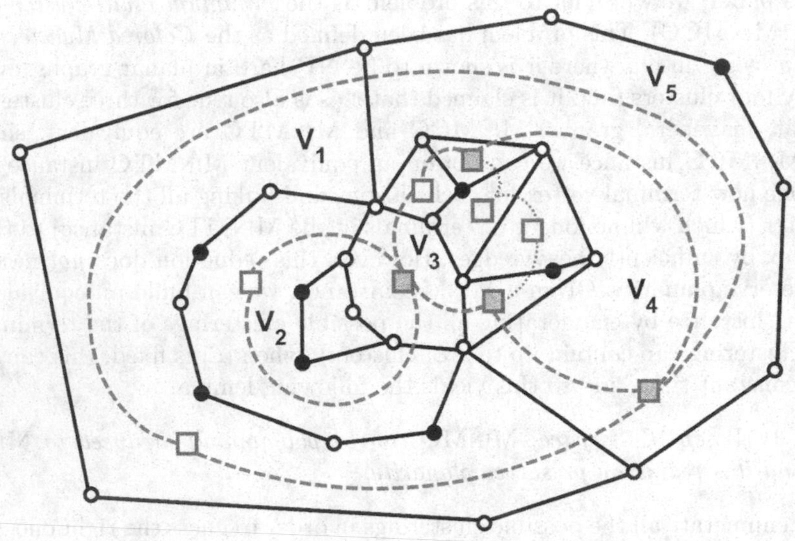


**Fig. 1.** A multi-cluster cut in a planar graph with five clusters. The edges of the initial (primal) graph are in plain lines, the non-terminal vertices are the white round vertices, the terminals are the black round vertices, the dual vertices are the square vertices, and the dual edges associated with the multi-cluster cut $\mathcal{C}$ are in dashed and dotted lines. (The edges of $\mathcal{C}_1$ are in dashed lines, and the four grey square vertices are the joint-vertices of $\mathcal{C}$.)

Given a MinMCC instance $I = (G = (V, E), w, \mathcal{T} = \{\mathcal{T}_1, \ldots, \mathcal{T}_p\})$ and an optimal multi-cluster cut $C$ for $I$, we denote by $C^*$ the edge set dual to $C$, and, for each $i$, by $V_i$ the vertices of the connected component of $G' = (V, E \setminus C)$ containing the terminals in $\mathcal{T}_i$, and by $C_i$ the set of edges such that $C_i \subseteq C$ and $C_i$ has exactly one endpoint in $V_i$. We define a *joint-vertex* as a dual vertex (a vertex of the dual graph $G^*$ of $G$) of degree at least 3 in $C^*$. Note that each $C_i$ corresponds to a set of (not necessarily simple) cycles in $G^*$. Let us assume for now that each $C_i$ corresponds to only one cycle.

If the edges in the embedding of the dual graph are viewed as curves in the plane (the dual vertices being intersections between curves), then the dual image of each $C_i$ will be a closed curve $\mathcal{C}_i$ (the union of all the $\mathcal{C}_i$'s, i.e., the geometric representation of $C^*$, will be denoted by $\mathcal{C}$); if this closed curve is simple (this may not be the case, see below), then, by the Jordan curve theorem, the faces of $G^*$ associated with all the terminals in $\mathcal{T}_i$ are *inside* this curve, and the faces of $G^*$ associated with all the terminals in $\bigcup_{j \neq i} \mathcal{T}_j$ are *outside* this curve (which simply means that the edges associated with $\mathcal{C}_i$ isolate the terminals in $\mathcal{T}_i$ from all the other terminals). When $C_i \subset \mathbb{R}^2$ is not simple (as this is the case for $C_1$ in Figure 1), i.e., when $C_i$ self-intersects in one or more points of the plane, the situation is a bit more complex: in this case, by a simple corollary of the Jordan curve theorem, $\mathbb{R}^2 \setminus C_i$ contains more than two connected regions (a connected region of $\mathbb{R}^2 \setminus C_i$ being a region of $\mathbb{R}^2$ such that any two points of this region can be linked by a curve without crossing $C_i$), and one of these connected regions is unbounded (it is called the *unbounded region*), while all the other ones are bounded. The only bounded region of $\mathbb{R}^2 \setminus C_i$ (and all the faces it contains) that is adjacent to the unbounded region is called the *inside* of $C_i$ (it is unique since $V_i$ is connected), and every other bounded region of $\mathbb{R}^2 \setminus C_i$ is called an *inner region* of $C_i$ (although it does belong to the outside of $C_i$, and not to its inside).

Notice that, if some $C_i^*$ contains more than one cycles (either simple or not), then either this means that there is one cycle $\bar{C}_i^*$ contained in $C_i^*$, corresponding to a closed curve $\bar{C}_i$ in $G^*$, such that any other cycle contained in $C_i^*$ lies inside $\bar{C}_i$, or this means that $V_i$ is the only component in contact with the infinite face. (In the first case, note that there is at least one other $C_j^*$ for some $j \neq i$ that lies inside the closed curve corresponding to each cycle in $C_i^* \setminus \bar{C}_i^*$.) So, we have:

**Lemma 2.** *For each $i$, if $C_i$ is a closed curve, then the faces associated with $\mathcal{T}_i$ are inside $C_i$, while the faces associated with $\mathcal{T}_j$ are outside $C_i$, for each $j \neq i$.*

We also need to define *homotopic curves.* Roughly speaking, given a set $\mathcal{O}$ of $\mu$ obstacles (typically, faces) $\mathcal{O}_1, \ldots, \mathcal{O}_\mu$ in the plane, two simple curves $C_1, C_2$ in $\mathbb{R}^2 \setminus \mathcal{O}$ sharing the same endpoints (or two closed curves) are said to be *homotopic with respect to* $\mathcal{O}$ if $C_1$ can be continuously deformed into $C_2$ in $\mathbb{R}^2 \setminus \mathcal{O}$. We can also say that $C_1$ is homotopic to $C_2$ with respect to $\mathcal{O}$, or alternatively that $C_1$ and $C_2$ belong to the same homotopy class. In the present setting, the curves we will consider are the ones that are associated with (i.e., that are the dual images of) the $C_i$'s (or parts of them); the set of obstacles $\mathcal{O}$ we will consider is the set of faces associated with the terminals. Then, the following lemma is easy to see:

**Lemma 3.** *Two simple closed curves having the same faces of $\mathcal{O}$ in their insides and the same faces of $\mathcal{O}$ in their outsides are homotopic with respect to $\mathcal{O}$.*

Finally, let us notice that the number of vertices in $G^*$ is bounded by $2|V| - 4$, since it is equal to the number of faces $f_G$ of $G$. Indeed, $G$ is a simple, loopless and connected planar graph, and hence each of its faces contains at least three vertices and edges: this implies that $2|E| \geq 3f_G$, which, combined with Euler's formula $|V| + f_G - |E| = 2$, yields $f_G \leq 2|V| - 4$. However, we still have to bound the number of joint-vertices in $C^*$. To this end, the following lemma will be useful in the next section:

**Lemma 4.** *The number of joint-vertices in $C^*$ is at most $2p - 4$.*

*Proof.* This can be shown by a simple application of Euler's formula. Consider the subgraph of $G^*$ induced by $C^*$. In this subgraph, there is no vertex of degree 1, and we contract any vertex of degree 2 in this subgraph (this does not modify the number of vertices of degree at least 3) in order to obtain the graph $G_C^*$. The number of faces in $G_C^*$ is $p$, since each cluster in $\{\mathcal{T}_1, \ldots, \mathcal{T}_p\}$ induces exactly one connected component in $G$. We remove loops (and associated faces) as well as multiple edges (and associated faces) from $G_C^*$: each time we remove such an edge, we remove one face. If we denote by $m_C$ and $f_C$ the number of edges and faces in $G_C^*$, and by $n_C, m'_C, f'_C, \kappa_C$ the number of vertices, edges, faces, and connected components in this updated (simple) graph, respectively, then by Euler's formula we have $n_C + f'_C - m'_C = 1 + \kappa_C$, i.e., $n_C + f_C - m_C = 1 + \kappa_C$. (Note that $n_C$ is the number of joint-vertices we have to consider.) Any vertex in $G_C^*$ has degree at least 3, and hence $2m_C \geq 3n_C$. Since $\kappa_C \geq 1$, we have $n_C + f_C - m_C \geq 2$, i.e., $n_C \geq m_C - f_C + 2 \geq 3n_C/2 - f_C + 2$, and this yields $n_C/2 \leq f_C - 2$, i.e., $n_C \leq 2f_C - 4 = 2p - 4$. □

A similar result was presented in [19, Theorem 5], using the notion of *component graph* (in which there is a vertex for each component $V_i$ and a single edge between any two vertices if the corresponding components share at least one edge); however, a joint-vertex may actually *not* induce a face in the component graph (see the joint-vertex belonging to $C_2^*$ in Figure 1 for instance), since this graph is simple by definition, and hence this proof was incomplete.

# 4 Description and Proof of the Algorithm

## 4.1 A Structural Description of Optimal Solutions

Dahlhaus et al. [9], and later Hartvigsen [13], gave structural descriptions of optimal planar multiterminal cuts (one is based on the notion of *topology* and on minimum spanning trees computation, and the other is based on links between optimal planar multiterminal cuts and Gomory-Hu cut collections). However, it is not clear whether these structural results could be extended to optimal planar multi-cluster cuts; in fact, it seems that they cannot. Here, we give a new and somewhat simpler structural description of optimal planar multiterminal cuts (although it may imply enumerating more elements than in the approaches described by Dahlhaus et al. and Hartvigsen), that is also valid for optimal planar multi-cluster cuts.

We use the definitions and notations from the previous section. Let $\mathcal{F}$ (resp. $\mathcal{F}_i$) be the faces of $G^*$ associated with the terminals in $\mathcal{T}$ (resp. in $\mathcal{T}_i$), and let $C$ be any multi-cluster cut that partitions the plane into $p$ connected regions (each one containing one cluster), such as a minimum multi-cluster cut (for instance). Let us now consider $C_i$ for some $i$, and assume that the dual image $\mathcal{C}_i$ of $C_i$ consists of only one closed curve. This curve goes through a certain number of joint-vertices: let us call them $\omega_1, \ldots, \omega_{q_i}$, in clockwise order (with $\omega_1 = \omega_{q_i}$). Recall that, by definition, the curve $\mathcal{C}_i$ intersects other $\mathcal{C}_j$'s *only* at joint-vertices. Assume that $q_i \geq 2$. Then, we have:

**Lemma 5.** *Let $V_i$ be a connected component of $G' = (V, E \setminus C)$, let $\mathcal{C}_i$ be the associated curve in $G^*$, and let $\omega_1, \ldots, \omega_{q_i}$ be the joint-vertices $\mathcal{C}_i$ goes through.*

*Then, $\mathcal{C}' = (\mathcal{C} \setminus \mathcal{C}_i) \cup \mathcal{C}'_i$ is also a valid multi-cluster cut for $I$, where $\mathcal{C}'_i$ is any cycle in $G^*$ going through $\omega_1, \ldots, \omega_{q_i}$, and such that the faces associated with $\mathcal{T}_i$ are inside $\mathcal{C}'$, while the faces associated with $\mathcal{T}_j$ are outside $\mathcal{C}'$ for each $j \neq i$.*

*Proof.* Assume that one such $\mathcal{C}'$ is not a multi-cluster cut. Consider any path $\mu_{a,b}$ in $G' = (V, E \setminus C')$ between two terminal vertices $t_a \in \mathcal{T}_j$ and $t_b \in \mathcal{T}_{j'}$ for some $j \neq j'$. We cannot have $j = i$ or $j' = i$, by the definition of $\mathcal{C}'_i$. Moreover, since $C$ is a multi-cluster cut, we know that $\mu_{a,b}$ contains at least one edge in $C_i$, say $uv$. Choose an edge dual to such an edge in $C_i$, and assume without loss of generality that this dual edge belongs to the curve $\mathcal{C}_i[\omega_1, \omega_2]$, defined as the part of $\mathcal{C}_i$ linking $\omega_1$ and $\omega_2$. From Lemma 3, $\mathcal{C}'_i$ is homotopic to $\mathcal{C}_i$ with respect to $\mathcal{F}$. Hence, $\mathcal{C}_i$ can be continuously deformed into $\mathcal{C}'_i$ in $\mathbb{R}^2 \setminus \mathcal{F}$. In particular, since $\mathcal{C}'_i$ goes through $\omega_1$ and $\omega_2$, it contains some curve $\mathcal{C}'_i[\omega_1, \omega_2]$ homotopic to $\mathcal{C}_i[\omega_1, \omega_2]$ with respect to $\mathcal{F}$. Hence, the inside of the closed curve $\mathcal{C}_i[\omega_1, \omega_2] \cup \mathcal{C}'_i[\omega_1, \omega_2]$ contains neither $t_a$ nor $t_b$ (since $i, j, j'$ are all distinct). We claim the following :

**Claim 1.** *$\mu_{a,b}$ must "intersect" (i.e. have an edge in common with) $\mathcal{C}'_i[\omega_1, \omega_2]$ at least once.*

*Proof.* Since $\mathcal{C}_i[\omega_1, \omega_2] \cup \mathcal{C}'_i[\omega_1, \omega_2]$ is a closed (but not necessarily simple) curve, the edge dual to any edge on its boundary either belongs to both $\mathcal{C}_i[\omega_1, \omega_2]$ and $\mathcal{C}'_i[\omega_1, \omega_2]$ (which is clearly not the case for $uv$, otherwise we are done), or has one endpoint inside $\mathcal{C}_i[\omega_1, \omega_2] \cup \mathcal{C}'_i[\omega_1, \omega_2]$ and one endpoint outside $\mathcal{C}_i[\omega_1, \omega_2] \cup \mathcal{C}'_i[\omega_1, \omega_2]$ (so, this is the case for $uv$).

Now, assume that $\mu_{a,b}$ has $t \geq 1$ (for some $t$) edges in common with $\mathcal{C}_i[\omega_1, \omega_2]$ (none of them is of the first type described above, otherwise we are done). If $\mu_{a,b}$ crosses $\mathcal{C}'_i[\omega_1, \omega_2]$, then we are done. Assume otherwise. $\omega_1$ and $\omega_2$ being two consecutive joint-vertices in $\mathcal{C}_i$, then by definition each of these $t$ edges has one endpoint in $V_i$ and the other one in $V_l$ for some $l$ (the same $l$ for all these edges). In particular, the vertices inside $\mathcal{C}_i[\omega_1, \omega_2] \cup \mathcal{C}'_i[\omega_1, \omega_2]$ that are incident to edges in $\mathcal{C}_i[\omega_1, \omega_2]$ all belong to the same connected component of $(V, E \setminus C)$ (either $V_i$ or $V_l$). Hence, each time $\mu_{a,b}$ "crosses" $\mathcal{C}_i[\omega_1, \omega_2]$, it "changes side" (going for instance from $V_i$ to $V_l$, then from $V_l$ to $V_i$, then again from $V_i$ to $V_l$, etc.). If it crosses $\mathcal{C}_i[\omega_1, \omega_2]$ an even number of times (the first edge crossed being $uv$ and the last one $u'v'$ for instance), then $u$ and $v'$ either both belong to $V_i$ or both belong to $V_l$ (i.e., belong to the same connected component of $(V, E \setminus C)$). So, instead, we can find a new path $\mu'_{a,b}$ from $t_a$ to $t_b$ that does not cross $\mathcal{C}_i[\omega_1, \omega_2]$ at all, by replacing the part of $\mu_{a,b}$ going from $u$ to $v'$ by a path from $u$ to $v'$ using vertices of $V_i$ (or $V_l$) only; this yields a contradiction. By the same argument, we can show that if $\mu_{a,b}$ crosses $\mathcal{C}_i[\omega_1, \omega_2]$ an odd number of times (the first edge crossed being $uv$ and the last one $u'v'$ for instance; note that $u'v'$ may be $uv$), then $v'$ is inside $\mathcal{C}_i[\omega_1, \omega_2] \cup \mathcal{C}'_i[\omega_1, \omega_2]$. Since the part of $\mu_{a,b}$ going from $v'$ to $t_b$ crosses neither $\mathcal{C}_i[\omega_1, \omega_2]$ (by definition) nor $\mathcal{C}'_i[\omega_1, \omega_2]$ (by assumption), and since neither $t_a$ nor $t_b$ are inside $\mathcal{C}_i[\omega_1, \omega_2] \cup \mathcal{C}'_i[\omega_1, \omega_2]$, this yields a contradiction. Thus, $\mu_{a,b}$ must cross $\mathcal{C}'_i[\omega_1, \omega_2]$. □

From this claim, $\mathcal{C}'$ intersects any path between two terminals lying in different clusters: it contradicts the fact that $\mathcal{C}'$ is not a multi-cluster cut. □

We can then use this lemma to show that, if $q_i \geq 2$:

**Corollary 1.** *Let $C$ be a minimum multi-cluster cut in a graph $G = (V, E)$, let $V_i$ be a connected component of $G' = (V, E \setminus C)$, let $\mathcal{C}_i$ be the associated curve in $G^*$, and let $\omega_1, \ldots, \omega_{q_i}$ be the joint-vertices $\mathcal{C}_i$ goes through. Then, $\mathcal{C}_i$ is a shortest cycle in $G^*$, that is homotopic to any cycle $\Gamma$ in $G^*$ going through $\omega_1, \ldots, \omega_{q_i}$ and being such that the faces in $\mathcal{F}_i$ are inside $\Gamma$, while the faces in $\mathcal{F}_j$ are outside $\Gamma$ for each $j \neq i$.*

*Proof.* Assume that $\mathcal{C}_i$ is not such a shortest cycle. Then, we can replace $\mathcal{C}_i$ by a shortest cycle $\Gamma^*$ in $G^*$ going through $\omega_1, \ldots, \omega_{q_i}$, and such that the faces in $\mathcal{F}_i$ are inside $\Gamma^*$, while the faces in $\mathcal{F}_j$ are outside $\Gamma^*$ for each $j \neq i$. From Lemma 5, $\mathcal{C}' = (\mathcal{C} \setminus \mathcal{C}_i) \cup \Gamma^*$ is also a valid multi-cluster cut for $I$. Moreover, $\Gamma^*$ is strictly shorter than $\mathcal{C}_i$ (since from Lemmas 2 and 3 they are homotopic with respect to $\mathcal{F}$), and hence $\mathcal{C}'$ is a strictly better solution than $\mathcal{C}$: a contradiction. □

### 4.2 Algorithmic Aspects

From Subsection 4.1, we can construct $\mathcal{C}$ in an iterative way, by first "guessing" (i.e., enumerating) all the joint-vertices, then computing each $\mathcal{C}_i$ corresponding to a single closed curve one after the other, and finally removing the vertices inside it, and go on. (We assume without loss of generality that we look for an optimal solution having the maximum number of joint-vertices among the ones with $p$ clusters, and this implies that we cannot create "new" joint-vertices when computing each $\mathcal{C}_i$.) Hence, we have to guess an $i$ for which $\mathcal{C}_i$ corresponds to a single closed curve, compute $\mathcal{C}_i$ and remove it, and then go on by identifying another $i$ for which the part of $\mathcal{C}_i$ lying in the remaining graph (i.e., after removing the previous component) corresponds to a single closed curve, until there remains only one component. We can do this by enumerating all the possible sets of inclusions between the $C_i$'s (i.e., for each $i$ and $j \neq i$, whether there is one cycle $\bar{C}_i$ contained in $C_i$, that corresponds to a closed curve $\bar{\mathcal{C}}_i$ in the dual graph, and such that $C_j$ lies inside $\bar{\mathcal{C}}_i$; or whether there is one cycle $\bar{C}_j$ contained in $C_j$, that corresponds to a closed curve $\bar{\mathcal{C}}_j$ in the dual graph, and such that $C_i$ lies inside $\bar{\mathcal{C}}_j$; or finally whether none lies inside the other). Since the number of $C_i$'s is $p$ and since $p$ is fixed, this can be done in constant time.

In order to compute $\mathcal{C}_i$ for each $i$, we must first "guess" which joint-vertices $\mathcal{C}_i$ goes through (from Lemma 4, the maximum number of joint-vertices is $2p - 4$, so guessing them requires to try all the possible ways of choosing at most $2p - 4$ vertices among $2|V| - 4$, which implies that the running time will depend on $p$), and then we can apply Corollary 1 and find a shortest cycle homotopic to some predefined curve in $G^*$ (keeping in mind that $\mathcal{C}_i$ may go through no joint-vertex; in this case, we only need to compute a minimum cut separating $\mathcal{T}_i$ from $\mathcal{T}_j$, for all $j \neq i$). (If needed, we can reduce the computation of a shortest cycle to the computation of a shortest path, by "guessing" the first edge of this path.) Finding a shortest homotopic path or cycle can be hard, if we require that it must be elementary; however, this property is not needed in our case. (And, indeed, some $\mathcal{C}_i$'s may be non simple cycles, such as $\mathcal{C}_1$ in Figure 1.) We can compute a shortest homotopic path or cycle using for instance the algorithms given in [17, Proposition 1] or in [8].

Finally, we have two last points to address. First, we must ensure that the shortest cycles or paths we compute go through predetermined joint-vertices. Second, we need to be able to generate all the possible predefined curves that the shortest paths we compute can be homotopic to. We now describe the strategy we use to deal with both points at the same time. Each time a given $\mathcal{C}_i$ goes through a given joint-vertex, this means that some vertices of the primal face associated with this joint-vertex belong to $V_i$. Actually, we even know that, on each face associated with a joint-vertex $\mathcal{C}_i$ goes through, there are at most $h_i + 1$ sets of consecutive vertices (called *intervals*) that belong to $V_i$, where $h_i \leq p$ is the number of inner regions of $\mathcal{C}_i$. Therefore, to the joint-vertices associated with a given $\mathcal{C}_i$ corresponds a set $B_i$ of distinct vertices of $V_i$ lying on the primal faces associated with these joint-vertices. The best way to encode this set $B_i$ is to include two vertices of each interval. For a given interval lying on the face associated with a given joint-vertex, call $a$ and $b$ the two vertices of this interval. Then, the vertices in $B_i$ associated with that interval are all the vertices of this face encountered while traveling clockwise from $a$ to $b$ on this face. Let us denote by $\mathcal{B}_i$ the set of dual faces associated with the vertices in $B_i$. By definition, each face associated with a joint-vertex contains at least two vertices belonging to two different $B_i$'s, thus from Lemma 3, for each $i$, $\mathcal{C}_i$ is homotopic, with respect to the faces in $\mathcal{F}$ and $\bigcup_j \mathcal{B}_j$, to any closed curve being such that the faces in $\mathcal{F}_i \cup \mathcal{B}_i$ are inside it, and the faces in $\mathcal{F}_j \cup \mathcal{B}_j$, for each $j \neq i$, are outside it. More generally, any closed curve goes through the same joint-vertices as $\mathcal{C}_i$, if this curve is such that the faces in $\mathcal{B}_i$ belong to its inside, and the faces in $\mathcal{B}_j$ belong to its outside, for each $j \neq i$.

Since for each $i$ the inside of $\mathcal{C}_i$ is a connected region, i.e., the subgraph of $G$ induced by $V_i$ is connected, we also know that in $G' = (V, E \setminus C)$ all the vertices in $B_i$, as well as all the terminals in $\mathcal{T}_i$, are connected together. This implies that, for each $i$, we can construct a closed curve $\mathcal{C}'_i$ homotopic to $\mathcal{C}_i$ by choosing some tree spanning both $\mathcal{T}_i$ and $B_i$, and then removing the edges having exactly one endpoint in the $i$th of these spanning trees (these trees span vertices in distinct connected components of $G'$, and hence have to be vertex-disjoint). (For each $i$, $\mathcal{C}'_i$ goes through the same joint-vertices as $\mathcal{C}_i$, and $\mathcal{C}'_i$ and $\mathcal{C}_i$ are indeed homotopic with respect to the faces in $\mathcal{F}$, since $\mathcal{T}_i$ is the only cluster that belongs to the inside of $\mathcal{C}'_i$, i.e., $\mathcal{C}'_i$ isolates $\mathcal{T}_i$ from $\mathcal{T}_j$, for all $j \neq i$.) In practice, we have to "guess" $B_i$ for each $i$ (which, as mentioned above, can be done by enumerating at most two vertices of $G$ for each interval), making sure that the $B_i$'s define a partition of the vertex set of the faces associated with all the joint-vertices, and then construct $p$ vertex-disjoint trees (each one spanning $\mathcal{T}_i$ and $B_i$ for some $i$), and finally remove the edges isolating each tree from the rest of the graph. For each combination of $B_i$'s, finding such vertex-disjoint trees can be done in polynomial time (since the graph is planar, $\sum_{i=1}^{p} |\mathcal{T}_i|$ is fixed, and the number of mandatory vertices that the $p$ trees must span lie on at most $\sum_{i=1}^{p} |\mathcal{T}_i| + (2p-4)$ faces), using for instance the algorithm given in [17, Theorem 4].

So, our algorithm for planar MinMCC is as follows:

1. For each possible clustering of the terminals, for each possible set of inclusions between the $C_i$'s, for each possible combination of joint-vertices, and for each possible choice of the $B_i$'s do:
   (a) Compute $p$ vertex-disjoint trees, each spanning $\mathcal{T}_i$ and $B_i$ for some $i$, and construct the curves $\mathcal{C}'_i$ by removing, for each $i$, each edge incident to exactly one vertex of the $i$th tree;
   (b) For each $i$ except the last one (in the order given by the current set of inclusions, starting from a $\mathcal{C}_i$ including no other $\mathcal{C}_j$ for $j \neq i$), compute a shortest cycle homotopic to $\mathcal{C}'_i$ with respect to $\mathcal{F}$ and $\bigcup_j \mathcal{B}_j$; then, remove the vertices of the connected component of $G$ lying inside this cycle.
2. Output the best feasible solution found.

We already explained why all steps run in polynomial time, and it should be clear from our above discussion that this algorithm is correct. This yields:

**Theorem 1.** MinMCC *can be solved in polynomial time in planar graphs, if the sum of the sizes of the clusters is fixed.*

Therefore, we can finally state:

**Corollary 2.** MinMC *can be solved in polynomial time in planar graphs, if the number of source-sink pairs is fixed.*

**Acknowledgements.** The author thanks Sylvie Poirier for her help, and Éric Colin de Verdière for fruitful discussions on shortest homotopic paths.

## References

1. Bateni, M., Hajiaghayi, M., Klein, P., Mathieu, C.: A polynomial-time approximation scheme for planar multiway cut. In: 23th SODA (2012)
2. Bentz, C.: Résolution exacte et approchée de problèmes de multiflot entier et de multicoupe: algorithmes et complexité. PhD Thesis, CNAM, Paris (2006) (in French)
3. Bentz, C.: On the complexity of the multicut problem in bounded tree-width graphs and digraphs. Discrete Applied Mathematics 156, 1908–1917 (2008)
4. Bentz, C.: A simple algorithm for multicuts in planar graphs with outer terminals. Discrete Applied Mathematics 157, 1959–1964 (2009)
5. Bentz, C.: New results on planar and directed multicuts. In: EUROCOMB 2009 (2009); Electronic Notes in Discrete Mathematics, vol. 34, pp. 207–211 (2009)
6. Călinescu, G., Fernandes, C.G., Reed, B.: Multicuts in unweighted graphs and digraphs with bounded degree and bounded tree-width. Journal of Algorithms 48, 333–359 (2003)
7. Cheung, K., Harvey, K.: Revisiting a simple algorithm for the planar multiterminal cut problem. Operations Research Letters 38, 334–336 (2010)

8. Colin de Verdière, E., Erickson, J.: Tightening non-simple paths and cycles on surfaces. In: 17th SODA, pp. 192–201 (2006)
9. Dahlhaus, E., Johnson, D.S., Papadimitriou, C.H., Seymour, P.D., Yannakakis, M.: The complexity of multiterminal cuts. SIAM J. on Computing 23, 864–894 (1994)
10. Downey, R.G., Fellows, M.R.: Parameterized Complexity. Springer, New York (1999)
11. Garg, N., Vazirani, V.V., Yannakakis, M.: Primal-dual approximation algorithms for integral flow and multicut in trees. Algorithmica 18, 3–20 (1997)
12. Guo, J., Hüffner, F., Kenar, E., Niedermeier, R., Uhlmann, J.: Complexity and exact algorithms for vertex multicut in interval and bounded treewidth graphs. European Journal of Operational Research 186, 542–553 (2008)
13. Hartvigsen, D.: The planar multiterminal cut problem. Discrete Applied Mathematics 85, 203–222 (1998)
14. Klein, P.N., Marx, D.: Solving PLANAR $k$-TERMINAL CUT in $O(n^{c\sqrt{k}})$ Time. In: Czumaj, A., Mehlhorn, K., Pitts, A., Wattenhofer, R. (eds.) ICALP 2012, Part I. LNCS, vol. 7391, pp. 569–580. Springer, Heidelberg (2012)
15. Marx, D.: A Tight Lower Bound for Planar Multiway Cut with Fixed Number of Terminals. In: Czumaj, A., Mehlhorn, K., Pitts, A., Wattenhofer, R. (eds.) ICALP 2012, Part I. LNCS, vol. 7391, pp. 677–688. Springer, Heidelberg (2012)
16. Schrijver, A.: Homotopic routing methods. In: Korte, B., Lovasz, L., Prömel, H.J., Schrijver, A. (eds.) Paths, Flows and VLSI-Layout. Algorithms and Combinatorics, vol. 9, pp. 329–371. Springer, Berlin (1990)
17. Schrijver, A.: Disjoint Homotopic Paths and Trees in a Planar Graph. Discrete & Computational Geometry 6, 527–574 (1991)
18. Yannakakis, M., Kanellakis, P., Cosmadakis, S., Papadimitriou, C.: Cutting and Partitioning a Graph After a Fixed Pattern. In: Díaz, J. (ed.) ICALP 1983. LNCS, vol. 154, pp. 712–722. Springer, Heidelberg (1983)
19. Yeh, W.-C.: A simple algorithm for the planar multiway cut problem. Journal of Algorithms 39, 68–77 (2001)

# Instance Compression for the Polynomial Hierarchy and beyond

Chiranjit Chakraborty and Rahul Santhanam

School of Informatics, University of Edinburgh, UK
C.Chakraborty@sms.ed.ac.uk, rsanthan@inf.ed.ac.uk

**Abstract.** We define instance compressibility ([5,7] ) for parametric problems in PH and PSPACE. We observe that the problem $\Sigma_i$CIRCUITSAT of deciding satisfiability of a quantified Boolean circuit with $i-1$ alternations of quantifiers starting with an existential quantifier is complete for parametric problems in the class $\Sigma_i^p$ with respect to $W$-reductions, and that analogously the problem QBCSAT (Quantified Boolean Circuit Satisfiability) is complete for parametric problems in PSPACE with respect to $W$-reductions. We show the following results about these problems:

1. If CIRCUITSAT is non-uniformly compressible within NP, then $\Sigma_i$CIRCUITSAT is non-uniformly compressible within NP, for any $i \geq 1$.
2. If QBCSAT is non-uniformly compressible (or even if satisfiability of quantified Boolean $CNF$ formulae is non-uniformly compressible), then PSPACE $\subseteq$ NP/poly and PH collapses to the third level.

Next, we define Succinct Interactive Proof (Succinct IP) and by adapting the proof of IP = PSPACE ([4,2]), we show that QBFORMULASAT (Quantified Boolean Formula Satisfiability) is in Succinct IP. On the contrary if QBFORMULASAT has Succinct PCPs ([11]), Polynomial Hierarchy (PH) collapses.

## 1 Introduction

An NP problem is said to be instance compressible if there is a polynomial-time reduction mapping instances of size $m$ and parameter $n$ to instances of size $poly(n)$ (possibly of a different problem). The notion of instance compressibility for NP problems was defined by Harnik and Naor ([5]) motivated by applications in cryptography. This notion of compression is basically same as the notion of polynomial kernelizability in parametrized complexity ([7,14,8]), which is motivated by algorithmic applications. Fortnow and Santhanam showed ([11], Theorem 3.1) that the compressibility of the satisfiability problem for Boolean formulae (even non-uniformly) is unlikely, since it would imply that the Polynomial Hierarchy (PH) collapses. Since then, there's been a very active stream of research extending this negative result to other problems in NP ([7,17] etc.). Instance compressibility is a useful notion for complexity theory as well - Buhrman and Hitchcock ([6]) use it to study the question of whether NP has sub-exponentially-sparse complete sets.

Given different possibilities of application of this notion, it is a natural question whether we can extend it to other complexity classes, such as PH and PSPACE. Our

D.M. Thilikos and G.J. Woeginger (Eds.): IPEC 2012, LNCS 7535, pp. 120–134, 2012.

first contribution here is to define such an extension. The key to defining instance compressibility for NP problems is that there is a notion of "witness" for instances of NP problems, and in general the witness size can be much smaller than the instance size. We observe that the characterization of PH and PSPACE using alternating time Turing machines yields a natural notion of "guess size" - namely the total number of non-deterministic or co-non-deterministic bits used during the computation. We use this characterization to extend the definition of compressibility to parametric problems in PH and PSPACE in a natural way.

Some proposals ([8,9]) have already been made in the parametrized complexity setting for defining in general the parametrized complexity analogue of a classical complexity class. Our definition (Section 2) seems similar in spirit, but all the problems we consider *are* in fact fixed-parameter tractable. What we are interested in is whether they are instance-compressible, or equivalently whether they have polynomial-size kernels.

One of our main motivations is to provide a structural theory of compressibility, analogous to the theory in the classical setting. Intuitively, instance compressibility provides a different, more relaxed notion of "solvability" than the traditional notion. So it is interesting to study what kinds of analogues to classical results hold for the new notion. The result of Fortnow and Santhanam ([11]) can be thought of as an analogue of the Karp-Lipton theorem ([13], Theorem 6.1), since non-uniform compressibility is a weakening of the notion of non-uniform solvability. Other well-known theorems in the classical setting are that NP has polynomial-size circuits iff all of PH does, as well as the Karp-Lipton theorem for PSPACE ([13], Theorem 4.1). The main results we prove here are analogues of these results for instance compressibility.

Our first main result is, if the language CIRCUITSAT is non-uniformly compressible within NP (i.e., the reduction is to an NP problem), then so is the language $\Sigma_i$CIRCUITSAT, which is in some sense complete for parametric problems in the class $\Sigma_i^p$. Note that we need a stronger assumption here compared to that in the Fortnow-Santhanam result ([11]): they need only to assume that SAT is compressible. This reflects the fact that the proof is more involved - it relies on the Fortnow-Santhanam result ([11]) as well as on the techniques used in the classical case. In addition, the code used by the hypothetical compression for CIRCUITSAT shows up not just in the resulting compression *algorithm* for $\Sigma_i$CIRCUITSAT, but also in the *instance* generated - this is why we need to work with circuits, as they can simulate any polynomial-time computation. This ability to interpret code as data is essential to our proof. We give more intuition about the proof in Section 3, where the detailed proof can also be found. We also observe that under the assumption of $\Sigma_3$CIRCUITSAT being compressible (we make no assumption about the complexity of the set we are reducing to), all of the PH is compressible as well.

Our second main result is that if QBCNFSAT is non-uniformly compressible, the Polynomial Hierarchy collapses to the third level. The proof of this is easier and an adaptation of the Fortnow-Santhanam technique ([11]) to PSPACE. Here we consider an "OR" version of the problem as they do, and derive the collapse of the hierarchy from the assumption that the OR version is compressible. In the case of NP, showing that compressing the OR version is at least as easy as compressing SAT is easier as there are no quantifiers; however, this is not the case for PSPACE and this is where we need to work a little harder.

Our third result is an analogue of the IP = PSPACE ([4,2]) result in the parametric world. We define the class Succinct IP, which consists of parametric problems with interactive protocols where the total amount of communication is polynomial in the size of the parameter. We observe that the traditional proof of IP = PSPACE ([4,2]) can be adapted to show that the problem of determining whether a quantified Boolean formula is valid, has succinct interactive proofs. This demonstrates a difference between succinctness in an interactive setting and succinctness in a non-interactive setting - it is shown in [11] that if SAT has succinct probabilistically checkable proofs, then PH collapses.

There are many open problems in the compressibility theory for NP, such as, whether there are any unlikely consequences of SAT being probabilistically compressible, and whether the problem AND-SAT is deterministically compressible. Our hope is that extending the theory to larger classes such as PH and PSPACE will give us more "room" to work with. Besides, if we manage to settle these questions for the larger classes, the techniques can be translated back to NP.

## 2 Some Complexity Theory Notions

**Definition 1.** ***Parametric problem:*** *A parametric problem is a subset of* $\{ \langle x, 1^n \rangle \mid x \in \{0,1\}^*, n \in \mathbb{N} \}$. *The term n is known as the parameter of the problem.*

**NP problems in parametric form:** Now consider some popular NP languages in parametric form.
**SAT** = $\{\langle \varphi, 1^n \rangle \mid \varphi$ is a satisfiable formula in $CNF$, and $n$ is the number of variables in $\varphi\}$.
**VC** = $\{\langle G, 1^{k\, log(m)} \rangle \mid G$ has a vertex cover of size at most $k$, where $m = |G|\}$.
**CLIQUE** = $\{\langle G, 1^{k\, log(m)} \rangle \mid G$ has a clique of size at least $k$, where $m = |G|\}$.
**DOMINATINGSET** = $\{\langle G, 1^{k\, log(m)} \rangle \mid G$ has a dominating set of size at most $k$, where $m = |G|\}$.
**OR-SAT** = $\{\langle \{\varphi_i\}, 1^n \rangle \mid$ At least one $\varphi_i$ is satisfiable, and each $\varphi_i$ is of bit-length at most $n\}$.

For the parametric problems above in NP, the parameter can be interpreted as the *witness size* for some natural $NTM$ deciding the language. For example in SAT, the number of variables, which captures the witness of satisfiability problem, is taken as the parameter. Note that in the definitions of the CLIQUE, VC and DOMINATINGSET problems, the parameter is $k\, log(m)$ rather than $k$ as in the typical parametrized setting. This is because, here $k\, log(m)$ bits will be required to represent the witness. We say that a parametric problem is in NP if there is a polynomial-time $NTM$ solving it.

**Definition 2.** ***Compression of parametric problem:*** *Suppose here L is a parametric problem. L is said to be compressible within a complexity class A if there is a polynomial p(.), and a polynomial-time computable function* $f$: $\{0,1\}^* \to \{0,1\}^*$, *such that for each* $x \in \{0,1\}^*$ *and* $n \in \mathbb{N}$, $|f(\langle x, 1^n \rangle)| \le p(n)$ *and* $\langle x, 1^n \rangle \in L$ *iff* $f(\langle x, 1^n \rangle) \in L_A$ *for some problem* $L_A$ *in the complexity class A.*

We say the parametric problem $L$ is *compressible* in general, if there exists any such complexity class $A$ as mentioned above, for the problem $L$ to be compressed within.

**Definition 3.** ***Non-uniform Compression:*** *A parametric problem L is said to be compressible with advice s(., .) if the compression function is computable in deterministic polynomial time when given access to an advice string of size s($m$, $n$) which depends only on $m$ and $n$ but not on the actual instance. Here $m$ is the length of the parametric problem instance and $n$ is the parameter. L is non-uniformly compressible if $s$ is polynomially bounded in $m$ and $n$.*

In other words, we can say that the machine compressing the parametric problem in the preceding definition takes advice in case of *Non-uniform Compression.*

**Definition 4.** ***W-Reduction:*** *[5] Given parametric problems $L_1$ and $L_2$, $L_1$ $W$-reduces to $L_2$ (denoted $L_1 \leq_w L_2$ ) if there is a polynomial-time computable function $f$ and polynomials $p_1$ and $p_2$ such that:*
*1. $f(\langle x, 1^{n_1} \rangle)$ is of the form $\langle y, 1^{n_2} \rangle$ where $n_2 \leq p_2(n_1)$.*
*2. $f(\langle x, 1^{n_1} \rangle) \in L_2$ iff $\langle x, 1^{n_1} \rangle \in L_1$.*

The semantics of a $W$-reduction is that if $L_1$ $W$-reduces to $L_2$ , it is at least as hard to compress $L_2$ as it is to compress $L_1$ . If $L_1 \leq_w L_2$ and $L_2$ is compressible, then $L_1$ is compressible. One can prove that OR-SAT $\leq_w$ SAT ([18]).

As we have already mentioned, our primary objective is to extend the idea of compression to higher classes, namely Polynomial Hierarchy (PH) and PSPACE [15]. In our work, by a quantified Boolean formula, we mean a Boolean formula in prenex normal form where the quantifiers are in the beginning as follows, $\psi = Q_1\, x_1\, Q_2\, x_2 \ldots Q_n\, x_n\, \phi$, for any Boolean formula $\phi$. Similarly we can consider quantified Boolean circuits. Let us now consider some standard PH and PSPACE languages but in parametric form.

**CIRCUITSAT** $= \{\langle C, 1^n \rangle \mid$ *C is a satisfiable circuit, and n is the number of variables in C*$\}$

$\Sigma_i$**SAT** $= \{\langle \varphi, 1^n \rangle \mid$ *$\varphi$ is a satisfiable quantified Boolean formula in CNF with $i-1$ alternations where odd position quantifiers are $\exists$ and even position quantifiers are $\forall$, and $n = (n_1 + n_2 + \ldots + n_i)$ where $n_i$ is the number of the variables corresponding to $i^{th}$ quantifier*$\}$

$\Sigma_i$**CIRCUITSAT** $= \{\langle C, 1^n \rangle \mid$ *C is a satisfiable quantified circuit with $i-1$ alternations where odd position quantifiers are $\exists$ and even position quantifiers are $\forall$, and $n = (n_1 + n_2 + \ldots + n_i)$ where $n_i$ is the number of the variables corresponding to $i^{th}$ quantifier*$\}$

*Similarly we can define $\Pi_i$SAT and $\Pi_i$CIRCUITSAT in parametric form.*

**QBCNFSAT** $= \{\langle \varphi, 1^n \rangle \mid$ *$\varphi$ is a satisfiable quantified Boolean formula in CNF, and n is the number of variables*$\}$

**QBFORMULASAT** $= \{\langle \varphi, 1^n \rangle \mid$ *$\varphi$ is a satisfiable quantified Boolean formula (not necessarily in CNF), and n is the number of variables*$\}$

*If $\varphi$ is replaced by the circuit C, then similarly we can define* **QBCSAT**.

**OR-QBCNFSAT** $= \{\langle \{\varphi_i\}, 1^n \rangle \mid$ *Each $\varphi_i$ is a quantified Boolean formula in CNF and at least one $\varphi_i$ is satisfiable, and each $\varphi_i$ is of bit-length at most n*$\}$.

Now we can generalize. For any language L we can define, **OR-L** = $\{\langle \{x_i\}, 1^n \rangle \mid$ At least one $x_i \in$ L, and each $x_i$ is of bit-length at most $n\}$.

Here we would like to mention that the non-parametric versions of $\Sigma_i$CIRCUITSAT and $\Sigma_i$SAT are complete for the class $\Sigma_i^p$ according to Cook-Levin reduction, and similarly the non-parametric versions of QBCNFSAT, QBFORMULASAT and QBCSAT are complete for PSPACE.

We can define a parametric problem corresponding to any language $L$ in the class $\Sigma_i^p$, or more precisely to the $(i+1)$-ary polynomial-time computable relation $R$ defining $L$, as follows.

**Definition 5.** *For any $(i+1)$-ary polynomial-time computable relation R, we can define a parametric problem in $\Sigma_i^p$, $L_R = \{\langle x, 1^n \rangle \mid \exists\, u_1 \in \{0,1\}^{n_1}\, \forall\, u_2 \in \{0,1\}^{n_2} \ldots Q_i$ $u_i \in \{0,1\}^{n_i}$ $R\,(x, u_1, \ldots, u_i) = 1$ and $n = (n_1 + n_2 + \ldots + n_i)$ where $n_i$ is the parameter corresponding to $i^{th}$ quantifier$\}$*

We can do essentially the similar thing for any language $L \in$ PSPACE using the characterization of PSPACE as alternating polynomial time ([1], Corollary 3.6) as follows:

**Proposition 1.** *Any language $L$ is in* PSPACE *if and only if it is decidable by an Alternating Turing machine in polynomial time.*

Now we can define,

**Definition 6.** *For any binary polynomial-time computable relation R, we can define a parametric problem in* PSPACE, *$L_R = \{\langle x, 1^n \rangle \mid Q_1\, u_1 \in \{0,1\}^{n_1}\, Q_2\, u_2 \in \{0,1\}^{n_2} \ldots Q_i\, u_i \in \{0,1\}^{n_i}$ $R\,(x, \langle u_1, \ldots, u_i \rangle) = 1$ and $n = (n_1 + n_2 + \ldots + n_i)$ where all the Q variables denote $\exists$ or $\forall$ alternately, depending on whether its suffix is odd or even, $i$ is polynomially bounded with respect to the size of $x$ and $n_i$ is the parameter corresponding to $i^{th}$ quantifier$\}$*

So using the general definition of compression of any language in parametric form given above, we can define the compression for all the PH and PSPACE parametric problems where the "witness length" or "guess length" is the parameter of the problem.

**Proposition 2.** *$\Sigma_i$CIRCUITSAT is a complete parametric problem with respect to W-reduction for the class of parametric problems in $\Sigma_i^p$.*

*Proof.* Firstly we can observe that $\Sigma_i$CIRCUITSAT is among the parametric problems in the class $\Sigma_i^p$ as there is an Alternating Turing Machine accepting this language with $i-1$ alternations, starting with existential guesses. Let us now consider the parametric problem $L_R \in \Sigma_i^p$. So there exists a polynomial-time computable relation $R$ such that, $\langle x, 1^n \rangle \in L_R \Leftrightarrow \exists\, u_1 \in \{0,1\}^{n_1}\, \forall\, u_2 \in \{0,1\}^{n_2} \ldots Q_i\, u_i \in \{0,1\}^{n_i}$ $R\,(x, u_1, \ldots, u_i) = 1$, where $Q_i$ denotes $\exists$ or $\forall$ depending on whether $i$ is odd or even respectively. Here $n = (n_1 + n_2 + \ldots + n_i)$.

Now for the parametric problem $L_R$ the parameter is the number of guess bits used by $R$ which is $n$ in this case. We know that any polynomial time computable relation has uniform polynomial size circuits ([15], Theorem 6.7). Let $C_m$ be the circuit on inputs of length $m$ - we can generate $C_m$ from $1^m$ in polynomial time. Hence, $\langle x, 1^n \rangle \in L_R \Leftrightarrow \exists\, u_1 \in \{0,1\}^{n_1}\, \forall\, u_2 \in \{0,1\}^{n_2} \ldots Q_i\, u_i \in \{0,1\}^{n_i}$ $C_m\,(x, u_1, \ldots, u_i) = 1$, where $Q_i$ denotes $\exists$ or $\forall$ depending on whether $i$ is odd or even respectively. This gives a $W$-reduction from the parametric problem $L_R$ to $\Sigma_i$CIRCUITSAT, since the length of the parameter is preserved. □

A similar proposition holds for $\Pi_i$CIRCUITSAT as well. We can also show, using a similar proof, a completeness result for PSPACE as follows.

**Proposition 3.** QBCSAT *is a complete parametric problem for the class of parametric problems in* PSPACE *with respect to* $W$*-reduction.*

*Proof.* Firstly we can observe that QBCSAT is among the parametric problems in the class PSPACE as there is an Alternating Turing Machine accepting this language with at most $n$ alternations, where $n$ is the number of variables of the QBCSAT instance. Let us now consider the parametric problem $L_R \in$ PSPACE. So there exists a polynomial-time computable relation $R$ such that,
$\langle x, 1^n \rangle \in L_R \Leftrightarrow Q_1\, u_1 \in \{0,1\}^{n_1}\, Q_2\, u_2 \in \{0,1\}^{n_2} \ldots Q_i\, u_i \in \{0,1\}^{n_i}\, R\,(x, \langle u_1, \ldots, u_i \rangle) = 1$, where all the $Q$ variables denote $\exists$ or $\forall$ alternately, depending on whether its suffix is odd or even. $i$ is polynomially bounded with respect to the size of $x$. Here $n = (n_1 + n_2 + \ldots + n_i)$.

Now for the parametric problem $L_R$ the parameter is the number of guess bits used by $R$ which is $n$ in this case. We know that any polynomial time computable relation has uniform polynomial size circuits ([15], Theorem 6.7). Let $C_m$ be the circuit on inputs of length $m$ - we can generate $C_m$ from $1^m$ in polynomial time. Hence, $\langle x, 1^n \rangle \in L_R \Leftrightarrow Q_1\, u_1 \in \{0,1\}^{n_1}\, Q_2\, u_2 \in \{0,1\}^{n_2} \ldots Q_i\, u_i \in \{0,1\}^{n_i}\, C_m\,(x, \langle u_1, \ldots, u_i \rangle) = 1$, where all the $Q$ variables denote $\exists$ or $\forall$ alternately, depending on whether its suffix is odd or even. This gives a $W$-reduction from the parametric problem $L_R$ to QBCSAT, since the length of the parameter is preserved. □

We note that all the parametric problems we have defined so far are in fact fixed-parameter tractable, simply by using brute force search.

**Proposition 4.** QBCSAT *is solvable in time* $O(2^n poly(m))$ *by brute force enumeration.*

# 3 Instance Compression for Polynomial Hierarchy

## 3.1 Instance Compression in Second Level

In this section, we are going to show that non-uniform compression of CIRCUITSAT within NP implies the non-uniform compression of $\Sigma_2$CIRCUITSAT within NP as well. In the next subsection, essentially by using induction we show how to extend this to the entire Polynomial Hierarchy. We have used the following result by Fortnow and Santhanam ([11], Theorem 3.1):

**Theorem 1.** *If* OR-SAT *is compressible, then* CoNP $\subseteq$ NP/poly, *and hence* PH *collapses.*

The same technique actually shows that, any language L for which OR-L (section 2) is compressible, lies within CoNP/poly.

**Theorem 2.** *If* CIRCUITSAT *is non-uniformly compressible within* NP*, then* $\Sigma_2$CIRCUITSAT *is non-uniformly compressible within* NP.

*Proof.* Let us consider the parametric problem $\Sigma_2$CIRCUITSAT first. For the sake of convenience, we often omit the parameter when talking about an instance of this problem. According to the definition,

$$C \in \Sigma_2\text{CIRCUITSAT} \Leftrightarrow \exists u \in \{0,1\}^{n_1} \forall v \in \{0,1\}^{n_2} C(u,v) = 1 \quad (1)$$

Let $m$ be the length of the description of the circuit $C$ and $n = (n_1 + n_2)$ to be the number of variables of $C$.

Let us now fix a specific $u = u_1$. Now, we can define a new parametric problem $L^{'}$ as follows,

$$\langle C, u_1, 1^{n_2} \rangle \in L^{'} \Leftrightarrow \forall v \in \{0,1\}^{n_2} C(u_1, v) = 1 \quad (2)$$

It is clear from the above definition that $L^{'}$ is a parametric problem in CoNP (of instance size $\leq O(m + n_1)$) and any instance of $L^{'}$ can be polynomial-time reduced to an instance of CIRCUIT-UNSAT, say $C^{'}$ (because CIRCUIT-UNSAT, the parametric problem of all unsatisfiable circuits, is CoNP-Complete with respect to $W$-reduction). As shown in Proposition 2, the size of the witness will be preserved in this reduction.

$C \in \Sigma_2$CIRCUITSAT $\Leftrightarrow \exists u_1 \langle C, u_1 \rangle \in L^{'}$ and $\langle C, u_1 \rangle \in L^{'} \Leftrightarrow C^{'} \in$ CIRCUIT-UNSAT. Here the instance length $|C| = m$ and $|C^{'}| = poly(m)$. $poly(.)$ is denoting just an arbitrary polynomial function.

Let $g$ be the polynomial-time reduction used to obtain $C^{'}$ from $C$ and $u_1$. Namely, $C^{'} = g(C, u_1)$. If CIRCUITSAT is non-uniformly compressible within NP, using the same reduction we can non-uniformly compress CIRCUIT-UNSAT within CoNP. That means we can reduce a CIRCUIT-UNSAT instance into another CIRCUIT-UNSAT instance in polynomial time, as CIRCUIT-UNSAT is CoNP-complete with respect to $W$-reduction. Assume this polynomial time compression function be $f_1$ with polynomial size advice. So we will use $f_1$ to compress CIRCUIT-UNSAT instance $C^{'}$ to another CIRCUIT-UNSAT instance, say $C^{''}$, of size $poly(n_2)$.

Therefore, $C^{'} \in$ CIRCUIT-UNSAT $\Leftrightarrow C^{''} = f_1(C^{'}, w_1) = f_1(g(C, u_1), w_1) \in$ CIRCUIT-UNSAT, where $|C^{''}| = poly(n_2)$ and the string $w_1$ (of size at most $poly(m)$) is capturing the notion of polynomial size advice. Here the compression function $f_1$ is computable in polynomial (in $m$) time.

Now, if CIRCUITSAT is non-uniformly compressible within NP so is SAT as SAT is $W$-reducible to CIRCUITSAT. Now, OR-SAT is also non-uniformly compressible as OR-SAT $W$-reduces to SAT. It can be proved from Theorem 1 that if OR-SAT is non-uniformly compressible then CoNP $\subseteq$ NP/poly, as mentioned in the beginning of this section.

Now combining the statements in the above paragraph we can say that if CIRCUITSAT is non-uniformly compressible within NP then CoNP $\subseteq$ NP/poly. So we can now reduce our parametric problem in CoNP (here CIRCUIT-UNSAT) instance $C^{''}$ to a NP-complete parametric problem instance using polynomial size advice. As CIRCUITSAT is a NP-complete with respect to $W$-reduction, we can reduce $C^{''}$ to a CIRCUITSAT instance, say $C^{'''}$, using a polynomial time computable function $f_2$ with advice $w_2$. In the above procedure, the length of the instance definitely will not increase by more than a polynomial factor. So clearly $|C^{'''}| = poly(n_2)$.

So from the above arguments we can say that,
$C' \in$ CIRCUIT-UNSAT $\Leftrightarrow C''' = f_2(C'', w_2) = f_2(f_1(g(C, u_1), w_1), w_2) \in$ CIRCUIT-SAT, where $|C'''| = poly(n_2)$ and the string $w_2$ (of size at most $poly(n_2)$) is capturing the notion of polynomial size advice which arises in the proof of Theorem 1. Here the function $f_2$ is computable in polynomial (in $n_2$) time.

Now we define a new circuit $C_1$ as follows. $C_1$ is a non-deterministic circuit whose non-deterministic input is divided into two strings: $u$ of length $n_1$ and $v$ of length $poly(n_2)$. Given its non-deterministic input, $C_1$ first computes $C''' = f_2((f_1(g(C, u), w_1), w_2)$. This can be done in polynomial size in $m$ since the functions $f_2$, $f_1$ and $g$ are all polynomial-time computable and $C$, $w_1$ and $w_2$ are all fixed strings of size polynomial in $m$. It then uses its input $v$ as non-deterministic input to $C'''$ and checks if $v$ satisfies $C'''$. This can be done in polynomial-size since the computation of a polynomial-size circuit can be simulated in polynomial time. If so, it outputs 1, else it outputs 0. Now we have

$$C \in \Sigma_2\text{CIRCUITSAT} \Leftrightarrow \exists u \in \{0,1\}^{n_1} \exists v \in \{0,1\}^{n_2} C_1(u, v) = 1 \quad (3)$$

The key point is that we have reduced our original $\Sigma_2$CIRCUITSAT question to a CIRCUITSAT question, *without* a super-polynomial blow-up in the witness size. This allows us to apply the compressibility hypothesis again. Also, note that $C_1$ is computable from $C$ in polynomial size.

After that, using the compressibility assumption for CIRCUITSAT, we can non-uniformly compress $C_1$ to an instance $C_2$ of size $poly(n_1 + n_2)$ of a parametric problem in NP. Our final compression procedure just composes the procedures deriving $C_1$ from $C$ and $C_2$ from $C_1$, and since each of these can be implemented in polynomial size, our compression of the original $\Sigma_2$CIRCUITSAT instance is a valid non-uniform instance compression. Thus it is shown that if CIRCUITSAT is non-uniformly compressible within NP, $\Sigma_2$CIRCUITSAT is also non-uniformly compressible within NP. □

## 3.2 Instance Compression for Higher Levels

Now we are going to extend the idea for higher classes. It is not that difficult to see, if $\Sigma_2$CIRCUITSAT is non-uniformly compressible within NP, $\Pi_2$CIRCUITSAT is non-uniformly compressible within CoNP. We will use this in the following theorem.

**Theorem 3.** *If* CIRCUITSAT *is non-uniformly compressible within* NP, *then* $\Sigma_i$CIRCUITSAT *is non-uniformly compressible within* NP *for all* $i > 1$.

*Proof Outline.* We are going use induction here. Let us assume CIRCUITSAT is non-uniformly compressible within NP. To prove $\Sigma_i$CIRCUITSAT is compressible for all $i > 1$, the base case $i = 2$, directly follows from Theorem 2. Now suppose the statement is true for all $i \leq k$. We have to prove that the statement is true for $i = k + 1$ as well. So we are now assuming that $\Sigma_i$CIRCUITSAT is non-uniformly compressible within NP for all $i \leq k$ and going to prove that $\Sigma_{k+1}$CIRCUITSAT is also non-uniformly compressible within NP.

Now, fixing the first variable, $u_1$ to $u'$ of $\Sigma_{k+1}$CIRCUITSAT instance $C$ as before, we can define a new language similarly as we did in the proof of Theorem 2. Using similar argument we can introduce a circuit $C_1$ as well. The key point is, we can reduced our original $\Sigma_{k+1}$CIRCUITSAT instance to an instance of $\Sigma_k$CIRCUITSAT. Hence using the induction step, we can reduce it to a CIRCUITSAT instance, *without* a super-polynomial blow-up in the witness size. This allows us to apply the compressibility hypothesis again. Also, note that $C_1$ is computable from $C$ in polynomial size. Next, using the compressibility assumption for CIRCUITSAT, we can non-uniformly compress $C_1$ to an NP language instance $C_2$ of size $poly(n_1 + n^{'})$ i.e. $poly(n_1 + \ldots + n_{k+1})$. So using mathematical induction we can say if CIRCUITSAT is non-uniformly compressible within NP, $\Sigma_i$CIRCUITSAT is also non-uniformly compressible within NP for all $i > 1$. □

**Corollary 1.** *If* CIRCUITSAT *is compressible within* NP, $\Pi_i$CIRCUITSAT *is also non-uniformly compressible within* NP *for all* $i \geq 1$.

As $\Pi_i$CIRCUITSAT $W$-reduces to $\Sigma_{i+1}$CIRCUITSAT, the above Corollary is trivial. Theorems 2 and 3 require an assumption on non-uniform compressibility in NP. But we don't need this for compressibility of a problem higher in the hierarchy.

**Proposition 5.** *If* $\Sigma_3$CIRCUITSAT *is compressible, then* $\Sigma_i$CIRCUITSAT *is compressible for any* $i > 3$.

*Proof.* This proposition follows from the fact that $\Sigma_3$CIRCUITSAT being compressible implies that SAT is compressible. So, by the result of Fortnow and Santhanam (Theorem 1), PH collapses to $\Sigma_3^p$. As a result, every parametric problem in the class $\Sigma_i^p$ $W$-reduces to $\Sigma_3$CIRCUITSAT, as $\Sigma_3$CIRCUITSAT is complete for the class $\Sigma_3^p$ with respect to $W$-reduction. Hence, $\Sigma_3$CIRCUITSAT being compressible, $\Sigma_i$CIRCUITSAT is compressible for any $i > 3$. □

## 4 Instance Compression for PSPACE

In this section, we show that QBCNFSAT is unlikely to be compressible, even non-uniformly - compressibility of QBCNFSAT implies that PSPACE collapses to the third level of the Polynomial Hierarchy. The strategy we adopt is similar to that in Theorem 1 where it is shown that compressibility of SAT implies NP $\subseteq$ CoNP/poly. To show their result, they used the OR-SAT problem, which is $W$-reducible to SAT ([18]). Thus an incompressibility result for the OR-SAT problem translates directly to a corresponding result for SAT.

We similarly defined OR-QBCNFSAT problem in Section 2. But it is not that easy to show that OR-QBCNFSAT $W$-reduces to QBCNFSAT. There are a couple of different issues. First the quantifier patterns for the formulae $\{\phi_i\}, i = 1 \ldots m$ might all be different. This is easily taken care of, because we can assume quantifiers alternate between existential and universal - this just blows up the number of variables for any formula by a factor of at most 2. The more critical issue is that nothing as simple as the OR works for combining formulae. $\exists x \forall y \phi_1(x, y) \vee \exists x \forall y \phi_2(x, y)$ is not equivalent to $\exists x \forall y(\phi_1(x, y) \vee \phi_2(x, y))$. We are forced to adopt a different strategy as explained below. Later we have found similar strategy is used in [18], though it was in the context of OR-SAT, not OR-QBCNFSAT.

**Lemma 1.** OR-QBCNFSAT *is W-reducible to* QBCNFSAT

*Proof.* Let $\langle\{\phi_i\}, 1^n\rangle$ be an OR-QBCNFSAT instance of length $m$. Assume without loss of generality that each $\phi_i$ has exactly $n$ variables and that the quantifiers alternate starting with existential quantification over $x_1$, continuing with quantification over $x_2, x_3$ etc. We construct in polynomial time in $m$ an equivalent instance of QBCNFSAT with at most $poly(n)$ variables and of size $poly(m)$. We first take care of quantifications. The quantifier patterns for the formulae $\{\phi_i\}, i = 1 \ldots m$ might all be different. But we can assume quantifiers alternate between existential and universal - this just blows up the number of variables for any formula by a factor of at most 2. Then we check if the number of input formulae is greater than $2^n$ or not. If yes, we solve the original instance by brute force search and output either a trivial true formula or a trivial false formula depending on the result of the search. If not, then we define a new formula with $\lceil log(m) \rceil$ additional variables $y_1, y_2 \ldots y_k$. We identify each number between 1 and $m$ uniquely with a string in $\{0,1\}^k$. Now we define the formula $\psi_i$ corresponding to $\phi_i$ as follows. Let the string $w_i \in \{0,1\}^k$ correspond to the number $i$. Then $\psi_i = z_1 \wedge z_2 \ldots \wedge z_k \wedge \phi_i$, where $z_r = y_r$ if $w_r = 1$ and the complement of $y_r$ otherwise. The output formula $\psi$ starts with existential quantification over the $y$ variables followed by the standard pattern of quantification over the $x$ variables followed by the formula which is the OR of all $\psi_i$'s, $i = 1 \ldots m$. So $\psi$ will be as follows:
$\psi = \exists\, y_1 \,\exists\, y_2 \ldots \exists\, y_k \; Q_1\, x_1\, Q_2\, x_2 \ldots Q_n\, x_n\, (\psi_1 \vee \psi_2 \vee \ldots \vee \psi_m)$.
Where $Q_i$'s are the quantifications of the $x_i$'s as before. It is not that hard to check that $\psi$ is valid iff one of the $\phi_i$'s is. □

**Theorem 4.** *If* QBCNFSAT *is compressible, then* PSPACE $\subseteq$ NP/poly, *and hence* PSPACE = $\Sigma_3^p$.

*Proof.* Using Lemma 1, if QBCNFSAT is compressible, OR-QBCNFSAT is also compressible. From the proof of Theorem 1 we can say for any parametric problem L for which OR-L (section 2) is compressible, lies in CoNP/poly. Thus, since the parametric problem QBCNFSAT is PSPACE-complete and PSPACE is closed under complementation, a compression for OR-QBCNFSAT implies PSPACE is in NP/poly. Hence by the result of Yap [3], it follows that PH collapses to the third level. Combining this with the Karp-Lipton theorem for PSPACE ([13], Theorem 4.1), we have that PSPACE = $\Sigma_3^p$. □

## 5 Succinct IP and PSPACE

IP ([16,10]) is the class of problems solvable by an interactive proof system. An interactive proof system consists of two machines, a *Prover*, $P$, which presents a proof that a input string is a member of some language, and a *Verifier*, $V$, that checks that the presented proof is correct. Now we are extending this idea of IP to Succinct IP, where the total number of bits communicated between *prover* and the *verifier* is polynomially bounded in parameter length.

We define *Verifier* to be a function $V$ that computes its next transmission to the *Prover* from the message history sent so far. The function $V$ has three inputs:

(1) **Input String**, (2) **Random input** and (3) **Partial message history**

$m_1\#m_2\#\ldots\#m_i$ is used to represent the exchange of messages $m_1$ through $m_i$ between $P$ and $V$. The Verifier's output is either the next message $m_{i+1}$ in the sequence or *accept* or *reject*, designating the conclusion of the interaction. Thus $V$ has the function from $V$: $\Sigma^* \times \Sigma^* \times \Sigma^* \to \Sigma^* \cup \{$ accept, reject $\}$.

The *Prover* is a party with unlimited computational ability. We define it to be a function $P$ with two inputs:

(1) **Input String** and (2) **Partial message history**

The Prover's output is the next message to the Verifier. Formally, $P$: $\Sigma^* \times \Sigma^* \to \Sigma^*$. Next we define the interaction between Prover and the Verifier. For particular input string $w$ and random string $r$, we write $(V \leftrightarrow P)(w, r)$ = *accept* if a message sequence $m_1$ to $m_k$ exists for some $k$ whereby

1. for $0 \leq i < k$, where $i$ is an even number, $V(w, r, m_1\#m_2\#\ldots\#m_i) = m_{i+1}$;
2. $0 < i < k$, where $i$ is an odd number, $P(w, m_1\#m_2\#\ldots\#m_i) = m_{i+1}$; and
3. the final message $m_k$ in the message history is *accept*.

In the definition of the class Succinct IP, the lengths of the Verifier's random input and each of the messages exchanged are $p(n)$ for some polynomial $p$ that depends only on the Verifier. Here $n$ is the parameter length of input instance. Besides, total bits of messages exchanged is at most $p(n)$ as well.

**Succinct IP:** A parametric problem $L$ ($\subseteq \{\langle x, 1^n\rangle | x \in \{0,1\}^*, n \in \mathbb{N}\}$) is in Succinct IP if there exist some polynomial time function $V$ and arbitrary function $P$, with total $poly(n)$ many bits of messages communicated between them and for every function $\tilde{P}$ and string $w$,

1. $w \in L$ implies $\Pr[V \leftrightarrow P] \geq 2/3$, and
2. $w \notin L$ implies $\Pr[V \leftrightarrow \tilde{P}] \leq 1/3$.

Here $poly(n)$ denotes some polynomial that depends only on the Verifier and $n$ is the parameter length of input instance $w$.

We know that QBFORMULASAT is in IP, as IP = PSPACE ([4,2]). But we can even prove something more. Not only for QBCNFSAT, we can construct Succinct IP protocol for QBFORMULASAT as well. To prove that we are basically going to adapt the formal proof of the part, PSPACE $\subseteq$ IP ([4,2,12]).

**Proposition 6. QBFORMULASAT $\in$ SUCCINCT IP**

*Proof.* The key idea is to take an algebraic view of Boolean formulae by representing them as polynomials. We are considering the inputs are from some finite field $\mathbb{F}$. We can see that 0, 1 can be thought of both as truth values and as elements of $\mathbb{F}$. Thus we have the following correspondence between formulas and polynomials when the variables take 0/1 values:

$x \wedge y \leftrightarrow X . Y$

$\bar{x} \leftrightarrow 1 - X$

$x \vee y \leftrightarrow X{*}Y = 1 - (1 - X)(1 - Y)$

So, if there is a Boolean formula $\phi(x_1, x_2, \ldots, x_n)$ of length $m$, we can easily convert that into a polynomial $p$ of degree at most $m$ following the rules described above.

Let the given formula be,
$\Psi = Q_1\, x_1\, Q_2\, x_2\, Q_3\, x_3 \ldots\ Q_n\, x_n\, \phi(x_1, \ldots, x_n)$,
where the size of $\Psi$ is $m$. $\phi$ is any Boolean formula over $n$ variables.

To arithmetize $\Psi$ we introduce some new terms in quantification and rewrite the expression in the following manner:

$\Psi' = Q_1\, x_1\, R\, x_1\, Q_2\, x_2\, R\, x_1\, R\, x_2\, Q_3\, x_3\, R\, x_1\, R\, x_2\, R\, x_3 \ldots\ Q_n\, x_n\, R\, x_1\, R\, x_2 \ldots R\, x_n\, \phi(x_1, \ldots, x_n)$,

Here $R$ is introduced to enable linearize operation on the polynomial as explained later. We now rewrite this $\Psi'$ as follows : $\Psi' = S_1\, x_1\, S_2\, x_2\, S_3\, x_3 \ldots\ S_k\, x_k\, [\phi]$, where each $S_i \in \{\exists, \forall, R\}$. We are going to define $R$ very soon. We can see that value of $k$ can be at most $O(n^2)$.

For each $i \leq k$ we define the function $f_i$. We define $f_k(x_1, x_2, \ldots, x_n)$ to be the polynomial $p$ [i.e. $p(x_1, x_2, \ldots, x_n)$] obtained by arithmetization of $\phi$. For $i < k$ we define $f_i$ in terms of $f_{i+1}$:

$S_{i+1} = \forall$: $f_i(\ldots) = f_{i+1}(\ldots, 0).f_{i+1}(\ldots, 1)$;
$S_{i+1} = \exists$: $f_i(\ldots) = f_{i+1}(\ldots, 0) * f_{i+1}(\ldots, 1)$;
$S_{i+1} = R$: $f_i(\ldots, a) = (1-a) f_{i+1}(\ldots, 0) + a f_{i+1}(\ldots, 1)$.

Here we reorder the inputs of the functions in such a way that variable $y_{i+1}$ is always the last argument. If $S$ is $\exists$ or $\forall$, $f_i$ has one fewer input variable than $f_{i+1}$ does. But if $S$ is $R$, both of them have same number of arguments. To avoid complicated subscripts, we use "$\ldots$" which can be replaced by $a_1$ through $a_j$ for appropriate values of $j$ after the reordering of the inputs.

We can observe that operation $R$ on polynomial doesn't change their values for Boolean inputs. So $f_0()$ is still the truth value of $\Psi$. Now we can observe that these $Rx$ operation produces a result that is linear in $x$. We added $Rx_1\, Rx_2 \ldots Rx_i$ after $Q_i x_i$ in $\Psi'$ in order to reduce the degree of each variable to 1 prior to the squaring due to arithmetization of $Q_i$.

We are now ready to describe the protocol. Here $P$ is denoted to be the prover and $V$ to be the verifier as we always use.

**Phase 0:** [$P$ sends $f_0()$]
$P \rightarrow V$: $P$ sends $f_0()$ to $V$. $V$ checks that $f_0() = 1$ and *rejects* if not.
Progressing similarly,

**Phase $i$:** [$P$ persuades $V$ that $f_{i-1}(r_1, \ldots)$ is correct if $f_i(r_1, \ldots, r)$ is correct]
$P \rightarrow V$: $P$ sends the coefficients of $f_i(r_1, \ldots, z)$ as a polynomial in $z$. (Here $r_1 \ldots$ denotes a setting of the variables to the previously selected random values $r_1, r_2, \ldots$)
$V$ uses these coefficients to evaluate $f_i(r_1, \ldots, 0)$ and $f_i(r_1, \ldots, 1)$. Then it checks that the polynomial degree is at most 2 and that these identities hold:

$$f_{i-1}(r_1, \ldots) = \begin{cases} f_i(r_1, \ldots, 0).f_i(r_1, \ldots, 1) & \text{if } S_i = \forall \\ f_i(r_1, \ldots, 0) * f_i(r_1, \ldots, 1) & \text{if } S_i = \exists \end{cases}$$

and

$$f_{i-1}(r_1,\ldots,r) = (1-r)f_i(r_1,\ldots,0) + rf_i(r_1,\ldots,1) \text{ if } S_i = R$$

If either fails, $V$ *rejects*.

$V \rightarrow P$: $V$ picks a random Boolean value $r$ from $\mathbb{F}$ and sends it to $P$. If $S_i = R$, this $r$ replaces the previous $r$

Then it goes to phase $i$+1, where $P$ must persuade $V$ that $f_i(r_1,\ldots,r)$ is correct.

Progressing similarly,

**Phase $k$+1:** [$V$ checks directly that $f_k(r_1,\ldots,r_n)$ is correct]
$V$ evaluates $p(r_1,\ldots,r_n)$ to compare with the value $V$ has for $f_k(r_1,\ldots,r_n)$. If they are equal, $V$ *accepts*, otherwise $V$ *rejects*. That completes the description of the protocol.

Here polynomial $p$ is nothing but the arithmetization of the formula $\phi$, as we have already seen. It can be shown that the evaluation of this polynomial can be done in polynomial time.

For the evaluation of the polynomial $p$ for $r_1,\ldots,r_n$, we will consider $\phi$ and apply the arithmetization for the nodes individually. We will evaluate the nodes from lower level. Before we evaluate for any node, corresponding inputs are already evaluated and ready to use. Evaluation for each node will take constant amount of time. So total evaluation of $p$ for $r_1,\ldots,r_n$ through modified $\phi$ will take $poly(m)$ time.

Now we can try to prove that the probability of error is bounded within the limit. If the prover $P$ always returns the correct polynomial, it will always convince $V$ . If $P$ is not honest then we are going to prove that $V$ rejects with high probability:

$$Pr[V\ rejects] \geq (1 - d/|\mathbb{F}|)^k \tag{4}$$

where $d$ is the highest degree of the polynomial sent in each stage. We can see that value of $k$ can be at most $O(n^2)$. As the value of $d$ is 2 in our case, the right hand side of the above expression is at least (1 - $2k/|\mathbb{F}|$), which is very close to 1 for sufficiently large values of $|\mathbb{F}|$. It will be sufficient for us if $|\mathbb{F}|$ is bounded by a large enough polynomial in $n$.

Now we are going to see how the proof works when the proves is trying to cheat for "no" instance. In the first round, the prover $P$ should send $f_0()$ which must be 1. Then $P$ is supposed to return the polynomial $f_1$. If it indeed returns $f_1$ then since $f_1(0)$ + $f_1(1) \neq f_0()$ by assumption, $V$ will immediately reject (i.e., with probability 1). So assume that the prover returns some $s(X_1)$, different from $f_1(X_1)$. Since the degree $d$ non-zero polynomial $s(X_1)$ - $f_1(X_1)$ has at most $d$ roots, there are at most $d$ values $r$ such that $s(r) = f_1(r)$. Thus when $V$ picks a random $r$,

$$Pr_r[s(r) \neq f_1(r)] \geq (1 - d/|\mathbb{F}|) \tag{5}$$

Then the prover is left with an incorrect claim to prove in all the phases. So prover should lie continuously. If $P$ is lucky, $V$ will not understand the lie. To prove equation (4), we will use induction here. We assume the induction hypothesis to be true for $k-1$ steps, that is, the prover fails to prove this false claim with probability at least $\geq$ $(1 - d/|\mathbb{F}|)^{k-1}$. Base case is easy to see from equation (5). Thus we have,

$$Pr[V\ rejects] \geq (1 - d/|\mathbb{F}|).(1 - d/|\mathbb{F}|)^{k-1} = (1 - d/|\mathbb{F}|)^k \tag{6}$$

If $P$ is not lucky, as the verifier is evaluating $p()$ explicitly in the last stage, $V$ will anyway detect the lie.

Here in the description of the protocol, we can see that the degree of the polynomial at each stage is at most 2. So we need just constant number of coefficients for encoding such polynomials. coefficients are from the field $\mathbb{F}$ which is of size $poly(m)$. So $O(log(poly(m)))$ i.e. $O(poly(n))$ size messages are sent in any phase. Even, it will be sufficient for us if $|\mathbb{F}|$ is bounded by a large enough polynomial in $n$. Number of such phases are bounded by $(k{+}1)$ which is $O(n^2)$. So we have constructed a *Succinct* Interactive proof protocol for QBFORMULASAT. □

**Issue in finding Succinct IP protocol for QBCSAT:** In case of QBCSAT, similar arithmetization technique will give polynomial of degree much larger size, actually exponential in $m$. Now, to reduce the error, we have to use Field $\mathbb{F}$ of larger size, basically exponential in $m$. This will give us each coefficients of the polynomials exchanged between $prover$ and $verifier$ to be of size $log(e^{poly(m)})$, i.e. $poly(m)$, which means the protocol is not succinct.

## 6 Future Directions

There are various possible directions. Suppose CIRCUITSAT is compressible within a class $C$. Here we have considered $C$ to be the class NP and got some interesting results. For any general class $C$ we know from Theorem 1 that the immediate consequence is the collapse of PH at third level. But it is still not known how our results for compression at second level of Polynomial Hierarchy will be affected for compression into an arbitrary class $C$. Besides, one could try to work under the weaker assumption that SAT or OR-SAT or OR-CIRCUITSAT is compressible instead of CIRCUITSAT. We also don't know whether there are similar implications for probabilistic compression where we allow certain amount of error in compression. One could also try to find a Succinct IP protocol for QBCSAT to show Succinct IP = PSPACE or try to find some negative implications of such a protocol existing for QBCSAT.

## References


1. Chandra, A.K., Kozen, D.C., Stockmeyer, L.J.: Alternation. Journal of the ACM 28(1), 114–133 (1981)
2. Shamir, A.: IP = PSPACE. Journal of the ACM 39(4), 869–877 (1992)
3. Yap, C.K.: Some consequences of non-uniform conditions on uniform classes. Theoretical Computer Science 26, 287–300 (1983)
4. Lund, C., Fortnow, L., Karloff, H., Nisan, N.: Algebraic methods for interactive proof systems. Journal of the ACM 39(4), 859–868 (1992)
5. Harnik, D., Naor, M.: On the compressibility of NP instances and cryptographic applications. In: Proceedings of the 47th Annual IEEE Symposium on Foundations of Computer Science, pp. 719–728 (2006)
6. Buhrman, H., Hitchcock, J.M.: NP-Hard Sets are Exponentially Dense Unless NP is contained in coNP/poly. Elect. Colloq. Comput. Complex (ECCC) 15(022) (2008)
7. Bodlaender, H.L., Downey, R.G., Fellows, M.R., Hermelin, D.: On problems without polynomial kernels. J. Comput. Syst. Sci. 75(8), 423–434 (2009)

8. Flum, J., Grohe, M.: Parameterized Complexity Theory. Springer (2006)
9. Abrahamson, K.A., Downey, R.G., Fellows, M.R.: Fixed-parameter tractability and completeness IV: On completeness for W[P] and PSPACE analogs. Annals of pure and applied logic 73, 235–276 (1995)
10. Babai, L., Moran, S.: Arthur-Merlin games: a randomized proof system, and a hierarchy of complexity classes. Journal of Computer and System Sciences 36, 254–276 (1988)
11. Fortnow, L., Santhanam, R.: Infeasibility of instance compression and succinct PCPs for NP. Journal of Computer and System Sciences 77(1), 91–106 (2011); special issues celebrating Karp's Kyoto Prize
12. Sipser, M.: Introduction to the Theory of Computation. Course Technology, 2nd edn. (2005)
13. Karp, R.M., Lipton, R.J.: Some connections between nonuniform and uniform complexity classes. In: Proceedings of the Twelfth Annual ACM Symposium on Theory of Computing, pp. 302–309 (1980), doi:10.1145/800141.804678
14. Niedermeier, R.: Invitation to Fixed Parameter Algorithms. Oxford University Press (2006)
15. Arora, S., Barak, B.: Computational Complexity: A Modern Approach. Cambridge University Press (2009)
16. Goldwasser, S., Micali, S., Rackoff, C.: The Knowledge complexity of interactive proof-systems. In: Proceedings of 17th ACM Symposium on the Theory of Computation, Providence, Rhode Island, pp. 291–304 (1985)
17. Kratsch, S., Wahlström, M.: Preprocessing of Min Ones Problems: A Dichotomy. In: Abramsky, S., Gavoille, C., Kirchner, C., Meyer auf der Heide, F., Spirakis, P.G. (eds.) ICALP 2010, Part I. LNCS, vol. 6198, pp. 653–665. Springer, Heidelberg (2010)
18. Chen, Y., Flum, J., Muller, M.: Lower bounds for kernelizations. CRM Publications (November 2008)

# Polynomial Time and Parameterized Approximation Algorithms for Boxicity

Abhijin Adiga[1], Jasine Babu[2], and L. Sunil Chandran[2]

[1] Network Dynamics and Simulation Science Laboratory,
Virginia Bioinformatics Institute, Virginia Tech Blacksburg, VA 24061, USA
[2] Department of Computer Science and Automation,
Indian Institute of Science, Bangalore 560012, India
abhijin@vbi.vt.edu, {jasine,sunil}@csa.iisc.ernet.in

**Abstract.** The boxicity (cubicity) of a graph $G$, denoted by $box(G)$ (respectively $cub(G)$), is the minimum integer $k$ such that $G$ can be represented as the intersection graph of axis parallel boxes (cubes) in $\mathbb{R}^k$. The problem of computing boxicity (cubicity) is known to be inapproximable in polynomial time even for graph classes like bipartite, co-bipartite and split graphs, within an $O(n^{0.5-\epsilon})$ factor for any $\epsilon > 0$, unless $NP = ZPP$.

We prove that if a graph $G$ on $n$ vertices has a clique on $n-k$ vertices, then $box(G)$ can be computed in time $n^2 2^{O(k^2 \log k)}$. Using this fact, various FPT approximation algorithms for boxicity are derived. The parameter used is the vertex (or edge) edit distance of the input graph from certain graph families of bounded boxicity - like interval graphs and planar graphs. Using the same fact, we also derive an $O\left(\frac{n\sqrt{\log\log n}}{\sqrt{\log n}}\right)$ factor approximation algorithm for computing boxicity, which, to our knowledge, is the first $o(n)$ factor approximation algorithm for the problem. We also present an FPT approximation algorithm for computing the cubicity of graphs, with vertex cover number as the parameter.

**Keywords:** Boxicity, Parameterized Algorithm, Approximation Algorithm.

## 1 Introduction

Let $G(V, E)$ be a graph. If $I_1, I_2, \ldots, I_k$ are (unit) interval graphs on the vertex set $V$ such that $E(G) = E(I_1) \cap E(I_2) \cap \cdots \cap E(I_k)$, then $\{I_1, I_2, \ldots, I_k\}$ is called a box (cube) representation of $G$ of dimension $k$. The boxicity (cubicity) of an incomplete graph $G$, $box(G)$ (respectively $cub(G)$), is defined as the minimum integer $k$ such that $G$ has a box (cube) representation of dimension $k$. For a complete graph, it is defined to be zero. Equivalently, boxicity (cubicity) of $G$ is the minimum integer $k$ such that $G$ can be represented as the intersection graph of axis parallel boxes (cubes) in $\mathbb{R}^k$. Boxicity was introduced by Roberts [15] in 1969 for modeling problems in the social sciences and ecology. Box representations of low dimension are memory efficient for representing dense graphs. Some

D.M. Thilikos and G.J. Woeginger (Eds.): IPEC 2012, LNCS 7535, pp. 135–146, 2012.

well known NP-hard problems like the max-clique problem are polynomial time solvable, if low dimensional box representations are known [16].

Boxicity is a combinatorially well studied parameter and its bounds in terms of parameters like maximum degree [2] and tree-width [7] are known. Roberts [15] showed that for any graph $G$ on $n$ vertices, $box(G) \leq \lfloor \frac{n}{2} \rfloor$ and $cub(G) \leq \lfloor \frac{2n}{3} \rfloor$. In 1986, Thomassen [17] proved that the boxicity of planar graphs is at most 3. Boxicity is also closely related to other dimensional parameters of graphs like partial order dimension and threshold dimension [2,1,19].

Cozzens [8] proved that computing boxicity is NP-hard. In fact, determining whether a graph has boxicity 2 is itself NP-hard (see Yannakakis [19] and Kratochvíl [11]). Recently, Adiga et al. [1] proved that it is not possible to approximate boxicity within a factor of $O(n^{0.5-\epsilon})$ for any $\epsilon > 0$ in polynomial time, even for bipartite, co-bipartite and split graphs, unless $NP = ZPP$. In this work, we present $o(n)$ factor approximation algorithms for computing boxicity and cubicity - the first of their kind, to our knowledge.

Since NP-hard problems are often impractical to solve, it is natural to introduce parameters along with the input, and design algorithms which run in polynomial time for small values of the parameter. We say that a decision problem with input size $n$ and a parameter $k$ is Fixed Parameter Tractable (FPT) if the problem can be decided in time $f(k) \cdot n^{O(1)}$, for some computable function $f$. Often, a similar terminology is used in the case of optimization problems too. An FPT approximation algorithm is an approximation algorithm that runs in $f(k) \cdot n^{O(1)}$ time. For an introduction to parameterized complexity, please refer to [14].

The decision problem BOXICITY takes a graph on $n$ vertices and an integer $b$ as inputs and asks whether $box(G) \leq b$. The standard parameterization of this problem using boxicity itself as the parameter $k$ is meaningless since the problem is NP-hard even for $k = 2$. Parameterizations with vertex cover number (MVC), minimum feedback vertex set size (FVS) and max leaf number as parameters were studied by Adiga et al. [3]. With vertex cover number as the parameter $k$, they gave an algorithm which computes boxicity exactly in $2^{O(2^k k^2)}n$ time, and another algorithm which gives an additive one approximation for boxicity in $2^{O(k^2 \log k)}n$ time, where $n$ is the number of vertices in the graph. Using FVS as the parameter $k$, they gave a $2 + \frac{2}{box(G)}$ factor approximation algorithm to compute boxicity that runs in $2^{O(2^k k^2)}n^{O(1)}$ time. With max leaf number as the parameter $k$, they gave an additive two approximation algorithm for boxicity that runs in $2^{O(k^3 \log k)}n^{O(1)}$ time.

In this work, we consider vertex and edge edit distance from families of graphs of bounded boxicity as parameters. The notion of edit distance refers, in general, to the smallest number of some well-defined modifications to be applied to the input graph so that the resultant graph possesses some desired properties. Edit distance from graph classes is a well-studied problem in parameterized complexity [5,10,13,18].

Cai [6] introduced a framework for parameterizing problems with edit distance as the parameter. For a family $\mathcal{F}$ of graphs, and $k \geq 0$ an integer, the author used $\mathcal{F} + ke$ (respectively, $\mathcal{F} - ke$) to denote the family of graphs that can be converted to a graph in $\mathcal{F}$ by deleting (respectively, adding) at most $k$ edges, and $\mathcal{F} + kv$ to denote the family of graphs that can be converted to a graph in $\mathcal{F}$ by deleting at most $k$ vertices. Cai [6] considered the parameterized complexity of the vertex coloring problem on $\mathcal{F} - ke$, $\mathcal{F} + ke$ and $\mathcal{F} + kv$ for various families $\mathcal{F}$ of graphs, with $k$ as the parameter. This was further studied by Marx [12].

In the same framework, we consider the parameterized complexity of computing the boxicity of $\mathcal{F} + k_1e - k_2e$ and $\mathcal{F} + kv$ graphs for families $\mathcal{F}$ of bounded boxicity graphs, using $k_1 + k_2$ and $k$ as parameters. We will see that many relevant parameters for the boxicity problem, including MVC and FVS considered by Adiga et al. [3], are special cases of our parameters. We provide an improved FPT algorithm with the parameter FVS and give FPT approximation algorithms with some parameters smaller than MVC. With the parameter max leaf number, our method achieves the same result as obtained in Adiga et al. [3]. (See corollaries 1-7 for more details.)

We also give a factor-2 FPT approximation algorithm for cubicity, using vertex cover number as the parameter. This can be improved to a $(1+\epsilon)$ factor algorithm for any $\epsilon > 0$, by sacrificing more on the running time.

## 2 Prerequisites

In this section, we give some basic facts necessary for the later part of the paper. For a vertex $v \in V$ of a graph $G$, we use $N_G(v)$ to denote the set of neighbors of $v$ in $G$. We use $G[S]$ to denote the induced subgraph of $G(V, E)$ on the vertex set $S \subseteq V$. Let $I$ be an interval representation of an interval graph $G(V, E)$. We use $l_v(I)$ and $r_v(I)$ respectively to denote the left and right end points of the interval corresponding to $v \in V$ in $I$. The interval is denoted as $[l_v(I), r_v(I)]$.

**Lemma 1 (Roberts [15]).** *Let $G(V, E)$ be any graph. For any $x \in V$, $box(G) \leq 1 + box(G \setminus \{x\})$.*

The following lemmas are easy to prove.

**Lemma 2.** *Let $G(V, E)$ be a graph. Let $S \subseteq V$ be such that $\forall v \in V \setminus S$ and $u \in V$, $(u, v) \in E$. If a box representation $\mathcal{B}_S$ of $G[S]$ is known, then, in $O(n^2)$ time we can construct a box representation $\mathcal{B}$ of $G$ of dimension $|\mathcal{B}_S|$. Moreover, $box(G) = box(G[S])$.*

**Lemma 3.** *Let $G(V, E)$ be a graph and let $A \subseteq V$. Let $G_1(V, E_1)$ be a supergraph of $G$ with $E_1 = E \cup \{(x, y) \mid x, y \in A,\ x \neq y\}$. If a box representation $\mathcal{B}$ of $G$ is known, then in $O(n^2)$ time we can construct a box representation $\mathcal{B}_1$ of $G_1$ of dimension $2 \cdot |\mathcal{B}|$. In particular, $box(G_1) \leq 2 \cdot box(G)$.*

We know that there are at most $2^{O(nb \log n)}$ distinct $b$-dimensional box representations of a graph $G$ on $n$ vertices and all these can be enumerated in time $2^{O(nb \log n)}$ [3, Proposition 1]. In linear time, it is also possible to check whether a given graph is a unit interval graph and if so, generate a unit interval representation of it [4]. Hence, a similar result holds for cubicity as well.

**Proposition 1.** *Let $G(V, E)$ be a graph on $n$ vertices of boxicity (cubicity) $b$. Then an optimal box (cube) representation of $G$ can be computed in $2^{O(nb \log n)}$ time.*

If $S \subseteq V$ induces a clique in $G$, then it is easy to see that the intersection of all the intervals in $I$ corresponding to vertices of $S$ is nonempty. This property is referred to as the *Helly property of intervals* and we refer to this common region of intervals as the *Helly region* of the clique $S$.

**Definition 1.** *Let $G(V, E)$ be a graph in which $S \subseteq V$ induces a clique in $G$. Let $H(V, E')$ be an interval supergraph of $G$. Let $p$ be a point on the real line. If $H$ has an interval representation $I$ satisfying the following conditions:*

*(1) $p$ belongs to the Helly region of $S$ in $I$.*
*(2) The end points of intervals corresponding to vertices of $V \setminus S$ are all distinct in $I$.*
*(3) For each $v \in S$,*

$$l_v(I) = \min\left(p, \min_{u \in N_G(v) \cap (V \setminus S)} r_u(I)\right) \text{ and}$$

$$r_v(I) = \max\left(p, \max_{u \in N_G(v) \cap (V \setminus S)} l_u(I)\right)$$

*then we call $I$ a nice interval representation of $H$ with respect to $S$ and $p$. If $H$ has a nice interval representation with respect to clique $S$ and some point $p$, then $H$ is called a nice interval supergraph of $G$ with respect to clique $S$.*

**Lemma 4.** *Let $G$ be a graph on $n$ vertices, with its vertices arbitrarily labeled as $1, 2, \ldots, n$. If $G$ contains a clique of size $n - k$ or more, then :*

*(a) A subset $A \subseteq V$ such that $|A| \leq k$ and $G[V \setminus A]$ is a clique, can be computed in $O(n2^k)$ time.*
*(b) There are at most $2^{O(k \log k)}$ nice interval supergraphs of $G$ with respect to the clique $V \setminus A$. These can be enumerated in $n^2 2^{O(k \log k)}$ time.*
*(c) If $G$ has a box representation $\mathcal{B}$ of dimension $b$, then it has a box representation $\mathcal{B}'$ of the same dimension, in which $\forall I \in \mathcal{B}'$, $I$ is a nice interval supergraph of $G$ with respect to the clique $V \setminus A$.*
*(d) By construction, vertices of the nice interval supergraphs obtained in (b) and (c) retain their original labels as in $G$.*

*Proof.* (a) We know that, if $G$ contains a clique of size $n - k$ or more, then the complement graph $\overline{G}$ has a vertex cover of size at most $k$. We can compute a minimum vertex cover $A$ of $\overline{G}$ in $O(n2^k)$ time [14]. We have $|A| \leq k$ and $G[V \setminus A]$ is a clique because $V \setminus A$ is an independent set in $\overline{G}$.

(b) Let $H$ be any nice interval supergraph of $G$ with respect to $V \setminus A$. Let $I$ be a nice interval representation of $H$ with respect to $V \setminus A$ and a point $p$. Let $P$ be the set of end points (both left and right) of the intervals corresponding to vertices of $A$ in $H$. Clearly $|P| = 2|A| \leq 2k$. The order of end points of vertices of $A$ in $I$ from left to right corresponds to a permutation of elements of $P$ and therefore, there are at most $(2k)!$ possibilities for this ordering. Moreover, note that the points of $P$ divide the real line into $|P|+1$ regions and that $p$ can belong to any of these regions. From the definition of nice interval representation, it is clear that, once the point $p$ and the end points of vertices of $A$ are fixed, the end points of vertices in $V \setminus A$ get automatically decided.

Thus, to enumerate every nice interval supergraph $H$ of $G$ with respect to clique $V \setminus A$, it is enough to enumerate all the $(2k)! = 2^{O(k \log k)}$ permutations of elements of $P$ and consider $|P| + 1 \leq 2k + 1$ possible placements of $p$ in each of them. Some of these orderings may not produce an interval supergraph of $G$ though. In $O(k^2)$ time, we can check whether the resultant graph is an interval supergraph of $G$ and output the interval representation in $O(n)$ time. The number of supergraphs enumerated is only $(2k+1)2^{O(k \log k)} = 2^{O(k \log k)}$.

(c) Let $\mathcal{B} = \{I_1, I_2, \ldots, I_b\}$ be a box representation of $G$. Without loss of generality, we can assume that all $2|V|$ interval end points are distinct in $I_i$, for $1 \leq i \leq b$. (Otherwise, we can always alter the end points locally and make them distinct.) Let $p_i \in \mathbb{R}$ be a point belonging to the Helly region corresponding to $V \setminus A$ in $I_i$. For $1 \leq i \leq b$, let $I'_i$ be the interval graph defined by the interval assignments given below. Vertices of $I'_i$ are assigned their original labels as in $I_i$.

$$[l_v(I'_i), r_v(I'_i)] = \begin{cases} [l_v(I_i), r_v(I_i)] & \text{if } v \in A, \\ [l'_v(i), r'_v(i)] & \text{if } v \in V \setminus A. \end{cases}$$

where $l'_v(i) = \min\left(p_i, \min_{u \in N_G(v) \cap A} r_u(I_i)\right)$ and $r'_v(i) = \max\left(p_i, \max_{u \in N_G(v) \cap A} l_u(I_i)\right)$.

*Claim.* $\mathcal{B}' = \{I'_1, I'_2, \ldots, I'_b\}$ is a box representation of $G$ such that $\forall I'_i \in \mathcal{B}'$, $I'_i$ is a nice interval supergraph of $G$ with respect to clique $V \setminus A$.

*Proof.* Consider any $I'_i \in \mathcal{B}'$. For $u, v \in A$, intervals corresponding to $u$ and $v$ are the same in both $I_i$ and $I'_i$. If $(u, v) \in E(G)$, with $u, v \in A$, then the intervals corresponding to $u$ and $v$ intersect in $I'_i$ because they were intersecting in $I_i$. For any $(u, v) \in E(G)$, with $u \in A$ and $v \in V \setminus A$, the interval of $v$ intersects the interval of $u$ in $I'_i$, by the definition of $[l'_v(i), r'_v(i)]$ Vertices of $V \setminus A$ share the common point $p_i$. Thus, $I'_i$ is an interval supergraph of $G$. It is easy to see that $I'_i$ is a nice interval supergraph of $G$ with respect to clique $V \setminus A$ and point $p_i$.

Since $\mathcal{B}$ is a valid box representation of $G$, for each $(u, v) \notin E(G)$, $\exists I_i \in \mathcal{B}$ such that $(u, v) \notin E(I_i)$. Observe that for any vertex $v \in V$, the interval of $v$ in $I_i$ contains the interval of $v$ in $I'_i$. Therefore, if $(u, v) \notin E(I_i)$, then $(u, v) \notin E(I'_i)$ too. Thus, $\mathcal{B}'$ is also a valid box representation of $G$. □

(d) Since vertices of $G$ are labeled initially, we just need to retain the same labeling during the definition and construction of nice interval supergraphs of $G$. (We have included this obvious fact in the statement of the lemma, just to give better clarity.) □

## 3 Boxicity of Graphs with Large Cliques

One of the central ideas in this work is the following theorem about computing the boxicity of graphs which contain very large cliques. Using this theorem, in Section 4 we derive $o(n)$ factor approximation algorithms for computing the boxicity and cubicity of graphs. Further, it is used in Section 5 to derive parameterized approximation algorithm for the boxicity problem parameterized by vertex edit distance from a family of graphs of bounded boxicity.

**Theorem 1.** *Let $G$ be a graph on $n$ vertices, containing a clique of size $n-k$ or more. Then, $box(G) \leq k$ and an optimal box representation of $G$ can be found in time $n^2 2^{O(k^2 \log k)}$.*

*Proof.* Let $G(V, E)$ be a graph on $n$ vertices containing a clique of size $n-k$ or more. Arbitrarily label the vertices of $G$ as $1, 2, \ldots, n$. Using part (a) of Lemma 4, we can compute in $O(n2^k)$ time, $A \subseteq V$ such that $|A| \leq k$ and $G[V \setminus A]$ is a clique. It is easy to infer from Lemma 1 that $box(G) \leq box(G \setminus A) + |A| = k$, since $box(G \setminus A) = 0$ by definition.

From part (c) of Lemma 4, we get that, if $box(G) = b$, then there exists a box representation $\mathcal{B}' = \{I'_1, I'_2, \ldots, I'_b\}$ of $G$ in which each $I'_i$ is a nice interval supergraph of $G$ with respect to clique $V \setminus A$. We call such a representation a nice box representation of $G$ with respect to clique $V \setminus A$. To construct a nice box representation of $G$ with respect to clique $V \setminus A$ and of dimension $d$, we choose $d$ of the $2^{O(k \log k)}$ nice interval supergraphs of $G$ with respect to clique $V \setminus A$ (guaranteed by part (b) of Lemma 4) and check if this gives a valid box representation of $G$. This validation is straightforward because vertices in supergraphs being considered retain their original labels as in $G$ by part (d) of Lemma 4. All possible nice box representations of dimension $d$ can be computed and validated in $n^2 2^{O(k \cdot d \log k)}$ time. We might have to repeat this process for $1 \leq d \leq b$ in that order, to obtain an optimal box representation. Hence the total time required to compute an optimal box representation of $G$ is $bn^2 2^{O(k \cdot b \log k)}$, which is $n^2 2^{O(k^2 \log k)}$, because $b \leq k$ by the first part of this theorem. □

*Remark 1.* Theorem 1 gives an FPT algorithm for computing the boxicity of $G$, with the parameter $k = MVC(\overline{G})$, where $\overline{G}$ is the graph complement of $G$.

## 4 Approximation Algorithms for Boxicity and Cubicity

In this section, we use Theorem 1 and derive $o(n)$ factor approximation algorithms for boxicity and cubicity. Let $G(V, E)$ be the given graph with $|V| = n$. Without loss of generality, we can assume that $G$ is connected. Let $k = \frac{\sqrt{\log n}}{\sqrt{\log \log n}}$ and $t = \lceil \frac{n}{k} \rceil$. The algorithm proceeds by defining $t$ supergraphs of $G$ and computing their optimal box representations. Let the vertex set $V$ be partitioned arbitrarily into $t$ sets $V_1, V_2, \ldots, V_t$ where $|V_i| \leq k$, for each $1 \leq i \leq t$. We define supergraphs $G_1, G_2, \ldots, G_t$ of $G$ with $G_i(V, E_i)$ defined by setting $E_i = E \cup \{(x, y) | x, y \in V \setminus V_i\}$, for $1 \leq i \leq t$.

**Lemma 5.** *Let $G_i$ be as defined above, for $1 \leq i \leq t$. An optimal box representation $\mathcal{B}_i$ of $G_i$ can be computed in $n^{O(1)}$ time, where $n = |V|$.*

*Proof.* Noting that $G[V \setminus V_i]$ is a clique and $|V_i| \leq k = \frac{\sqrt{\log n}}{\sqrt{\log\log n}}$, by Theorem 1, we can compute an optimal box representation $\mathcal{B}_i$ of $G_i$ in $n^2 2^{O(k^2 \log k)} = n^{O(1)}$ time, where $n = |V|$. □

**Lemma 6.** *Let $\mathcal{B}_i$ be as computed above, for $1 \leq i \leq t$. Then, $\mathcal{B} = \bigcup_{1 \leq i \leq t} \mathcal{B}_i$ is a valid box representation of $G$ such that $|\mathcal{B}| \leq t' \cdot box(G)$, where $t'$ is $O\left(\frac{n\sqrt{\log\log n}}{\sqrt{\log n}}\right)$. The box representation $\mathcal{B}$ is computable in $n^{O(1)}$ time.*

*Proof.* We can compute optimal box representations $\mathcal{B}_i$ of $G_i$, for $1 \leq i \leq t = \left\lceil \frac{n\sqrt{\log\log n}}{\sqrt{\log n}} \right\rceil$ as explained in Lemma 5 in total $n^{O(1)}$ time. Observe that $E(G) = E(G_1) \cap E(G_2) \cap \cdots \cap E(G_t)$. Therefore, it is a trivial observation that the union $\mathcal{B} = \bigcup_{1 \leq i \leq t} \mathcal{B}_i$ gives us a valid box representation of $G$.

We will prove that this representation gives the approximation ratio as required. By Lemma 3 we have, $|\mathcal{B}_i| = box(G_i) \leq 2 \cdot box(G)$. Hence, $|\mathcal{B}| = \sum_{i=1}^{t} |\mathcal{B}_i| \leq 2t \cdot box(G)$. Substituting $t = \left\lceil \frac{n\sqrt{\log\log n}}{\sqrt{\log n}} \right\rceil$ in this inequality gives the approximation ratio as required. □

The box representation $\mathcal{B}$ obtained from Lemma 6 can be extended to a cube representation $\mathcal{C}$ of $G$ as stated in the following lemma. We omit its proof due to space constraints.

**Lemma 7.** *A cube representation $\mathcal{C}$ of $G$, such that $|\mathcal{C}| \leq t' \cdot cub(G)$, where $t'$ is $O\left(\frac{n(\log\log n)^{\frac{3}{2}}}{\sqrt{\log n}}\right)$, can be computed in $n^{O(1)}$ time.*

Combining Lemma 6 and Lemma 7, we get the follwing theorem which gives $o(n)$ factor approximation algorithms for computing boxicity and cubicity.

**Theorem 2.** *Let $G(V, E)$ be a graph on $n$ vertices. Then a box representation $\mathcal{B}$ of $G$, such that $|\mathcal{B}| \leq t \cdot box(G)$, where $t$ is $O\left(\frac{n\sqrt{\log\log n}}{\sqrt{\log n}}\right)$, can be computed in polynomial time. Further, a cube representation $\mathcal{C}$ of $G$, such that $|\mathcal{C}| \leq t' \cdot cub(G)$, where $t'$ is $O\left(\frac{n(\log\log n)^{\frac{3}{2}}}{\sqrt{\log n}}\right)$, can also be computed in polynomial time.*

## 5 Computing the Boxicity of Graphs with Edit Distances as the Parameter

In this section we give parameterized approximation algorithms for the boxicity problem parameterized by various vertex(edit) distance parameters. A subset $S \subseteq V$ such that $|S| \leq k$ is called a **modulator** for an $\mathcal{F} + kv$ graph $G(V, E)$

if $G \setminus S \in \mathcal{F}$. Similarly, a set $E_k$ of pairs of vertices such that $|E_k| \leq k$ is called a modulator for an $\mathcal{F} - ke$ graph $G(V, E)$ if $G'(V, E \cup E_k) \in \mathcal{F}$. Modulators for graphs in $\mathcal{F}+ke$ and $\mathcal{F}+k_1e-k_2e$ are defined in a similar manner. The following theorem is gives us a parameterized algorithm for computing the boxicity of $\mathcal{F} + kv$ graphs.

**Theorem 3.** *Let $\mathcal{F}$ be a family of graphs such that $\forall G' \in \mathcal{F}$, $box(G') \leq b$. Let $T(n)$ denote the time required to compute a $b$-dimensional box representation of a graph belonging to $\mathcal{F}$ on $n$ vertices. Let $G$ be an $\mathcal{F}+kv$ graph on $n$ vertices. Given a modulator of $G$, a box representation $\mathcal{B}$ of $G$, such that $|\mathcal{B}| \leq 2 \cdot box(G) + b$ can be computed in time $T(n-k) + n^2 2^{O(k^2 \log k)}$.*

*Proof.* Let $\mathcal{F}$ be the family of graphs of boxicity at most $b$. Let $G(V, E)$ be an $\mathcal{F} + kv$ graph on $n$ vertices, with a modulator $S_k$ on $k$ vertices such that $G' = G \setminus S_k \in \mathcal{F}$. We define two supergraphs of $G$, namely $H_1(V, E_1)$ and $H_2(V, E_2)$ such that $E = E_1 \cap E_2$ with $box(H_1) \leq 2 \cdot box(G)$, $box(H_2) \leq b$ and their required valid box representations are computable within the time specified in the theorem. It is easy to see that the union of valid box representations of $H_1$ and $H_2$ will be a valid box representation $\mathcal{B}$ of $G$ and hence $|\mathcal{B}| \leq box(H_1) + box(H_2) \leq 2 \cdot box(G) + b$. This will complete our proof of Theorem 3.

We define $H_1$ to be the graph obtained from $G$ by making $V \setminus S_k$ a clique on $n-k$ vertices, without altering other adjacencies in $G$. Formally, $E_1 = E \cup \{(x, y) \mid x, y \in V \setminus S_k, x \neq y\}$. Using Theorem 1, we can get an optimal box representation $\mathcal{B}_1$ of $H_1$ in $n^2 2^{O(k^2 \log k)}$ time. By Lemma 3, $|\mathcal{B}_1| \leq 2 \cdot box(G)$.

We define $H_2$ to be the graph obtained from $G$ by making each vertex in $S_k$ adjacent to every other vertex in the graph and leaving other adjacencies in $G$ unaltered. Formally, $E_2 = E \cup \{(x, y) \mid x \in S_k, y \in V, x \neq y\}$. Let $\mathcal{B}'$ be a box representation of $G'$ of dimension at most $b$ (computed in time $T(n-k)$). Then, $\mathcal{B}'$ is a box representation of $H_2[V \setminus S_k]$ as well, because $H_2[V \setminus S_k] = G'$. By Lemma 2, $box(H_2) = box(H_2[V \setminus S_k])$ and a box representation $\mathcal{B}_2$ of $H_2$ of dimension at most $|\mathcal{B}'| \leq b$ can be produced in $O(n^2)$ time.

Since $G = H_1 \cap H_2$, $\mathcal{B} = \mathcal{B}_1 \cup \mathcal{B}_2$ is a valid box representation of $G$, of dimension at most $2 \cdot box(G) + b$. All computations were done in $T(n-k) + n^2 2^{O(k^2 \log k)}$ time. □

Using a similar method, we also get a parameterized approximation algorithm for computing the boxicity of $\mathcal{F} + k_1e - k_2e$ graphs.

**Theorem 4.** *Let $\mathcal{F}$ be a family of graphs such that $\forall G' \in \mathcal{F}$, $box(G') \leq b$. Let $T(n)$ denote the time required to compute a $b$-dimensional box representation of a graph belonging to $\mathcal{F}$ on $n$ vertices. Let $G$ be an $\mathcal{F}+k_1e-k_2e$ graph on $n$ vertices and let $k = k_1 + k_2$. Given a modulator of $G$, a box representation $\mathcal{B}$ of $G$, such that $|\mathcal{B}| \leq box(G) + 2b$, can be computed in time $T(n) + O(n^2) + 2^{O(k^2 \log k)}$.*

*Proof.* Let $\mathcal{F}$ be the family of graphs of boxicity at most $b$. Let $G(V, E)$ be an $\mathcal{F} + k_1e - k_2e$ graph on $n$ vertices, where $k_1 + k_2 = k$. Let $E_{k_1} \cup E_{k_2}$ be a modulator of $G$ such that $|E_{k_1}| = k_1$, $|E_{k_2}| = k_2$ and $G'(V, (E \cup E_{k_2}) \setminus E_{k_1}) \in \mathcal{F}$. Let $T \subseteq V(G)$ be the set of vertices incident with edges in $E_{k_1} \cup E_{k_2}$.

As in the proof of Theorem 3, we define two supergraphs of $G$, namely $H_1(V, E_1)$ and $H_2(V, E_2)$ such that $E = E_1 \cap E_2$ with $box(H_1) \leq 2b$, $box(H_2) \leq box(G)$ and their required valid box representations are computable within the time specified in the theorem. As earlier, the union of valid box representations of $H_1$ and $H_2$ will be a valid box representation of $\mathcal{B}$ of $G$ and hence $|\mathcal{B}| \leq box(H_1) + box(H_2) \leq 2b + box(G)$. This will complete our proof of Theorem 4.

Let $H_1(V, E_1)$ be the graph obtained from $G'$ by making $T$ a clique, without altering other adjacencies in $G'$. Formally, $E_1 = E' \cup \{(x, y) | x, y \in T, x \neq y\}$. Let $\mathcal{B}'$ be a box representation of $G'$ of dimension at most $b$ computed in time $T(n)$. From the box representation $\mathcal{B}'$ of $G'$, in $O(n^2)$ time we can construct (by Lemma 3) a box representation $\mathcal{B}_1$ of $H_1$ with dimension $2b$.

Let $H_2(V, E_2)$ be the graph obtained from $G$ by making each vertex in $V \setminus T$ adjacent to every other vertex in the graph and leaving other adjacencies in $G$ unaltered. Formally, $E_2 = E \cup \{(x, y) | x \in V \setminus T, y \in V, x \neq y\}$. Clearly, $|T| \leq 2k$ and therefore, using the construction in Proposition 1, an optimal box representation $\mathcal{B}_T$ of $H_2[T]$ can be computed in $2^{O(k^2 \log k)}$ time. By Lemma 2, $box(H_2) = box(H_2[T])$ and a box representation $\mathcal{B}_2$ of $H_2$ of dimension $box(H_2[T])$ can be computed from the box representation $\mathcal{B}_T$ of $H_2[T]$ in $O(n^2)$ time. Observe that $H_2[T] = G[T]$. Therefore, $|\mathcal{B}_2| = box(G[T]) \leq box(G)$, because $G[T]$ is an induced subgraph of $G$.

Since $G = H_1 \cap H_2$, $\mathcal{B} = \mathcal{B}_1 \cup \mathcal{B}_2$ is a valid box representation of $G$, of dimension at most $box(G) + 2b$. All computations were done in $T(n) + O(n^2) + 2^{O(k^2 \log k)}$ time. □

*Remark 2.* Though in Theorem 3 and Theorem 4 we assumed that a modulator of $G$ for $\mathcal{F}$ is given, in several important special cases (as in the case of corollaries discussed below), the modulator for $\mathcal{F}$ can be computed from $G$ in FPT time. Moreover, in those cases, $T(n)$ is a polynomial in $n$. Thus, the algorithms given by Theorem 3 and Theorem 4 turns out to be FPT approximation algorithms for boxicity.

*Corollaries of Theorem 3 :* FPT approximation algorithms for computing boxicity with various parameters of interest result as consequences of Theorem 3. It is easy to see that these parameters are special cases of the vertex edit distance parameter. Detailed proofs of these corollaries are omitted due to space constraints. The general procedure is :

(i) Use known FPT algorithms to compute the parameter of interest and obtain the modulator $S_k$ for the corresponding family $\mathcal{F}$.

(ii) Compute a low dimensional box representation for the graph $G' = (G \setminus S_k) \in \mathcal{F}$, in polynomial time.

(iii) Use our algorithm of Theorem 3 to get the FPT approximation algorithm for computing boxicity with the parameter of interest.

**Corollary 1.** *FVS as the parameter : If $FVS(G) \leq k$, we get a $\left(2 + \frac{2}{box(G)}\right)$ factor approximation for boxicity with FVS as the parameter k, which runs in time $2^{O(k^2 \log k)} n^{O(1)}$.*

Note that, for the boxicity problem parameterized by FVS, the algorithm in Adiga et al. [3] gave the same approximation factor but with running time $2^{O(2^k k^2)} n^{O(1)}$. Our algorithm gives a better running time.

**Corollary 2.** *Proper Interval Vertex Deletion number (PIVD) as the parameter : The minimum number of vertices to be deleted from the graph G, so that the resultant graph is a proper interval graph, is called $PIVD(G)$. If $PIVD(G) \leq k$, we get a $2 + \frac{1}{box(G)}$ factor approximation for boxicity with PIVD as the parameter k, which runs in time $2^{O(k^2 \log k)} n^{O(1)}$.*

It is easy to see that $PIVD(G) \leq MVC(G)$. Hence, $PIVD(G)$ is a better parameter than the parameter $MVC(G)$ discussed in Adiga et al. [3]. Our algorithm has the same running time as the additive one approximation algorithm with $MVC(G)$ as the parameter, discussed in Adiga et al. [3].

**Corollary 3.** *Planar Vertex Deletion number (PVD) as the parameter : The minimum number of vertices to be deleted from G to make it a planar graph, is called the planar vertex deletion number of G. If $PVD(G) \leq k$, we get an FPT algorithm for boxicity, giving a $\left(2 + \frac{3}{box(G)}\right)$ factor approximation for boxicity using planar vertex deletion number as the parameter.*

*Corollaries of Theorem 4 :* Theorem 4 also gives us FPT approximation algorithms for computing boxicity with various parameters of interest.

**Corollary 4.** *Interval Completion number as the parameter : The minimum number of edges to be added to a graph G, so that the resultant graph is an interval graph, is called the interval completion number of G. If the interval completion number G is at most k, we get an FPT algorithm that achieves an additive 2 approximation for box(G) which runs in time $2^{O(k^2 \log k)} n^{O(1)}$.*

**Corollary 5.** *Proper Interval Edge Deletion number (PIED) as the parameter : The minimum number of edges to be deleted from the graph G, so that the resultant graph is a proper interval graph, is called $PIED(G)$. If $PIED(G)$ is at most k, we get an FPT algorithm that achieves an additive 2 approximation for box(G), with $PIED(G)$ as the parameter k, which runs in time $2^{O(k^2 \log k)} n^{O(1)}$.*

**Corollary 6.** *Planar Edge Deletion number (PED) as the parameter : The minimum number of edges to be deleted from G so that the resultant graph is planar is called $PED(G)$. If $PED(G) \leq k$, we get an FPT algorithm that gives an additive 6 approximation for box(G) with $PED(G)$ as the parameter.*

**Corollary 7.** *Max Leaf number (ML) as the parameter : The number of the maximum possible leaves in any spanning tree of a graph G is called $ML(G)$. If $ML(G) \leq k$, we get an FPT algorithm that achieves an additive 2 approximation for box(G) which runs in time $2^{O(k^3 \log k)} n^{O(1)}$, the running time and approximation ratio being the same as in Adiga et al. [3].*

## 6 An FPT Approximation Algorithm for Cubicity

Fellows et al. [9, Corollary 5] proved an existential result that for every fixed pair of integers $k$ and $b$, there is an $f(k) \cdot n$ time algorithm which determines whether a given graph $G$ on $n$ vertices and $MVC(G) \leq k$ has cubicity at most $b$. In the theorem below, we derive a FPT approximation algorithm, for computing the cubicity of graphs, using their vertex cover number as the parameter. Our algorithm is constructive.

**Theorem 5.** *Let $G$ be a graph on $n$ vertices. A cube representation of $G$ which is of dimension at most $2 \cdot cub(G)$ can be computed in time $2^{O(2^k k^2)} n^{O(1)}$, where $k = MVC(G)$. By allowing a larger running time of $2^{O(g(k,\epsilon))} n^{O(1)}$, we can achieve a $(1+\epsilon)$ approximation factor, for any $\epsilon > 0$, where $g(k,\epsilon) = \frac{1}{\epsilon} k^3 2^{\frac{4k}{\epsilon}}$.*

*Proof.* Due to space constraints, we give only an outline of the proof of the 2 factor approximation algorithm here. Let $G(V,E)$ be a graph on $n$ vertices. As in the previous sections, we define two supergraphs of $G$, namely $H_1(V,E_1)$ and $H_2(V,E_2)$ such that $E = E_1 \cap E_2$ with $cub(H_1) \leq cub(G)$ and $cub(H_2) \leq cub(G)$.

Let $S \subseteq V$ be a vertex cover of $G$ of cardinality $k$. First we define an equivalence relation on the vertices of the independent set $V \setminus S$ such that vertices $u$ and $v$ are in the same equivalence class if and only if $N_G(u) = N_G(v)$. Let $A_1, A_2, \ldots, A_t$ be the equivalence classes. We define $H_1$ to be the graph obtained from $G$ by making each $A_i$ into a clique and leaving other adjacencies as they are in $G$. Formally, $E_1 = E \cup \{(u,v) \mid u \neq v \text{ and } u,v \text{ belong to the same } A_i, \text{ for some } 1 \leq i \leq t\}$.

For each $A_i$, let us consider the mapping $n_{A_i} : A_i \mapsto \{1, 2, \cdots, |A_i|\}$, where $n_{A_i}(v)$ is the unique number representing $v \in A_i$. (Note that if $u \in A_i$ and $v \in A_j$, where $i \neq j$, then, $n_{A_i}(u)$ and $n_{A_j}(v)$ could potentially be the same.) Let $s = \max_{1 \leq i \leq t} |A_i|$. We define one more partitioning of the independent set $V \setminus S$ into equivalence classes $B_1, B_2, \ldots, B_s$ such that for $1 \leq i \leq s$, $B_i = \{v \mid n_{A_j}(v) = i,$ for some $1 \leq j \leq t\}$. We define $H_2$ to be the graph obtained from $G$ by making each $B_i$ into a clique, and making each vertex in $S$ adjacent to every other vertex in $V$. Formally, $E_2 = \{(u,v) \mid u \neq v \text{ and } u \in S, v \in V\} \cup \{(u,v) \mid u \neq v \text{ and } u,v$ belong to the same $B_i$, for some $1 \leq i \leq s\}$.

The following observations complete the proof : We have $E = E_1 \cap E_2$ with $cub(H_1)$, $cub(H_2) \leq cub(G)$. An optimal cube representation of $H_1$ can be constructed in $2^{O(2^k k^2)} n^{O(1)}$ time and that of $H_2$ can be constructed in $n^{O(1)}$ time. □

## 7 Conclusions and Open Problems

Among the several parameters giving FPT approximations for boxicity, we know the existence of exact FPT algorithms with parameter $MVC(G)$ only. The FPT status of the problem with other parameters is still open. Our FPT approximation algorithms for boxicity are dependent on the fact that intervals can be of different lengths. Hence, we do not know of a direct way of producing similar FPT approximation algorithms for cubicity. It will be interesting to investigate the possibility of FPT algorithms or approximations for cubicity, with parameters smaller than $MVC(G)$. We have presented $o(n)$ factor approximation algorithms for computing the boxicity and cubicity of graphs. The known hardness results only rule out the possibility of $O(n^{0.5-\epsilon})$-factor, for any $\epsilon > 0$. It is interesting to see whether it is possible to improve this hardness result to $O(n^{1-\epsilon})$-factor, or to get better approximation algorithms.

## References

1. Adiga, A., Bhowmick, D., Chandran, L.S.: The hardness of approximating the boxicity, cubicity and threshold dimension of a graph. Discrete Appl. Math. 158, 1719–1726 (2010)
2. Adiga, A., Bhowmick, D., Chandran, L.S.: Boxicity and poset dimension. SIAM J. Discrete Math. 25(4), 1687–1698 (2011)
3. Adiga, A., Chitnis, R., Saurabh, S.: Parameterized Algorithms for Boxicity. In: Cheong, O., Chwa, K.-Y., Park, K. (eds.) ISAAC 2010, Part I. LNCS, vol. 6506, pp. 366–377. Springer, Heidelberg (2010)
4. Booth, K.S., Lueker, G.S.: Testing for the consecutive ones property, interval graphs, and graph planarity using pq-tree algorithms. J. Comput. Syst. Sci. 13(3), 335–379 (1976)
5. Cai, L.: Fixed-parameter tractability of graph modification problems for hereditary properties. Inf. Process. Lett. 58(4), 171–176 (1996)
6. Cai, L.: Parameterized complexity of vertex colouring. Discrete Applied Mathematics 127(3), 415–429 (2003)
7. Chandran, L.S., Sivadasan, N.: Boxicity and treewidth. J. Comb. Theory Ser. B 97, 733–744 (2007)
8. Cozzens, M.B.: Higher and multi-dimensional analogues of interval graphs. Ph.D. thesis, Department of Mathematics, Rutgers University, New Brunswick, NJ (1981)
9. Fellows, M.R., Hermelin, D., Rosamond, F.A.: Well-Quasi-Orders in Subclasses of Bounded Treewidth Graphs. In: Chen, J., Fomin, F.V. (eds.) IWPEC 2009. LNCS, vol. 5917, pp. 149–160. Springer, Heidelberg (2009)
10. Grohe, M.: Computing crossing numbers in quadratic time. J. Comput. Syst. Sci. 68, 285–302 (2004)
11. Kratochvíl, J.: A special planar satisfiability problem and a consequence of its NP-completeness. Discrete Appl. Math. 52(3), 233–252 (1994)
12. Marx, D.: Parameterized coloring problems on chordal graphs. Theor. Comput. Sci. 351(3), 407–424 (2006)
13. Marx, D., Schlotter, I.: Obtaining a planar graph by vertex deletion. Algorithmica 62(3-4), 807–822 (2012)
14. Niedermeier, R.: Invitation to fixed-parameter algorithms (2002)
15. Roberts, F.S.: On the boxicity and cubicity of a graph. In: Recent Progresses in Combinatorics, pp. 301–310. Academic Press, New York (1969)
16. Rosgen, B., Stewart, L.: Complexity results on graphs with few cliques. Discrete Mathematics and Theoretical Computer Science 9, 127–136 (2007)
17. Thomassen, C.: Interval representations of planar graphs. J. Comb. Theory Ser. B 40, 9–20 (1986)
18. Villanger, Y., Heggernes, P., Paul, C., Telle, J.A.: Interval completion is fixed parameter tractable. SIAM J. Comput. 38(5), 2007–2020 (2008)
19. Yannakakis, M.: The complexity of the partial order dimension problem. SIAM J. Alg. Disc. Meth. 3(3), 351–358 (1982)

# Homomorphic Hashing for Sparse Coefficient Extraction

Petteri Kaski[1], Mikko Koivisto[2], and Jesper Nederlof[3]

[1] Helsinki Institute for Information Technology HIIT,
Department of Information and Computer Science, Aalto University, Finland
petteri.kaski@aalto.fi

[2] Helsinki Institute for Information Technology HIIT,
Department of Computer Science, University of Helsinki, Finland
mkhkoivi@cs.helsinki.fi

[3] Utrecht University, Utrecht, The Netherlands
j.nederlof@uu.nl

**Abstract.** We study classes of Dynamic Programming (DP) algorithms which, due to their algebraic definitions, are closely related to coefficient extraction methods. DP algorithms can easily be modified to exploit sparseness in the DP table through memorization. Coefficient extraction techniques on the other hand are both space-efficient and parallelisable, but no tools have been available to exploit sparseness. We investigate the systematic use of homomorphic hash functions to combine the best of these methods and obtain improved space-efficient algorithms for problems including LINEAR SAT, SET PARTITION and SUBSET SUM. Our algorithms run in time proportional to the number of nonzero entries of the last segment of the DP table, which presents a strict improvement over sparse DP. The last property also gives an improved algorithm for CNF SAT and SET COVER with sparse projections.

## 1 Introduction

*Coefficient extraction* can be seen as a general method for designing algorithms, recently in particular in the area of exact algorithms for various NP-hard problems [2,3,13,15,17,24] (cf. [7,26] for an introduction to exact algorithms). The approach of the method is the following (see also [14]):

1. Define a variable (the so-called coefficient) whose value (almost) immediately gives the solution of the problem to be solved,
2. Show that the variable can be expressed by a relatively small formula or circuit over a (cleverly chosen) large algebraic object like a ring or field,
3. Show how to perform operations in the algebraic object relatively efficiently.

In a typical application of the method, the first two steps are derived from an existing Dynamic Programming (DP) algorithm, and the third step deploys a carefully selected algebraic isomorphism, such as the discrete Fourier transform

D.M. Thilikos and G.J. Woeginger (Eds.): IPEC 2012, LNCS 7535, pp. 147–158, 2012.

to extract the desired solution/coefficient. Algorithms based on coefficient extraction have two key advantages over DP algorithms; namely, they are space-efficient and they parallelise well (see, for example, [15]).

Yet, DP has an advantage if the problem instance is *sparse*. By this we mean that the number of candidate/partial solutions that need to be considered during DP is small, that is, most entries in the DP table are not used at all. In such a case we can readily adjust the DP algorithm to take this into account through *memorization* so that both the running time and space usage become proportional to the number of partial solutions considered. Unfortunately, it is difficult to parallelise or lower the space usage of memorization. Coefficient extraction algorithms relying on interpolation of sparse polynomials [16] improve over memorization by scaling proportionally only to the number of *candidate* solutions, but their space usage is still not satisfactory (see also [26]).

This paper aims at obtaining what is essentially the best of both worlds, by investigating the systematic use of homomorphisms to "hash down" circuit-based coefficient extraction algorithms so that the domain of coefficient extraction – and hence the running time – matches or improves that of memorization-based DP algorithms, while providing space-efficiency and efficient parallelisation. The key idea is to take an existing algebraic circuit for coefficient extraction (over a sparsely populated algebraic domain such as a ring or field), and transform the circuit into a circuit over a smaller domain by a homomorphic hash function, and only then perform the actual coefficient extraction. Because the function is homomorphic, by hashing the values at the input gates and evaluating the circuit, the output evaluates to the hash of the original output value. Because the function is a hash function, the coefficient to be extracted collides with other coefficients only with negligible probability in the smaller domain, and coefficient extraction can be successfully used on the new (hashed-down) circuit. We call this approach *homomorphic hashing.*

**Our and Previous Results**

We study sparse DP/coefficient extraction in three domains: (a) the univariate polynomial ring $\mathbb{Z}[x]$ in Section 3, (b) the group algebra $\mathbb{F}[\mathbb{Z}_2^n]$ where $\mathbb{F}$ is a field of odd characteristic in Section 4 and (c) the Möbius algebra of the subset lattice in Section 5. The subject of sparse DP or coefficient extraction is highly motivated and well-studied [5,6,16,27]. In [16], a sparse polynomial interpolation algorithm using exponential space was already given for (a) and (b); our algorithms improve these to polynomial space. In [13] a polynomial-space algorithm for finding a small multilinear monomial in $\mathbb{F}_2[\mathbb{Z}_2^n]$ was given. In [15] a study of settings (a) and (b) was initiated, but sparsity was not addressed. Our main technical contribution occurs with (c) and hashing down to the "Solomon algebra" of a poset.

Our methods work for general arithmetic circuits similarly as in [13,15,16], and most of our algorithms work for counting variants as well. But, for concreteness, we will work here with specific decision problems. Although we mainly give improvements for sparse variants of these problems, we feel the results will be useful to deal with the general case as well (as we will see in Section 4).

*Subset Sum.* The SUBSET SUM problem is the following: given a vector $\boldsymbol{a} = (a_1, \ldots, a_n)$ and integer $t$, determine whether there exists a subset $X \subseteq [n]$ such that $\sum_{e \in X} a_e = t$. It is known to be solvable $\mathcal{O}^\star(2^{n/2})$ time and $\mathcal{O}^\star(2^{n/4})$ space [11,21], and solving it faster, or even in $\mathcal{O}^\star(1.99^n)$ time and polynomial space are interesting open questions [26]. Recently, a polynomial space algorithm using $\mathcal{O}^\star(t)$ time was given in [15]. Standard sparse DP gives an $\mathcal{O}^\star(S)$ time and $\mathcal{O}^\star(S)$ space algorithm. As a first "warm-up" application of our technique, we improve this to polynomial space as follows. The proofs of claims marked with a "†" are relegated to the full version in order to meet the page limit.

**Theorem 1 (†).** *Any instance* $(\boldsymbol{a}, t)$ *of the* SUBSET SUM *problem can be solved (a) in* $\mathcal{O}^\star(S)$ *expected time and polynomial space, and (b) in* $\mathcal{O}^\star(S^2)$ *time and polynomial space, where* $S = |\{\sum_{e \in X} a_e : X \subseteq [n]\}|$ *is the number of distinct sums.*

Informally stated, our algorithms hash the instances by working modulo randomly chosen prime numbers and apply the algorithm of [15]. While interesting on their own, these results may be useful in resolving the above open questions when combined with other techniques.

*Linear Sat.* The LINEAR SAT problem is defined as follows: given a matrix $\boldsymbol{A} \in \mathbb{Z}_2^{n \times m}$, vectors $\boldsymbol{b} \in \mathbb{Z}_2^m$ and $\boldsymbol{\omega} \in \mathbb{N}^n$, and an integer $t = n^{\mathcal{O}(1)}$, determine whether there is a vector $\boldsymbol{x} \in \mathbb{Z}_2^n$ such that $\boldsymbol{x}\boldsymbol{A} = \boldsymbol{b}$ and $\boldsymbol{\omega}\boldsymbol{x}^T \leq t$. Variants of LINEAR SAT have been studied, perhaps most notably in [10], where approximability was studied; Fixed Parameter Tractability was studied in [1,4]. Here, it was also quoted from [10] that (a variant of) LINEAR SAT is "as basic as satisfiability".

It can be observed that using the approach from [11], LINEAR SAT can be solved in $\mathcal{O}(2^{n/2}m)$ time and $\mathcal{O}(2^{n/2}m)$ space. Also, using standard "sparse dynamic programming", it can be solved in $\mathcal{O}^\star(2^{\mathrm{rk}(\boldsymbol{A})})$ time and $\mathcal{O}^\star(2^{\mathrm{rk}(\boldsymbol{A})})$ space, where $\mathrm{rk}(\boldsymbol{A})$ is the rank of $\boldsymbol{A}$. We give algorithms using about the same amount of time but only polynomial space:

**Theorem 2.** *Every instance* $(\boldsymbol{A}, \boldsymbol{b}, \boldsymbol{\omega}, t)$ *of* LINEAR SAT *can be solved by algorithms with constant one-sided error probability in (a)* $\mathcal{O}^\star(2^{\mathrm{rk}(\boldsymbol{A})})$ *time and polynomial space, and (b)* $\mathcal{O}^\star(2^{n/2})$ *time and polynomial space.*

The first algorithm hashes the input down using a random linear map and afterwards determines the answer using the Walsh-Hadamard transform. The second algorithm uses a Win/Win approach, combining the first algorithm with the fact that an $\boldsymbol{A}$ with high rank can be solved with a complementary algorithm.

*Satisfiability.* The CNF-SAT problem is defined as follows: given a CNF-formula $\phi = C_1 \wedge C_2 \wedge \ldots \wedge C_m$ over $n$ variables, determine whether $\phi$ is satisfiable. There are many interesting open questions related to this problem, a major one being whether it can be solved in time $\mathcal{O}^\star((2-\epsilon)^n)$ (the 'Strong Exponential Time Hypothesis' [12] states this is not possible), and another being whether satisfying assignments can be counted in time $\mathcal{O}^\star((2-\epsilon)^n)$ for some $\epsilon > 0$ (e.g. [23]).

A *prefix assignment* is an assignment of 0/1 values to the variables $v_1, \ldots, v_i$ for some $1 \leq i \leq n$. A *projection* (*prefix projection*) of a CNF-formula is a subset $\pi \subseteq [m]$ such that there exists an assignment (prefix assignment) of the variables such that for every $1 \leq j \leq m$ it satisfies $C_j$ if and only if $j \in \pi$. An algorithm for CNF-SAT running in time linear in the number of *prefix* projections can be obtained by standard sparse DP. However, it is sensible to ask about complexity of CNF-SAT if the number of projections is small. We give a positive answer:

**Theorem 3.** *Satisfiability of a formula $\phi = C_1 \wedge \ldots \wedge C_m$ can be determined in $\mathcal{O}^\star(P^2)$ time and $\mathcal{O}^\star(P)$ space, where $P = |\{\pi \subseteq [m] : \pi \text{ is a projection of } \phi\}|$.*

We are not aware of previous work that studies instances with few projections, but find it a natural parameter. For example, it is easy to see that hitting formulas[1] have only $m$ (and hence the minimum number of) projections, and that formulas having a strong backdoor set[2] of size $k$ have at most $2^k m$ projections. The formula with $2^n$ (and hence the maximum number of) projections is the one with a singleton clause for every variable. Naturally, there are more interesting cases and upper bounds for special classes of formula's, but to not lose focus from our main contribution we shall not discuss more structural properties of projections.

Underlying Theorem 3 is our main technical contribution (Theorem 15) that enables us to circumvent partial projections and access projections directly, namely homomorphic hashing from the Möbius algebra of the lattice of subsets of $[m]$ to the Solomon algebra of a poset. We think our result opens up a fresh technical perspective that may contribute towards solving the above mentioned and related questions. A full proof of Theorem 15 is given in the full version; we give a specialized, more direct proof of Theorem 3 and another application to SET COVER in Section 5.

## 2 Notation and Preliminaries

Lower-case boldface characters refer to vectors, while capital boldface letters refer to matrices, $\boldsymbol{I}$ being the identity matrix. The rank of a matrix $\boldsymbol{A}$ is denoted by $\mathrm{rk}(\boldsymbol{A})$. If $R$ and $S$ are sets, and $S$ is finite, denote by $R^S$ the set of all $|S|$-dimensional vectors with values in $R$, indexed by elements of $S$, that is, if $\boldsymbol{v} \in R^S$, then for every $e \in S$ we have $v_e \in R$. We denote by $\mathbb{Z}$ and $\mathbb{N}$ the set of integers and non-negative integers, respectively, and by $\mathbb{Z}_p$ the field of integers modulo a prime $p$. An arbitrary field is denoted by $\mathbb{F}$.

For a logical proposition $P$, we use Iverson's bracket notation $[P]$ to denote a 1 if $P$ is true and a 0 if $P$ is false. For a function $h : A \to B$ and $b \in B$, the preimage $h^{-1}(b)$ is defined as the set $\{a \in A : h(a) = b\}$. For an integer $n$ and $A \subseteq \{1, \ldots, n\}$, denote by $\chi(A) \in \mathbb{Z}_2^n$ the characteristic vector of $A$. Sometimes

[1] Every pair of clauses have a conflicting literal [18], also called "semi-complete" [1].

[2] $k$ variables such that each assignment of them leaves a hitting formula (from [25], see also e.g. [8]).

we will state running times of algorithms with the $\mathcal{O}^\star$ notation, which suppresses any factor polynomial in the input size.

For a ring $R$ and a finite set $S$, we write $R^S$ for the ring consisting of the set $R^S$ (the set of all vectors over $R$ with coordinates indexed by elements of $S$) equipped with coordinate-wise addition $+$ and multiplication $\circ$ (the *Hadamard product*), that is, for $\boldsymbol{a}, \boldsymbol{b} \in R^S$ and $\boldsymbol{a}+\boldsymbol{b} = \boldsymbol{c}$, $\boldsymbol{a} \circ \boldsymbol{b} = \boldsymbol{d}$ we set $a_z + b_z = c_z$ and $a_z b_z = d_z$ for each $z \in S$, where $+$ and the juxtaposition denote addition and multiplication in $R$, respectively. The inner-product $\boldsymbol{a}, \boldsymbol{b} \in R^S$ is denoted by $\boldsymbol{a}^T \cdot \boldsymbol{b}$. For $\boldsymbol{v} \in R^S$ denote by $\mathrm{supp}(\boldsymbol{v}) \subseteq S$ the *support* of $\boldsymbol{v}$, that is, $\mathrm{supp}(\boldsymbol{v}) = \{z \in S : v_z \neq 0\}$, where 0 is the additive identity element of $R$. A vector $\boldsymbol{v}$ is called a *singleton* if $|\mathrm{supp}(\boldsymbol{v})| = 1$. We denote by $\langle z \to w \rangle$ the singleton with value $w$ on index $z$, that is, $\langle z \to w \rangle_y = w[y = z]$ for all $y \in S$.

If $R$ is a ring and $(S, \cdot)$ is a finite semigroup, denote by $R[S]$ the ring consisting of the set $R^S$ equipped with coordinate-wise addition and multiplication defined by the convolution operator $*$, where for $\boldsymbol{a}, \boldsymbol{b} \in R^S$, $\boldsymbol{a} * \boldsymbol{b} = \boldsymbol{c}$ we set $c_z = \sum_{x \cdot y = z} a_x b_y$ for every $z \in S$.

If $R, S$ are rings with operations $(+, *)$ and $(\oplus, \circledast)$ respectively, a *homomorphism from $R$ to $S$* is a function $h : R \to S$ such that $h(e_1 + e_2) = h(e_1) \oplus h(e_2)$ and $h(e_1 * e_2) = h(e_1) \circledast h(e_2)$ for every $e_1, e_2 \in R$.

**Observation 4.** *Let $R$ be a ring, and let $(S, \cdot)$ and $(T, \odot)$ be finite semigroups. Suppose $\varphi : S \to T$ such that for every $x, y \in S$ we have $\varphi(x \cdot y) = \varphi(x) \odot \varphi(y)$. Then the function $h : R[S] \to R[T]$ defined by $h : \boldsymbol{a} \mapsto \boldsymbol{b}$ where $b_z = \sum_{y \in \varphi^{-1}(z)} a_y$ for all $z \in T$ is a homomorphism.*

A *circuit* $C$ over a ring $R$ is a labeled directed acyclic graph $D = (V, A)$ where the elements of $V$ are called *gates* and $D$ has a unique sink called the *output gate of* $C$. All sources of $C$ are called *input gates* and are labeled with elements from $R$. All gates with non-zero in-degree are labeled as either an *addition* or a *multiplication gate.* (If multiplication in $R$ is not commutative, the in-arcs of each multiplication gate are also ordered.) Every gate $g$ of $C$ can be associated with a ring element in the following natural way: If $g$ is an input gate, we associate the label of $g$ with $g$. If $g$ is an addition gate we associate the ring element $e_1 + \ldots + e_d$ with $g$, and if $g$ is a multiplication gate we associate the ring element $e_1 * \ldots * e_d$ with $g$ where $e_1, \ldots, e_d$ are the ring elements associated with the $d$ in-neighbors of $g$, and $+$ and $*$ are the operations of the ring $R$.

Suppose the ground set of $R$ is of the type $A^B$ where $A, B$ are sets. Then $C$ is said to have *singleton inputs* if the label of every input-gate of $C$ is a singleton vector of $R$.

**Definition 5.** *Let $R^1$ and $R^2$ be rings, let $h : R^1 \to R^2$ be a homomorphism, and suppose that $C$ is a circuit over $R$. Then, the circuit $h(C)$ over $R^2$* obtained *by applying $h$ to $C$ is defined as the circuit obtained from $C$ by replacing for every input gate the label $l$ by $h(l)$.*

Note that the following is immediate from the definition of a homomorphism:

**Observation 6.** *Suppose $C$ is a circuit over a ring $R^1$ with output $v \in R^1$. Then the circuit over $R^2$ obtained by applying a homomorphism $h : R^1 \to R^2$ to $C$ outputs $h(v) \in R^2$.*

## 3 Homomorphic Hashing for Subset Sum

In this section we will study the Subset Sum problem and prove Theorem 1. As mentioned in the introduction, it should be noted that this merely serves as an illustration of how similar problems can be tackled as well since the same method applies to the more general sparse polynomial interpolation problem. However, to avoid a repeat of the analysis of [15], we have chosen to restrict ourselves to the Subset Sum problem. Our central contribution over [15] is that we take advantage of sparsity. Given an integer $p \in \mathbb{N}$, let $c^p : \mathbb{N}^n \to \mathbb{N}^p$ be defined by

$$c^p(\boldsymbol{a})_j = \left|\left\{X \subseteq [n] : \sum_{e \in X} a_e \equiv j \pmod p\right\}\right| \quad \text{for every } j \in \mathbb{Z}_p \text{ and } \boldsymbol{a} \in \mathbb{N}^n.$$

We also use the shorthand $\boldsymbol{c}(\boldsymbol{a}) = \boldsymbol{c}^\infty(\boldsymbol{a})$. We use the following corollary from [15] and two results on primes:

**Corollary 7 (†, [15]).** *Given an instance $(\boldsymbol{a}, t)$ of* Subset Sum *and an integer $p$, $c^p(\boldsymbol{a})_t$ can be computed in $\mathcal{O}^\star(p)$ time and $\mathcal{O}^\star(1)$ space.*

**Theorem 8 ([20]).** *If $55 < u$, the number of primes at most $u$ is at least $\frac{u}{\ln u + 2}$.*

**Lemma 9 (†, Folklore).** *There exists an algorithm* `pickprime`$(u)$ *running in* polylog$(u)$ *time that, given integer $u \geq 2$ as input, outputs either a prime chosen uniformly at random from the set of primes at most $u$ or* `notfound`. *Moreover, the probability that the output is* `notfound` *is at most $\frac{1}{e}$.*

We will run a data reduction procedure similar to the one of Claim 2.7 in [9], before applying the algorithm of Corollary 7. The idea of the data reduction procedure is to work modulo a prime of size roughly $|\text{supp}(c(\boldsymbol{a}))|$ or larger:

**Lemma 10.** *Let $S \geq |\text{supp}(c(\boldsymbol{a}))|$ and let $\beta$ be an upper bound on the number of bits needed to represent the integers, i.e. $2^\beta > \max\{t, \max_i a_i\}$. Then for sufficiently large $\beta$ and $n$, $\text{Prob}_p[c(\boldsymbol{a})_t = c^p(\boldsymbol{a})_t] \geq \frac{1}{2}$, where the probability is taken uniformly over all primes $p \leq S\beta n(\log \beta)(\log n)$.*

*Proof.* Suppose $c(\boldsymbol{a})_t \neq c^p(\boldsymbol{a})_t$. Then there exists an integer $u \in \text{supp}(c(\boldsymbol{a}))$ such that $u \neq t$ and $u \equiv t \pmod p$. This implies that $p$ is a divisor of $|t - u|$, so let us bound the probability of this event. Since $|t - u| \leq 2^\beta n$, it has at most $\beta + \log n$ distinct prime divisors. Let $\gamma = S\beta n(\log \beta)(\log n)$. By Theorem 8 we have that $\text{Prob}_p[p \text{ divides } |t - u|]$ is at most

$$\frac{\beta + \log n}{\frac{\gamma}{\log \gamma + 2}} \leq \frac{\beta + \log n}{\frac{\gamma}{3(n + \log \beta)}} \leq \frac{3(n + \log \beta)(\beta + \log n)}{\gamma} \leq \frac{1}{2S}$$

for sufficiently high $\beta$ and $n$, where we use that $S \leq 2^n$ in the second inequality. Applying the union bound over the at most $S$ elements of $\text{supp}(c(\boldsymbol{a}))$, the event that there exists a $u \in \text{supp}(c(\boldsymbol{a}))$ with $u \neq t$ and $u \equiv t \pmod{p}$ occurs with probability at most $\frac{1}{2}$. □

Now we give an algorithm for the case where $S$ is known. The proof of Theorem 1, given in the full version, merely adds self-reduction arguments.

**Theorem 11.** *There exists an algorithm that, given an instance $(\boldsymbol{a}, t)$ of the* Subset Sum *problem and an integer $S \geq |\text{supp}(c(\boldsymbol{a}))|$ as input, outputs a non-negative integer $x$ in $\mathcal{O}^\star(S)$ time and polynomial space such that (i) $x = 0$ implies $c(\boldsymbol{a})_t = 0$ and (ii)* $\text{Prob}[c(\boldsymbol{a})_t = x] \geq \frac{1}{4}$.

*Proof.* The algorithm is: First, obtain prime $p = \texttt{pickprime}(S\beta n(\log \beta)(\log n))$ using Lemma 9. Second, compute and output $c^p(\boldsymbol{a})_t$ using Corollary 7. Condition (i) holds since $c^p(\boldsymbol{a})_t = 0$ implies $c(\boldsymbol{a})_t = 0$ for any $p, t$. Moreover, condition (ii) follows from Lemma 10 and Lemma 9 since $\frac{1}{2}(1 - \frac{1}{e}) \geq \frac{1}{4}$. The time and space bounds are met by Corollary 7 because $p = \mathcal{O}^\star(S)$. □

## 4 Homomorphic Hashing for Linear Satisfiability

In this section we assume that $\mathbb{F}$ is a field of non-even characteristic and that addition and multiplication refer to operations in $\mathbb{F}$. We prove the following general result, having Theorem 2(a) as a special case.

**Theorem 12.** *There exists a randomized algorithm that, given as input (i) a circuit $C$ with singleton inputs over $\mathbb{F}[\mathbb{Z}_2^n]$, (ii) an integer $S \geq |\text{supp}(\boldsymbol{v})|$, and (iii) an element $\boldsymbol{t} \in \mathbb{Z}_2^n$, outputs the coefficient $v_{\boldsymbol{t}} \in \mathbb{F}$ with probability at least $\frac{1}{2}$, where $\boldsymbol{v} \in \mathbb{F}[\mathbb{Z}_2^n]$ is the output of $C$. The algorithm uses $\mathcal{O}^\star(S)$ time, $\mathcal{O}^\star(S)$ arithmetic operations in $\mathbb{F}$, and storage for $\mathcal{O}^\star(1)$ bits and elements of $\mathbb{F}$.*

*Proof.* Consider Algorithm 1. Let us first analyse the complexity of this algorithm: Steps 1 and 2 can be performed in time polynomial in the input. Step 3 also be done in polynomial time since it amounts to relabeling all input gates with $h(\boldsymbol{e})$ where $\boldsymbol{e}$ was the old label. Indeed, we know that $\boldsymbol{e} \in \mathbb{F}[\mathbb{Z}_2^n]$ is a singleton $\langle \boldsymbol{y} \to v \rangle$, so $h(\boldsymbol{e})$ is the singleton $\langle \boldsymbol{y}\boldsymbol{H} \to v \rangle$ and this can be computed in polynomial time. Step 4 takes $\mathcal{O}^\star(S)$ operations and calls to `sub`, so for the complexity bound it remains to show that a call to `sub` runs in polynomial time. Step 5 can be implemented in polynomial time similar to Step 3 since the singleton $\boldsymbol{e} = \langle \boldsymbol{y} \to v \rangle$ is mapped to $(-1)^{\boldsymbol{x}\boldsymbol{y}^T} v$. Finally, the direct evaluation of $C_2$ uses $|C_2|$ operations in $\mathbb{F}$. Hence the algorithm meets the time bound, and also the space bound is immediate. The fact that `hashZ2` returns $v_{\boldsymbol{t}}$ with probability at least $\frac{1}{2}$ is a direct consequence of the following two claims, where $\boldsymbol{w}$ denotes the output of $C_1$.

**Claim 1** (†). $\text{Prob}_{\boldsymbol{H}}[v_{\boldsymbol{t}} = w_{\boldsymbol{t}\boldsymbol{H}}] \geq \frac{1}{2}$.

**Claim 2** (†). *Algorithm* `hashZ2` *returns* $w_{\boldsymbol{t}\boldsymbol{H}}$.

□

**Algorithm** hashZ2
1: Let $s = \lceil \log S \rceil + 1$.
2: Choose a matrix $\boldsymbol{H} \in \mathbb{Z}_2^{s \times n}$ uniformly at random from the set of all $s \times n$ matrices with binary entries.
3: Let $h : \mathbb{F}[\mathbb{Z}_2^n] \to \mathbb{F}[\mathbb{Z}_2^s]$ be the homomorphism defined by $h(\boldsymbol{a}) = \boldsymbol{b}$ where $b_{\boldsymbol{x}} = \sum_{\boldsymbol{y} \in \mathbb{Z}_2^n : \boldsymbol{y}\boldsymbol{H} = \boldsymbol{x}} a_{\boldsymbol{y}}$ for all $\boldsymbol{x} \in \mathbb{Z}_2^s$. Apply $h$ to $C$ to obtain the circuit $C_1$.
4: **return** $\displaystyle \frac{1}{2^s} \sum_{\boldsymbol{x} \in \mathbb{Z}_2^s} (-1)^{(\boldsymbol{t}\boldsymbol{H})\boldsymbol{x}^T}$ $\texttt{sub}(C_1, \boldsymbol{x})$.

**Algorithm** sub$(C_1, \boldsymbol{x})$
5: Let $\varphi : \mathbb{F}[\mathbb{Z}_2^s] \to \mathbb{F}$ be the homomorphism defined by $\varphi(\boldsymbol{w}) = \sum_{\boldsymbol{y} \in \mathbb{Z}_2^s} (-1)^{\boldsymbol{x}\boldsymbol{y}^T} w_{\boldsymbol{y}}$ for all $\boldsymbol{w} \in \mathbb{F}[\mathbb{Z}_2^s]$. Apply $\varphi$ to $C_1$ to obtain the circuit $C_2$.
6: Evaluate $C_2$ and return the output.

**Algorithm 1:** Homomorphic hashing for Theorem 12

*Proof (of Theorem 2(a)).* For $1 \leq i \leq n$ and $0 \leq w \leq t$ denote by $\boldsymbol{A}^{(i)}$ the $i$th row of $\boldsymbol{A}$ and define $\boldsymbol{f}[i, w] \in \mathbb{Q}[\mathbb{Z}_2^m]$ by

$$\boldsymbol{f}[i, w] = \begin{cases} \langle \boldsymbol{0} \to 1 \rangle & \text{if } i = w = 0, \\ 0 & \text{if } i = 0 \wedge w \neq 0, \\ \boldsymbol{f}[i-1, w] + \boldsymbol{f}[i-1, w - \omega_i] * \left\langle \boldsymbol{A}^{(i)} \to 1 \right\rangle & \text{otherwise.} \end{cases} \tag{1}$$

It is easy to see that for every $1 \leq i \leq n$, $0 \leq w \leq t$, and $\boldsymbol{y} \in \mathbb{Z}_2^m$, the value $f[i, w]_{\boldsymbol{y}}$ is the number of $\boldsymbol{x} \in \mathbb{Z}_2^i$ such that $\tilde{\boldsymbol{\omega}}\boldsymbol{x}^T = w$ and $\boldsymbol{x}\widetilde{\boldsymbol{A}} = \boldsymbol{y}$ where $\tilde{\boldsymbol{\omega}}$ and $\widetilde{\boldsymbol{A}}$ are obtained by truncating $\boldsymbol{\omega}$ and $\boldsymbol{A}$ to the first $i$ rows. Hence, we let $C$ be the circuit implementing (1) and let its output be $\boldsymbol{v} = \sum_{w=0}^{t} \boldsymbol{f}[n, w]$. Thus, $v_{\boldsymbol{b}}$ is the number of $\boldsymbol{x} \in \mathbb{Z}_2^n$ with $\boldsymbol{x}\boldsymbol{A} = \boldsymbol{b}$ and $\boldsymbol{x}\boldsymbol{\omega}^T \leq t$.

Also, $|\mathrm{supp}(\boldsymbol{v})| \leq 2^{\mathrm{rk}(\boldsymbol{A})}$ since any element of the support of $\boldsymbol{v}$ is a sum of rows of $\boldsymbol{A}$ and hence in the row-space of $\boldsymbol{A}$, which has size at most $2^{\mathrm{rk}(\boldsymbol{A})}$. To apply Theorem 12, let $\mathbb{F} = \mathbb{Q}$ and observe that the computations are in fact carried out over integers bounded in absolute value poly-exponentially in $n$ and hence the operations in the base field can also be executed polynomial in $n$. The theorem follows from Theorem 12. □

To establish Theorem 2(b), let us first see how to exploit a high linear rank of the matrix $\boldsymbol{A}$ in an instance of LINEAR SAT. By permuting the rows of $\boldsymbol{A}$ as necessary, we can assume that the first $\mathrm{rk}(\boldsymbol{A})$ rows of $\boldsymbol{A}$ are linearly independent. We can now partition $\boldsymbol{x}$ into $\boldsymbol{x} = (\boldsymbol{y}, \boldsymbol{z})$, where $\boldsymbol{y}$ has length $\mathrm{rk}(\boldsymbol{A})$ and $\boldsymbol{z}$ has length $n - \mathrm{rk}(\boldsymbol{A})$. There are $2^{n-\mathrm{rk}(\boldsymbol{A})}$ choices for $\boldsymbol{z}$, each of which by linear independence has at most one corresponding $\boldsymbol{y}$ such that $\boldsymbol{x}\boldsymbol{A} = \boldsymbol{b}$. Given $\boldsymbol{z}$, we can determine the corresponding $\boldsymbol{y}$ (if any) in polynomial time by Gaussian elimination. Thus, we have:

**Observation 13.** LINEAR SAT *can be solved in* $\mathcal{O}^\star(2^{n-\mathrm{rk}(\boldsymbol{A})})$ *time and polynomial space.*

This enables a "Win/Win approach" where we distinguish between low and high ranks, and use an appropriate algorithm in each case.

*Proof (of Theorem 2(b)).* Compute $\mathrm{rk}(\boldsymbol{A})$. If $\mathrm{rk}(\boldsymbol{A}) \geq n/2$, run the algorithm of Observation 13. Otherwise, run the algorithm implied by Theorem 2(a). □

**Set Partition.** We now give a very similar application to the SET PARTITION problem: given an integer $t$ and a set family $\mathcal{F} \subseteq 2^U$ where $|\mathcal{F}| = n$, $|U| = m$, determine whether there is a subfamily $\mathcal{P} \subseteq \mathcal{F}$ with $|\mathcal{P}| \leq t$ such that $\bigcup_{S \in \mathcal{P}} S = U$ and $\sum_{S \in \mathcal{P}} |S| = |U|$.

The *incidence matrix* of a set system $(U, \mathcal{F})$ is the $|U| \times |\mathcal{F}|$ matrix $\boldsymbol{A}$ whose entries $A_{e,S} = [e \in S]$ are indexed by $e \in U$ and $S \in \mathcal{F}$.

**Theorem 14 (†).** *There exist algorithms that given an instance $(U, \mathcal{F}, t)$ of* SET PARTITION *output the number of set partitions of size at most $t$ with probability at least $\frac{1}{2}$, and use (a) $\mathcal{O}^\star(2^{\mathrm{rk}(\boldsymbol{A})})$ time and polynomial space, and (b) $(2^{\mathrm{rk}(\boldsymbol{A})} + n)m^{\mathcal{O}(1)}$ time and space, where $\boldsymbol{A}$ is the incidence matrix of $(U, \mathcal{F})$.*

## 5 Homomorphic Hashing for the Union Product

In this section our objective is to mimic the approach of the previous section for $\mathbb{N}[(2^U, \cup)]$, where $(2^U, \cup)$ is the semigroup defined by the set union $\cup$ operation on $2^U$, the power set of an $n$-element set $U$. The direct attempt to apply a homomorphic hashing function, unfortunately, fails. Indeed, let $h$ be an arbitrary homomorphism from $(2^U, \cup)$ to $(2^V, \cup)$ with $|V| < |U|$. Let $U = \{e_1, e_2, \ldots, e_n\}$ and consider the minimum value $1 \leq j \leq n - 1$ with $h(\{e_1, \ldots, e_j\}) = \cup_{i=1}^{j} h(\{e_i\}) = \cup_{i=1}^{j+1} h(\{e_i\}) = h(\{e_1, \ldots, e_{j+1}\})$; in particular, for $X = \{e_1, \ldots, e_j, e_{j+2}, \ldots, e_n\} \neq U$ we have $h(X) = h(U)$, which signals failure since we cannot isolate $X$ from $U$.

Instead, we use hashing to an algebraic structure based on a poset (the "Solomon algebra" of a poset due to [22]) that is obtained by the technique "Iterative Compression". This gives the following main result. For reasons of space we relegate a detailed proof to the full version; here we will give a simplified version of the proof in the special case of Theorem 3 in this section.

**Theorem 15 (†).** *Let and $|U| = n$. There are algorithms that, given a circuit $C$ with singleton inputs in $\mathbb{N}[(2^U, \cup)]$ outputting $\boldsymbol{v}$, compute*

*(a) a list with $v_X$ for every $X \in \mathrm{supp}(\boldsymbol{v})$ in $\mathcal{O}^\star(|\mathrm{supp}(\boldsymbol{v})|^2 n^{\mathcal{O}(1)})$ time,*
*(b) $v_U$ in time $\mathcal{O}^\star(2^{(1-\alpha/2)n} n^{\mathcal{O}(1)})$ if $0 < \alpha \leq 1/2$ such that $|\mathrm{supp}(\boldsymbol{v})| \leq 2^{(1-\alpha)n}$.*

The above result is stated for simplicity in the unit-cost model, that is, we assume that arithmetic operations on integers take constant time. For the more realistic log-cost model, where such operations are assumed to take time polynomial in the number of bits of the binary representation, we only mention here that our results also hold under some mild technical conditions. Let us first show that Theorem 3(a) indeed is a special case of Theorem 15:

*Proof (of Theorem 3).* Use the circuit over $\mathbb{N}[(2^{[m]}, \cup)]$ that implements

$$\boldsymbol{f} = (\langle V_1 \to 1\rangle + \langle \bar{V}_1 \to 1\rangle) * (\langle V_2 \to 1\rangle + \langle \bar{V}_2 \to 1\rangle) * \ldots * (\langle V_m \to 1\rangle + \langle \bar{V}_m \to 1\rangle),$$

where $V_i \subseteq [m]$ (respectively, $\bar{V}_i \subseteq [m]$) is the set of all indices of clauses that contain a positive (respectively, negative) literal of the variable $v_i$. Then use Theorem 15 to determine $f_{[m]}$, the number of satisfying assignments of $\phi$. □

Now we proceed with a self-contained proof Theorem 3. Given poset $(P, \leq)$, the *Möbius function* $\mu : P \times P \to \mathbb{N}$ of $P$ is defined for all $x, y \in P$ by

$$\mu(x,y) = \begin{cases} 1 & \text{if } x = y, \\ -\sum_{x \leq z < y} \mu(x,z) & \text{if } x < y, \\ 0 & \text{otherwise.} \end{cases} \tag{2}$$

The *zeta transform* $\boldsymbol{\zeta}$ and *Möbius transform* $\boldsymbol{\mu}$ are the $|P| \times |P|$ matrices defined by $\zeta_{x,y} = [x \leq y]$ and $\mu_{x,y} = \mu(x,y)$ for all $x, y \in P$. For a CNF-formula $\phi$ denote $\mathrm{supp}(\phi)$ for the set of all projections of $\phi$. Recall in Theorem 3 we are given a CNF-Formula $\phi = C^1 \wedge \ldots \wedge C^m$ over $n$ variables. For $i = 1, \ldots, m$ define $\phi_i = C_1 \wedge \ldots \wedge C_i$. Then we have the following easy observations:

1. $\mathrm{supp}(\phi_0) = \{\emptyset\}$,
2. $\mathrm{supp}(\phi_i) \subseteq \mathrm{supp}(\phi_{i-1}) \cup \{X \cup \{i\} : X \in \mathrm{supp}(\phi_{i-1})\}$ for every $i = 1, \ldots, m$,
3. $|\mathrm{supp}(\phi_{i-1})| \leq |\mathrm{supp}(\phi_i)|$ for every $i = 1, \ldots, m$.

Given the above lemma and observations, we will give an algoritm using a technique called *iterative compression* [19]. As we will see, by this technique it is sufficient to solve the following "compression problem":

**Lemma 16.** *Given a CNF-formula $\phi = C_1 \wedge \ldots \wedge C_m$ and a set family $\mathcal{F} \subseteq 2^{[m]}$ with* $\mathrm{supp}(\phi) \subseteq \mathcal{F}$, *the set* $\mathrm{supp}(\phi)$ *can be constructed in $O^\star(|\mathcal{F}|^2)$ time.*

*Proof.* In what follows $\boldsymbol{a} \in \{0,1\}^n$ refers to an assignment of values to the $n$ variables in $\phi$. Define $\boldsymbol{f} \in \mathbb{N}^{2^{[m]}}$ for all $X \subseteq [m]$ by

$$f_X = |\{\boldsymbol{a} \in \{0,1\}^n : \forall i \in [m] \text{ it holds that } \boldsymbol{a} \text{ satisfies } C_i \text{ iff } i \in X\}|.$$

It is easy to see that $\mathrm{supp}(\boldsymbol{f}) = \mathrm{supp}(\phi)$, so if we know $f_X$ for every $X \in \mathcal{F}$ we can construct $\mathrm{supp}(\phi)$ in $|\mathcal{F}|$ time. Towards this end, first note that for every $Y \subseteq [m]$, $(\boldsymbol{f\zeta})_Y$ equals

$$\sum_{\substack{X \in \mathrm{supp}(\boldsymbol{f}) \\ X \subseteq Y}} f(X) = \sum_{X \subseteq Y} f(X) = |\{\boldsymbol{a} \in \{0,1\}^n \mid \forall i : \boldsymbol{a} \text{ satisfies } C_i \text{ only if } i \in Y\}|.$$

Second, note that the last quantity can be computed in polynomial time: since every clause outside $Y$ must not be satisfied, each such clause forces the variables that occur in it to unique values; any other variables may be assigned to arbitrary values. That is, the count is 0 if the clauses outside $Y$ force at least one variable to conflicting values, otherwise the count is $2^a$ where $a$ is the number of variables that occur in none of the clauses outside $Y$.

Now the algorithm is the following: for every $X \in \mathcal{F}$ compute $(\boldsymbol{f\zeta})_X$ in polynomial time as discussed above. Then we can use algorithm `mobius` as described below to obtain $f_X$ for every $X \in \mathcal{F}$ since it follows that $\boldsymbol{f} = \texttt{mobius}((\mathcal{F}, \subseteq), \boldsymbol{f\zeta})$ from the definition of $\boldsymbol{\mu}$ and the fact that $\boldsymbol{\mu\zeta} = \boldsymbol{I}$. Algorithm `mobius` clearly runs in $\mathcal{O}^\star(|P|^2)$ time, so this procedure meets the claimed time bound.

**Algorithm** $\texttt{mobius}((P, \leq), \boldsymbol{w})$

Let $P = \{v_1, v_2 \ldots, v_{|P|}\}$ such that $v_j \leq v_i$ implies $j \leq i$.<br>
$\boldsymbol{z} \leftarrow \boldsymbol{w}$.<br>
**for** $i = 1, 2, \ldots, |P|$ **do**<br>
    **for** every $v_j \leq v_i$ **do**<br>
        $z_i = z_i - z_j$<br>
**return** $\boldsymbol{z}$.

□

*Proof (of Theorem 3, self-contained).* Recall that we already know that $\text{supp}(\phi_0) = \{\emptyset\}$. Now, for $i = 1, \ldots, m$ we set $\mathcal{F} = \text{supp}(\phi_{i-1}) \cup \{X \cup \{i\} : X \in \text{supp}(\phi_{i-1})\}$ and use $\mathcal{F}$ to obtain $\text{supp}(\phi_i)$ using Lemma 16. In the end we are given $\text{supp}(\phi_m)$ and since $\phi_m$ is exactly the original formula, the input is a yes-instance if and only if $[m] \in \text{supp}(\phi_m)$. The claimed running time follows from Observations 1 and 3 above and the running time of algorithm `mobius`. □

**Set Cover.** We will now give an application of Theorem 15(b) to Set Cover: Given a set family $\mathcal{F} \subseteq 2^U$ where $|U| = n$ and an integer $k$, find a subfamily $\mathcal{C} \subseteq \mathcal{F}$ such that $|\mathcal{C}| = k$ and $\bigcup_{S \in \mathcal{C}} S = U$.

**Theorem 17 (†).** *Given an instance of* Set Cover, *let* $0 < \alpha \leq 1/2$ *be the largest real such that* $|\{\bigcup_{S \in \mathcal{C}} S : \mathcal{C} \subseteq \mathcal{F}$ *and* $|\mathcal{C}| = k\}| \leq 2^{(1-\alpha)n}$. *Then the instance can be solved in* $\mathcal{O}^\star(2^{(1-\alpha/2)n} n^{\mathcal{O}(1)})$ *time.*

**Acknowledgements.** P.K. is supported by the Academy of Finland, Grants 252083 and 256287. M.K. is supported by the Academy of Finland, Grant 125637. J.N. is supported by the Nederlandse Organisatie voor Wetenschappelijk Onderzoek (NWO), project: 'Space and Time Efficient Structural Improvements of Dynamic Programming Algorithms'.

## References

1. Alon, N., Gutin, G., Kim, E.J., Szeider, S., Yeo, A.: Solving MAX-$r$-SAT above a tight lower bound. Algorithmica 61(3), 638–655 (2011)
2. Björklund, A.: Determinant sums for undirected Hamiltonicity. In: FOCS, pp. 173–182. IEEE Computer Society (2010)
3. Björklund, A., Husfeldt, T., Koivisto, M.: Set partitioning via inclusion-exclusion. SIAM J. Comput. 39(2), 546–563 (2009)
4. Crowston, R., Gutin, G., Jones, M., Yeo, A.: Lower bound for Max-r-Lin2 and its applications in algorithmics and graph theory. CoRR, abs/1104.1135 (2011)
5. Eppstein, D., Galil, Z., Giancarlo, R., Italiano, G.F.: Sparse dynamic programming I: linear cost functions. J. ACM 39, 519–545 (1992)

6. Eppstein, D., Galil, Z., Giancarlo, R., Italiano, G.F.: Sparse dynamic programming II: convex and concave cost functions. J. ACM 39(3), 546–567 (1992)
7. Fomin, F.V., Kratsch, D.: Exact Exponential Algorithms, 1st edn. Springer-Verlag New York, Inc., New York (2010)
8. Gaspers, S., Szeider, S.: Strong backdoors to nested satisfiability. CoRR, abs/1202.4331 (2012)
9. Harnik, D., Naor, M.: On the compressibility of NP instances and cryptographic applications. SIAM Journal on Computing 39(5), 1667–1713 (2010)
10. Håstad, J.: Some optimal inapproximability results. J. ACM 48, 798–859 (2001)
11. Horowitz, E., Sahni, S.: Computing partitions with applications to the knapsack problem. J. ACM 21, 277–292 (1974)
12. Impagliazzo, R., Paturi, R.: On the complexity of $k$-SAT. J. Comput. Syst. Sci. 62(2), 367–375 (2001)
13. Koutis, I., Williams, R.: Limits and Applications of Group Algebras for Parameterized Problems. In: Albers, S., Marchetti-Spaccamela, A., Matias, Y., Nikoletseas, S., Thomas, W. (eds.) ICALP 2009, Part I. LNCS, vol. 5555, pp. 653–664. Springer, Heidelberg (2009)
14. Lipton, R.J.: Beating Bellman for the knapsack problem (2010), `http://rjlipton.wordpress.com/2010/03/03/beating-bellman-for-the-knapsack-problem/`, `http://rjlipton.wordpress.com`
15. Lokshtanov, D., Nederlof, J.: Saving space by algebraization. In: Proceedings of the 42nd ACM Symposium on Theory of Computing, STOC 2010, pp. 321–330. ACM, New York (2010)
16. Mansour, Y.: Randomized interpolation and approximation of sparse polynomials. SIAM J. Comput. 24(2), 357–368 (1995)
17. Nederlof, J.: Fast Polynomial-Space Algorithms Using Möbius Inversion: Improving on Steiner Tree and Related Problems. In: Albers, S., Marchetti-Spaccamela, A., Matias, Y., Nikoletseas, S., Thomas, W. (eds.) ICALP 2009, Part I. LNCS, vol. 5555, pp. 713–725. Springer, Heidelberg (2009)
18. Nishimura, N., Ragde, P., Szeider, S.: Solving #sat using vertex covers. Acta Inf. 44(7), 509–523 (2007)
19. Reed, B.A., Smith, K., Vetta, A.: Finding odd cycle transversals. Oper. Res. Lett. 32(4), 299–301 (2004)
20. Rosser, B.: Explicit bounds for some functions of prime numbers. American Journal of Mathematics 63(1), 211–232 (1941)
21. Schroeppel, R., Shamir, A.: A $T = O(2^{n/2})$, $S = O(2^{n/4})$ algorithm for certain NP-complete problems. SIAM J. Comput. 10(3), 456–464 (1981)
22. Solomon, L.: The burnside algebra of a finite group. Journal of Combinatorial Theory 2(4), 603–615 (1967)
23. Traxler, P.: Exponential Time Complexity of SAT and Related Problems. PhD thesis, ETH Zürich (2010)
24. Williams, R.: Finding paths of length $k$ in $\mathcal{O}^\star(2^k)$ time. Inf. Process. Lett. 109(6), 315–318 (2009)
25. Williams, R., Gomes, C.P., Selman, B.: Backdoors to typical case complexity. In: IJCAI, pp. 1173–1178. Morgan Kaufmann (2003)
26. Woeginger, G.J.: Open problems around exact algorithms. Discrete Applied Mathematics 156(3), 397–405 (2008)
27. Zippel, R.: Probabilistic Algorithms for Sparse Polynomials. In: Ng, E.W. (ed.) EUROSAM 1979 and ISSAC 1979. LNCS, vol. 72, pp. 216–226. Springer, Heidelberg (1979)

# Fast Monotone Summation over Disjoint Sets*

Petteri Kaski[1], Mikko Koivisto[2], and Janne H. Korhonen[2]

[1] Helsinki Institute for Information Technology HIIT & Department of Information and Computer Science, Aalto University, Finland

[2] Helsinki Institute for Information Technology HIIT & Department of Computer Science, University of Helsinki, Finland

**Abstract.** We study the problem of computing an ensemble of multiple sums where the summands in each sum are indexed by subsets of size $p$ of an $n$-element ground set. More precisely, the task is to compute, for each subset of size $q$ of the ground set, the sum over the values of all subsets of size $p$ that are *disjoint* from the subset of size $q$. We present an arithmetic circuit that, without subtraction, solves the problem using $O((n^p + n^q) \log n)$ arithmetic gates, all monotone; for constant $p$, $q$ this is within the factor $\log n$ of the optimal. The circuit design is based on viewing the summation as a "set nucleation" task and using a tree-projection approach to implement the nucleation. Applications include improved algorithms for counting heaviest $k$-paths in a weighted graph, computing permanents of rectangular matrices, and dynamic feature selection in machine learning.

## 1 Introduction

**Weak Algebrisation.** Many hard combinatorial problems benefit from *algebrisation*, where the problem to be solved is cast in algebraic terms as the task of evaluating a particular expression or function over a suitably rich algebraic structure, such as a multivariate polynomial ring over a finite field. Recent advances in this direction include improved algorithms for the $k$-path [25], Hamiltonian path [4], $k$-coloring [9], Tutte polynomial [6], knapsack [21], and connectivity [14] problems. A key ingredient in all of these advances is the exploitation of an algebraic catalyst, such as the existence of additive inverses for inclusion–exclusion, or the existence of roots of unity for evaluation/interpolation, to obtain fast evaluation algorithms.

Such advances withstanding, it is a basic question whether the catalyst is *necessary* to obtain speedup. For example, fast algorithms for matrix multiplication [11,13] (and combinatorially related tasks such as finding a triangle in a graph [1,17]) rely on the assumption that the scalars have a ring structure, which prompts the question whether a weaker structure, such as a semiring without

* This research was supported in part by the Academy of Finland, Grants 252083 (P.K.), 256287 (P.K.), and 125637 (M.K.), and by the Helsinki Doctoral Programme in Computer Science - Advanced Computing and Intelligent Systems (J.K.).

D.M. Thilikos and G.J. Woeginger (Eds.): IPEC 2012, LNCS 7535, pp. 159–170, 2012.

additive inverses, would still enable fast multiplication. The answer to this particular question is known to be negative [18], but for many of the recent advances such an analysis has not been carried out. In particular, many of the recent algebrisations have significant combinatorial structure, which gives hope for *positive* results even if algebraic catalysts are lacking. The objective of this paper is to present one such positive result by deploying *combinatorial* tools.

**A Lemma of Valiant.** Our present study stems from a technical lemma of Valiant [22] encountered in the study of circuit complexity over a monotone versus a universal basis. More specifically, starting from $n$ variables $f_1, f_2, \ldots, f_n$, the objective is to use as few arithmetic operations as possible to compute the $n$ sums of variables where the $j$th sum $e_j$ includes all the other variables except the variable $f_j$, where $j = 1, 2, \ldots, n$.

If additive inverses are available, a solution using $O(n)$ arithmetic operations is immediate: first take the sum of all the $n$ variables, and then for $j = 1, 2, \ldots, n$ compute $e_j$ by subtracting the variable $f_j$.

Valiant [22] showed that $O(n)$ operations suffice also when additive inverses are *not* available; we display Valiant's elegant combinatorial solution for $n = 8$ below as an arithmetic circuit.

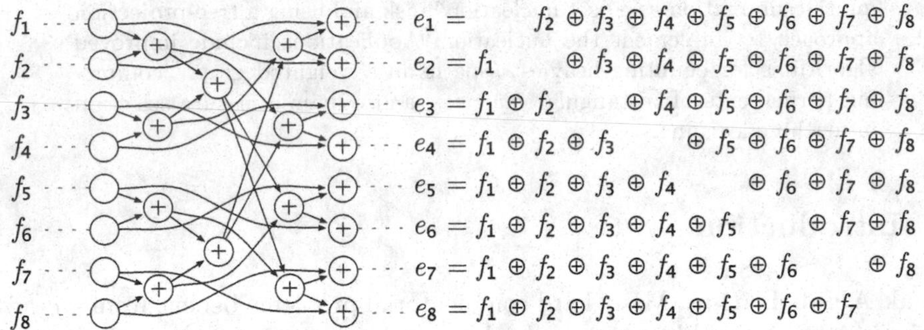


**Generalising to Higher Dimensions.** This paper generalises Valiant's lemma to higher dimensions using purely combinatorial tools. Accordingly, we assume that only very limited algebraic structure is available in the form of a commutative semigroup $(S, \oplus)$. That is, $\oplus$ satisfies the associative law $x \oplus (y \oplus z) = (x \oplus y) \oplus z$ and the commutative law $x \oplus y = y \oplus x$ for all $x, y, z \in S$, but nothing else is assumed.

By "higher dimensions" we refer to the input not consisting of $n$ values ("variables" in the example above) in $S$, but rather $\binom{n}{p}$ values $f(X) \in S$ indexed by the $p$-subsets $X$ of $[n] = \{1, 2, \ldots, n\}$. Accordingly, we also allow the output to have higher dimension. That is, given as input a function $f$ from the $p$-subsets $[n]$ to the set $S$, the task is to output the function $e$ defined for each $q$-subset $Y$ of $[n]$ by

$$e(Y) = \bigoplus_{X : X \cap Y = \emptyset} f(X), \tag{1}$$

where the sum is over all $p$-subsets $X$ of $[n]$ satisfying the intersection constraint. Let us call this problem $(p, q)$*-disjoint summation*.

In analogy with Valiant's solution for the case $p = q = 1$ depicted above, an algorithm that solves the $(p, q)$-disjoint summation problem can now be viewed as a circuit consisting of two types of gates: *input gates* indexed by $p$-subsets $X$ and *arithmetic gates* that perform the operation $\oplus$, with certain arithmetic gates designated as output gates indexed by $q$-subsets $Y$. We would like a circuit that has as few gates as possible. In particular, does there exist a circuit whose size for constant $p$, $q$ is within a logarithmic factor of the lower bound $\Theta(n^p + n^q)$?

**Main Result.** In this paper we answer the question in the affirmative. Specifically, we show that a circuit of size $O\big((n^p + n^q)\log n\big)$ exists to compute $e$ from $f$ over an arbitrary commutative semigroup $(S, \oplus)$, and moreover, there is an algorithm that constructs the circuit in time $O\big((p^2 + q^2)(n^p + n^q)\log^3 n\big)$. These bounds hold uniformly for all $p$, $q$. That is, the coefficient hidden by $O$-notation does not depend on $p$ and $q$.

From a technical perspective our main contribution is combinatorial and can be expressed as a solution to a specific *set nucleation* task. In such a task we start with a collection of "atomic compounds" (a collection of singleton sets), and the goal is to assemble a specified collection of "target compounds" (a collection of sets that are unions of the singletons). The assembly is to be executed by a straight-line program, where each operation in the program selects two *disjoint* sets in the collection and inserts their union into the collection. (Once a set is in the collection, it may be selected arbitrarily many times.) The assembly should be done in as few operations as possible.

Our main contribution can be viewed as a straight-line program of length $O\big((n^p + n^q)\log n\big)$ that assembles the collection $\{\{X : X \cap Y = \emptyset\} : Y\}$ starting from the collection $\{\{X\} : X\}$, where $X$ ranges over the $p$-subsets of $[n]$ and $Y$ ranges over the $q$-subsets of $[n]$. Valiant's lemma [22] in these terms provides an optimal solution of length $\Theta(n)$ for the specific case $p = q = 1$.

**Applications.** Many classical optimisation problems and counting problems can be algebrised over a commutative semigroup. A selection of applications will be reviewed in Sect. 3.

**Related Work.** "Nucleation" is implicit in the design of many fast algebraic algorithms, perhaps two of the most central are the fast Fourier transform of Cooley and Tukey [12] (as is witnessed by the butterfly circuit representation) and Yates's 1937 algorithm [26] for computing the product of a vector with the tensor product of $n$ matrices of size $2 \times 2$. The latter can in fact be directly used to obtain a nucleation process for $(p, q)$-disjoint summation, even if an inefficient one. (For an exposition of Yates's method we recommend Knuth [19, §4.6.4]; take $m_i = 2$ and $g_i(s_i, t_i) = [s_i = 0 \text{ or } t_i = 0]$ for $i = 1, 2, \ldots, n$ to extract the following nucleation process implicit in the algorithm.) For all $Z \subseteq [n]$ and $i \in \{0, 1, \ldots, n\}$, let

$$a_i(Z) = \{X \subseteq [n] : X \cap [n-i] = Z \cap [n-i],\ X \cap Z \setminus [n-i] = \emptyset\}. \quad (2)$$

Put otherwise, $a_i(Z)$ consists of $X$ that agree with $Z$ in the first $n - i$ elements of $[n]$ and are disjoint from $Z$ in the last $i$ elements of $[n]$. In particular, our objective is to assemble the sets $a_n(Y) = \{X : X \cap Y = \emptyset\}$ for each $Y \subseteq [n]$ starting from the singletons $a_0(X) = \{X\}$ for each $X \subseteq [n]$. The nucleation process given by Yates' algorithm is, for all $i = 1, 2, \ldots, n$ and $Z \subseteq [n]$, to set

$$a_i(Z) = \begin{cases} a_{i-1}(Z \setminus \{n+1-i\}) & \text{if } n+1-i \in Z, \\ a_{i-1}(Z \cup \{n+1-i\}) \cup a_{i-1}(Z) & \text{if } n+1-i \notin Z. \end{cases} \quad (3)$$

This results in $2^{n-1}n$ disjoint unions. If we restrict to the case $|Y| \leq q$ and $|X| \leq p$, then it suffices to consider only $Z$ with $|Z| \leq p+q$, which results in $O\big((p+q)\sum_{j=0}^{p+q}\binom{n}{j}\big)$ disjoint unions. Compared with our main result, this is not particularly efficient. In particular, our main result relies on "tree-projection" partitioning that enables a significant speedup over the "prefix-suffix" partitioning in (2) and (3).

We observe that "set nucleation" can also be viewed as a computational problem, where the output collection is given and the task is to decide whether there is a straight-line program of length at most $\ell$ that assembles the output using (disjoint) unions starting from singleton sets. This problem is known to be NP-complete even in the case where output sets have size 3 [15, Problem PO9]; moreover, the problem remains NP-complete if the unions are not required to be disjoint.

## 2 A Circuit for $(p, q)$-Disjoint Summation

**Nucleation of $p$-Subsets with a Perfect Binary Tree.** Looking at Valiant's circuit construction in the introduction, we observe that the left half of the circuit accumulates sums of variables (i.e., sums of 1-subsets of $[n]$) along what is a perfect binary tree. Our first objective is to develop a sufficient generalisation of this strategy to cover the setting where each summand is indexed by a $p$-subset of $[n]$ with $p \geq 1$.

Let us assume that $n = 2^b$ for a nonnegative integer $b$ so that we can identify the elements of $[n]$ with binary strings of length $b$. We can view each binary string of length $b$ as traversing a unique path starting from the root node of a perfect binary tree of height $b$ and ending at a unique leaf node. Similarly, we may identify any node at level $\ell$ of the tree by a binary string of length $\ell$, with $0 \leq \ell \leq b$. See Fig. 1(a) for an illustration. For $p = 1$ this correspondence suffices.

For $p > 1$, we are not studying individual binary strings of length $b$ (that is, individual elements of $[n]$), but rather $p$-subsets of such strings. In particular, we can identify each $p$-subset of $[n]$ with a $p$-subset of leaf nodes in the binary tree. To nucleate such subsets it will be useful to be able to "project" sets upward in the tree. This motivates the following definitions.

Let us write $\{0,1\}^\ell$ for the set of all binary strings of length $0 \leq \ell \leq b$. For $\ell = 0$, we write $\epsilon$ for the empty string. For a subset $X \subseteq \{0,1\}^b$, we define the *projection of $X$ to level $\ell$* as

$$X|_\ell = \left\{x \in \{0,1\}^\ell \colon \exists y \in \{0,1\}^{b-\ell} \text{ such that } xy \in X\right\}. \quad (4)$$

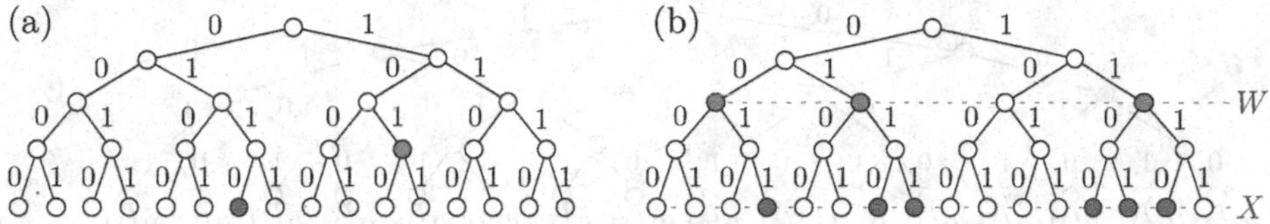


**Fig. 1.** Representing $\{0,1\}$-strings of length at most $b$ as nodes in a perfect binary tree of height $b$. Here $b = 4$. (a) Each string traces a unique path down from the root node, with the empty string $\epsilon$ corresponding to the root node. The nodes at level $0 \le \ell \le b$ correspond to the strings of length $\ell$. The red leaf node corresponds to 0110 and the blue node corresponds to 101. (b) A set of strings corresponds to a set of nodes in the tree. The set $X$ is displayed in red, the set $W$ in blue. The set $W$ is the projection of the set $X$ to level $\ell = 2$. Equivalently, $X|_\ell = W$.

That is, $X|_\ell$ is the set of length-$\ell$ prefixes of strings in $X$. Equivalently, in the binary tree we obtain $X|_\ell$ by lifting each element of $X$ to its ancestor on level-$\ell$ in the tree. See Fig. 1(b) for an illustration. For the empty set we define $\emptyset|_\ell = \emptyset$.

Let us now study a set family $\mathcal{F} \subseteq 2^{\{0,1\}^b}$. The intuition here is that each member of $\mathcal{F}$ is a summand, and $\mathcal{F}$ represents the sum of its members. A circuit design must assemble (nucleate) $\mathcal{F}$ by taking disjoint unions of carefully selected subfamilies. This motivates the following definitions.

For a level $0 \le \ell \le b$ and a string $W \subseteq \{0,1\}^\ell$ let us define *the subfamily of* $\mathcal{F}$ *that projects to* $W$ by

$$\mathcal{F}_W = \{X \in \mathcal{F} \colon X|_\ell = W\}\,. \tag{5}$$

That is, the family $\mathcal{F}_W$ consists of precisely those members $X \in \mathcal{F}$ that project to $W$. Again Fig. 1(b) provides an illustration: we select precisely those $X$ whose projection is $W$.

The following technical observations are now immediate. For each $0 \le \ell \le b$, if $\emptyset \in \mathcal{F}$, then we have

$$\mathcal{F}_\emptyset = \{\emptyset\}\,. \tag{6}$$

Similarly, for $\ell = 0$ we have

$$\mathcal{F}_{\{\epsilon\}} = \mathcal{F} \setminus \{\emptyset\}\,. \tag{7}$$

For $\ell = b$ we have for every $W \in \mathcal{F}$ that

$$\mathcal{F}_W = \{W\}\,. \tag{8}$$

Now let us restrict our study to the situation where the family $\mathcal{F} \subseteq 2^{\{0,1\}^b}$ contains only sets of size at most $p$. In particular, this is the case in our applications. For a set $U$ and an integer $p$, let us write $\binom{U}{p}$ for the family of all subsets of $U$ of size $p$, and $\binom{U}{\downarrow p}$ for the family of all subsets of $U$ with size at most $p$. Accordingly, for integers $0 \le k \le n$, let us use the shorthand $\binom{n}{\downarrow k} = \sum_{i=0}^{k} \binom{n}{i}$.

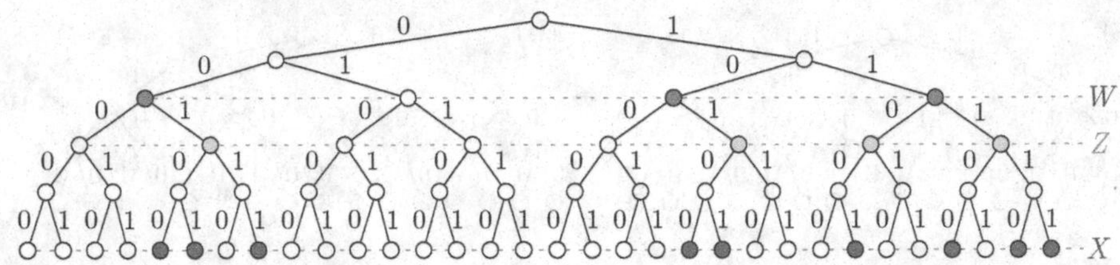


**Fig. 2.** Illustrating the proof of Lemma 1. Here $b = 5$. The set $X$ (indicated with red nodes) projects to level $\ell = 2$ to the set $W$ (indicated with blue nodes) and to level $\ell + 1 = 3$ to the set $Z$ (indicated with yellow nodes). Furtermore, the projection of $Z$ to level $\ell$ is $W$. Thus, each $X \in \mathcal{F}$ is included to $\mathcal{F}_W$ exactly from $\mathcal{F}_Z$ in Lemma 1.

The following lemma enables us to recursively nucleate any family $\mathcal{F} \subseteq \binom{\{0,1\}^b}{\downarrow p}$. In particular, we can nucleate the family $\mathcal{F}_W$ with $W$ in level $\ell$ using the families $\mathcal{F}_Z$ with $Z$ in level $\ell + 1$. Applied recursively, we obtain $\mathcal{F}$ by proceeding from the bottom up, that is, $\ell = b, b-1, \ldots, 1, 0$. The intuition underlying the lemma is illustrated in Fig. 2. We refer to the full version of this paper for the proof.

**Lemma 1.** *For all* $0 \leq \ell \leq b-1$, $\mathcal{F} \subseteq \binom{\{0,1\}^b}{\downarrow p}$, *and* $W \in \binom{\{0,1\}^\ell}{\downarrow p}$, *we have that the family* $\mathcal{F}_W$ *is a disjoint union* $\mathcal{F}_W = \bigcup \left\{ \mathcal{F}_Z : Z \in \binom{\{0,1\}^{\ell+1}}{\downarrow p}_W \right\}$.

**A Generalisation: $(p, q)$-Intersection Summation.** It will be convenient to study a minor generalisation of $(p, q)$-disjoint summation. Namely, instead of insisting on disjointness, we allow nonempty intersections to occur with "active" (or "avoided") $q$-subsets $A$, but require that elements in the intersection of each $p$-subset and each $A$ are "individualized." That is, our input is not given by associating a value $f(X) \in S$ to each set $X \in \binom{[n]}{\downarrow p}$, but is instead given by associating a value $g(I, X) \in S$ to each pair $(I, X)$ with $I \subseteq X \in \binom{[n]}{\downarrow p}$, where $I$ indicates the elements of $X$ that are "individualized." In particular, we may insist (by appending to $S$ a formal identity element if such an element does not already exist in $S$) that $g(I, X)$ vanishes unless $I$ is empty. This reduces $(p, q)$-disjoint summation to the following problem:

*Problem 1.* ($(p, q)$-**intersection summation**) Given as input a function $g$ that maps each pair $(I, X)$ with $I \subseteq X \in \binom{[n]}{\downarrow p}$ and $|I| \leq q$ to an element $g(I, X) \in S$, output the function $h\colon \binom{[n]}{\downarrow q} \to S$ defined for all $A \in \binom{[n]}{\downarrow q}$ by

$$h(A) = \bigoplus_{X \in \binom{[n]}{\downarrow p}} g(A \cap X, X)\,. \tag{9}$$

**The Circuit Construction.** We proceed to derive a recursion for the function $h$ using Lemma 1 to carry out nucleation of $p$-subsets. The recursion proceeds from the bottom up, that is, $\ell = b, b-1, \ldots, 1, 0$ in the binary tree representation. (Recall that we identify the elements of $[n]$ with the elements of $\{0,1\}^b$, where $n$ is a power of 2 with $n = 2^b$.) The intermediate functions $h_\ell$ computed by the recursion are "projections" of (9) using (5). In more precise terms, for $\ell = b, b-1, \ldots, 1, 0$, the function $h_\ell \colon \binom{\{0,1\}^b}{\downarrow q} \times \binom{\{0,1\}^\ell}{\downarrow p} \to S$ is defined for all $W \in \binom{\{0,1\}^\ell}{\downarrow p}$ and $A \in \binom{\{0,1\}^b}{\downarrow q}$ by

$$h_\ell(A, W) = \bigoplus_{X \in \binom{\{0,1\}^b}{\downarrow p}_W} g(A \cap X, X). \tag{10}$$

Let us now observe that we can indeed recover the function $h$ from the case $\ell = 0$. Indeed, for the empty string $\epsilon$, the empty set $\emptyset$ and every $A \in \binom{\{0,1\}^b}{\downarrow q}$ we have by (6) and (7) that

$$h(A) = h_0(A, \{\epsilon\}) \oplus h_0(A, \emptyset). \tag{11}$$

It remains to derive the recursion that gives us $h_0$. Here we require one more technical observation, which enables us to narrow down the intermediate values $h_\ell(A, W)$ that need to be computed to obtain $h_0$. In particular, we may discard the part of the active set $A$ that extends outside the "span" of $W$. This observation is the crux in deriving a succinct circuit design.

For $0 \le \ell \le b$ and $w \in \{0,1\}^\ell$, we define the *span* of $w$ by

$$\langle w \rangle = \{x \in \{0,1\}^b \colon \exists z \in \{0,1\}^{b-\ell} \text{ such that } wz = x\}.$$

In the binary tree, $\langle w \rangle$ consists of the leaf nodes in the subtree rooted at $w$. Let us extend this notation to subsets $W \subseteq \{0,1\}^\ell$ by $\langle W \rangle = \bigcup_{w \in W} \langle w \rangle$. The following lemma shows that it is sufficient to evaluate $h_\ell(A, W)$ only for $W \in \binom{\{0,1\}^\ell}{\downarrow p}$ and $A \in \binom{\{0,1\}^b}{\downarrow q}$ such that $A \subseteq \langle W \rangle$. We omit the proof; please refer to the full version of this paper for details.

**Lemma 2.** *For all* $0 \le \ell \le b$, $W \in \binom{\{0,1\}^\ell}{\downarrow p}$, *and* $A \in \binom{\{0,1\}^b}{\downarrow q}$, *we have*

$$h_\ell(A, W) = h_\ell(A \cap \langle W \rangle, W). \tag{12}$$

We are now ready to present the recursion for $\ell = b, b-1, \ldots, 1, 0$. The base case $\ell = b$ is obtained directly based on the values of $g$, because we have by (8) for all $W \in \binom{\{0,1\}^b}{\downarrow p}$ and $A \in \binom{\{0,1\}^b}{\downarrow q}$ with $A \subseteq W$ that

$$h_b(A, W) = g(A, W). \tag{13}$$

The following lemma gives the recursive step from $\ell + 1$ to $\ell$ by combining Lemma 1 and Lemma 2. Again, we defer the details of the proof to the full version of this paper.

**Lemma 3.** *For* $0 \le \ell \le b-1$, $W \in \binom{\{0,1\}^\ell}{\downarrow p}$, *and* $A \in \binom{\{0,1\}^b}{\downarrow q}$ *with* $A \subseteq \langle W \rangle$, *we have*

$$h_\ell(A,W) = \bigoplus_{Z \in \binom{\{0,1\}^{\ell+1}}{\downarrow p}_W} h_{\ell+1}(A \cap \langle Z \rangle, Z). \tag{14}$$

The recursion given by (13), (14), and (12) now defines an arithmetic circuit that solves $(p,q)$-intersection summation.

**Size of the circuit.** By (13), the number of input gates in the circuit is equal to the number of pairs $(I,X)$ with $I \subseteq X \in \binom{\{0,1\}^b}{\downarrow p}$ and $|X| \le q$, which is

$$\sum_{i=0}^{p} \sum_{j=0}^{q} \binom{2^b}{i}\binom{i}{j}. \tag{15}$$

To derive an expression for the number of $\oplus$-gates, we count for each $0 \le \ell \le b-1$ the number of pairs $(A,W)$ with $W \in \binom{\{0,1\}^\ell}{\downarrow p}$, $A \in \binom{\{0,1\}^b}{\downarrow q}$, and $A \subseteq \langle W \rangle$, and for each such pair $(A,W)$ we count the number of $\oplus$-gates in the subcircuit that computes the value $h_\ell(A,W)$ from the values of $h_{\ell+1}$ using (14).

First, we observe that for each $W \in \binom{\{0,1\}^\ell}{\downarrow p}$ we have $|\langle W \rangle| = 2^{b-\ell}|W|$. Thus, the number of pairs $(A,W)$ with $W \in \binom{\{0,1\}^\ell}{\downarrow p}$, $A \in \binom{\{0,1\}^b}{\downarrow q}$, and $A \subseteq \langle W \rangle$ is

$$\sum_{i=0}^{p} \sum_{j=0}^{q} \binom{2^\ell}{i}\binom{i2^{b-\ell}}{j}. \tag{16}$$

For each such pair $(A,W)$, the number of $\oplus$-gates for (14) is $\left|\binom{\{0,1\}^{\ell+1}}{\downarrow p}_W\right| - 1$.

**Lemma 4.** *For all* $0 \le \ell \le b-1$, $W \in \binom{\{0,1\}^\ell}{\downarrow p}$, *and* $|W| = i$, *we have*

$$\left|\binom{\{0,1\}^{\ell+1}}{\downarrow p}_W\right| = \sum_{k=0}^{p-i} \binom{i}{k} 2^{i-k}. \tag{17}$$

*Proof.* A set $Z \in \binom{\{0,1\}^{\ell+1}}{\downarrow p}_W$ can contain either one or both of the strings $w0$ and $w1$ for each $w \in W$. The set $Z$ may contain both elements for at most $p-i$ elements $w \in W$ because otherwise $|Z| > p$. Finally, for each $0 \le k \le p-i$, there are $\binom{i}{k}2^{i-k}$ ways to select a set $Z \in \binom{\{0,1\}^{\ell+1}}{\downarrow p}_W$ such that $Z$ contains $w0$ and $w1$ for exactly $k$ elements $w \in W$.

Finally, for each $A \in \binom{\{0,1\}^b}{\downarrow q}$ we require an $\oplus$-gate that is also designated as an output gate to implement (11). The number of these gates is

$$\sum_{j=0}^{q} \binom{2^b}{j}. \tag{18}$$

The total number of $\oplus$-gates in the circuit is obtained by combining (15), (16), (17), and (18). The number of $\oplus$-gates is thus

$$\sum_{i=0}^{p}\sum_{j=0}^{q}\binom{2^b}{i}\binom{i}{j}+\sum_{\ell=0}^{b-1}\sum_{i=0}^{p}\sum_{j=0}^{q}\binom{2^\ell}{i}\binom{i2^{b-\ell}}{j}\left(\sum_{k=0}^{p-i}\binom{i}{k}2^{i-k}-1\right)+\sum_{j=0}^{q}\binom{2^b}{j}$$
$$\leq\sum_{\ell=0}^{b}\sum_{i=0}^{p}\sum_{j=0}^{q}\binom{2^\ell}{i}\binom{i2^{b-\ell}}{j}3^i\leq\sum_{\ell=0}^{b}\sum_{i=0}^{p}\sum_{j=0}^{q}\frac{(2^\ell)^i}{i!}\frac{i^j(2^{b-\ell})^j}{j!}3^i$$
$$\leq\sum_{\ell=0}^{b}\sum_{i=0}^{p}\sum_{j=0}^{q}\frac{(2^\ell)^{\max(p,q)}}{i!}\frac{i^j(2^{m-\ell})^{\max(p,q)}}{j!}3^i$$
$$=n^{\max(p,q)}(1+\log_2 n)\sum_{i=0}^{p}\sum_{j=0}^{q}\frac{i^j3^i}{i!j!}\,.$$

The remaining double sum is bounded from above by a constant, and thus the circuit defined by (13), (14), and (12) has size $O((n^p+n^q)\log n)$, where the constant hidden by the $O$-notation does not depend on $p$ and $q$.

The circuit can be constructed in time $O\big((p^2+q^2)(n^p+n^q)\log^3 n\big)$. We omit the details.

## 3 Concluding Remarks and Applications

We have generalised Valiant's [22] observation that negation is powerless for computing simultaneously the $n$ different disjunctions of all but one of the given $n$ variables: now we know that, in our terminology, subtraction is powerless for $(p,q)$-disjoint summation for any constant $p$ and $q$. (Valiant proved this for $p=q=1$.) Interestingly, requiring $p$ and $q$ be constants turns out to be essential, namely, when subtraction is available, an inclusion–exclusion technique is known [5] to yield a circuit of size $O(p\binom{n}{\downarrow p}+q\binom{n}{\downarrow q})$, which, in terms of $p$ and $q$, is exponentially smaller than our bound $O\big((n^p+n^q)\log n\big)$. This gap highlights the difference of the algorithmic ideas behind the two results. Whether the gap can be improved to polynomial in $p$ and $q$ is an open question.

While we have dealed with the abstract notions of "monotone sums" or semigroup sums, in applications they most often materialise as maximisation or minimisation, as described in the next paragraphs. Also, in applications local terms are usually combined not only by one (monotone) operation but two different operations, such as "min" and "+". To facilitate the treatment of such applications, we extend the semigroup to a semiring $(S,\oplus,\odot)$ by introducing a product operation "$\odot$". Now the task is to evaluate

$$\bigoplus_{X,Y:X\cap Y=\emptyset} f(X)\odot g(Y)\,, \tag{19}$$

where $X$ and $Y$ run through all $p$-subsets and $q$-subsets of $[n]$, respectively, and $f$ and $g$ are given mappings to $S$. We immediately observe that the expression

(19) is equal to $\bigoplus_Y e(Y) \odot g(Y)$, where the sum is over all $q$-subsets of $[n]$ and $e$ is as in (1). Thus, by our main result, it can be evaluated using a circuit with $O((n^p + n^q)\log n)$ gates.

**Application to $k$-paths.** We apply the semiring formulation to the problem of counting the maximum-weight $k$-edge paths from vertex $s$ to vertex $t$ in a given edge-weighted graph with real weights, where we assume that we are only allowed to add and compare real numbers and these operations take constant time (cf. [24]). By straightforward Bellman–Held–Karp type dynamic programming [2,3,16] (or, even by brute force) we can solve the problem in $\binom{n}{\downarrow k} n^{O(1)}$ time. However, our main result gives an algorithm that runs in $n^{k/2+O(1)}$ time by solving the problem in halves: Guess a middle vertex $v$ and define $f_1(X)$ as the number of maximum-weight $k/2$-edge paths from $s$ to $v$ in the graph induced by the vertex set $X \cup \{v\}$; similarly define $g_1(X)$ for the $k/2$-edge paths from $v$ to $t$. Furthermore, define $f_2(X)$ and $g_2(X)$ as the respective maximum weights and put $f(X) = (f_1(X), f_2(X))$ and $g(X) = (g_1(X), g_2(X))$. These values can be computed for all vertex subsets $X$ of size $k/2$ in $\binom{n}{k/2} n^{O(1)}$ time. It remains to define the semiring operations in such a way that the expression (19) equals the desired number of $k$-edge paths; one can verify that the following definitions work correctly: $(c, w) \odot (c', w') = (c \cdot c', w + w')$ and

$$(c, w) \oplus (c', w') = \begin{cases} (c, w) & \text{if } w > w', \\ (c', w') & \text{if } w < w', \\ (c + c', w) & \text{if } w = w'. \end{cases}$$

Thus, the techniques of the present paper enable solving the problem essentially as fast as the fastest known algorithms for the special case of counting *all* the $k$-paths, for which quite different techniques relying on subtraction yield $\binom{n}{k/2} n^{O(1)}$ time bound [7]. On the other, for the more general problem of counting weighted subgraphs Vassilevska and Williams [23] give an algorithm whose running time, when applied to $k$-paths, is $O(n^{\omega k/3} + n^{2k/3+c})$, where $\omega < 2.3727$ is the exponent of matrix multiplication and $c$ is a constant; this of course would remain worse than our bound even if $\omega = 2$.

**Application to Matrix Permanent.** Consider the problem of computing the permanent of a $k \times n$ matrix $(a_{ij})$ over a *noncommutative semiring*, with $k \le n$ and even for simplicity, given by $\sum_\sigma a_{1\sigma(1)} a_{2\sigma(2)} \cdots a_{k\sigma(k)}$, where the sum is over all injective mappings $\sigma$ from $[k]$ to $[n]$. We observe that the expression (19) equals the permanent if we let $p = q = k/2 = \ell$ and define $f(X)$ as the sum of $a_{1\sigma(1)} a_{2\sigma(2)} \cdots a_{\ell\sigma(\ell)}$ over all injective mappings $\sigma$ from $\{1, 2, \ldots, \ell\}$ to $X$ and, similarly, $g(Y)$ as the sum of $a_{\ell+1\sigma(\ell+1)} a_{\ell+2\sigma(\ell+2)} \cdots a_{k\sigma(k)}$ over all injective mappings $\sigma$ from $\{\ell+1, \ell+2, \ldots, k\}$ to $Y$. Since the values $f(X)$ and $g(Y)$ for all relevant $X$ and $Y$ can be computed by dynamic programming in $\binom{n}{k/2} n^{O(1)}$ time, our main result yields the time bound $n^{k/2+O(1)}$ for computing the permanent.

Thus we improve significantly upon a Bellman–Held–Karp type dynamic programming algorithm that computes the permanent in $\binom{n}{\downarrow k} n^{O(1)}$ time, the best previous upper bound we are aware of for noncommutative semirings [8]. It should be noted, however, that essentally as fast algorithms are already known for *noncommutative rings* [8], and that faster, $2^k n^{O(1)}$ time, algorithms are known for *commutative semirings* [8,20].

**Application to Feature Selection.** The extensively studied feature selection problem in machine learning asks for a subset $X$ of a given set of available features $A$ so as to maximise some objective function $f(X)$. Often the size of $X$ can be bounded from above by some constant $k$, and sometimes the selection task needs to be solved repeatedly with the set of available features $A$ changing dynamically across, say, the set $[n]$ of all features. Such constraints take place in a recent work [10] on Bayesian network structure learning by branch and bound: the algorithm proceeds by forcing some features, $I$, to be included in $X$ and some other, $E$, to be excluded from $X$. Thus the key computational step becomes that of maximising $f(X)$ subject to $I \subseteq X \subseteq [n] \setminus E$ and $|X| \leq k$, which is repeated for varying $I$ and $E$. We observe that instead of computing the maximum every time from scratch, it pays off precompute a solution to $(p, q)$-disjoint summation for all $0 \leq p, q \leq k$, since this takes about the same time as a single step for $I = \emptyset$ and any fixed $E$. Indeed, in the scenario where the branch and bound search proceeds to exclude each and every subset of $k$ features in turn, but no larger subsets, such precomputation decreases the running time bound quite dramatically, from $O(n^{2k})$ to $O(n^k)$; typically, $n$ ranges from tens to some hundreds and $k$ from 2 to 7. Admitted, in practice, one can expect the search procedure match the said scenario only partially, and so the savings will be more modest yet significant.

**Acknowledgment.** We thank Jukka Suomela for useful discussions.

## References


1. Alon, N., Yuster, R., Zwick, U.: Finding and counting given length cycles. Algorithmica 17(3), 209–223 (1997)
2. Bellman, R.: Combinatorial processes and dynamic programming. In: Combinatorial Analysis. Proceedings of Symposia in Applied Mathematics, vol. 10, pp. 217–249. ACM (1960)
3. Bellman, R.: Dynamic programming treatment of the travelling salesman problem. J. ACM 9(1), 61–63 (1962)
4. Björklund, A.: Determinant sums for undirected Hamiltonicity. In: 51st Annual IEEE Symposium on Foundations of Computer Science (FOCS 2010), pp. 173–182. IEEE Computer Society, Los Alamitos (2010)
5. Björklund, A., Husfeldt, T., Kaski, P., Koivisto, M.: Fourier meets möbius: fast subset convolution (manuscript submitted for publication)
6. Björklund, A., Husfeldt, T., Kaski, P., Koivisto, M.: Computing the Tutte polynomial in vertex-exponential time. In: 49th Annual IEEE Symposium on Foundations of Computer Science (FOCS 2008), pp. 677–686. IEEE Computer Society, Los Alamitos (2008)
7. Björklund, A., Husfeldt, T., Kaski, P., Koivisto, M.: Counting Paths and Packings in Halves. In: Fiat, A., Sanders, P. (eds.) ESA 2009. LNCS, vol. 5757, pp. 578–586. Springer, Heidelberg (2009)

8. Björklund, A., Husfeldt, T., Kaski, P., Koivisto, M.: Evaluation of permanents in rings and semirings. Information Processing Letters 110(20), 867–870 (2010)
9. Björklund, A., Husfeldt, T., Koivisto, M.: Set partitioning via inclusion-exclusion. SIAM Journal on Computing 39(2), 546–563 (2009)
10. de Campos, C.P., Ji, Q.: Efficient structure learning of bayesian networks using constraints. Journal of Machine Learning Research 12, 663–689 (2011)
11. Cohn, H., Kleinberg, R., Szegedy, B., Umans, C.: Group-theoretic algorithms for matrix multiplication. In: 46th Annual IEEE Symposium on Foundations of Computer Science (FOCS 2005), pp. 379–388. IEEE Computer Society, Los Alamitos (2005)
12. Cooley, J.W., Tukey, J.W.: An algorithm for the machine calculation of complex Fourier series. Mathematics of Computation 19, 297–301 (1965)
13. Coppersmith, D., Winograd, S.: Matrix multiplication via arithmetic progressions. Journal of Symbolic Computation 9(3), 251–280 (1990)
14. Cygan, M., Nederlof, J., Pilipczuk, M., Pilipczuk, M., van Rooij, J.M.M., Wojtaszczyk, J.O.: Solving connectivity problems parameterized by treewidth in single exponential time (2011) (manuscript)
15. Garey, M.R., Johnson, D.S.: Computers and Intractability, A Guide to the Theory of NP-Completeness. W.H. Freeman and Company (1979)
16. Held, M., Karp, R.M.: A dynamic programming approach to sequencing problems. Journal of the Society for Industrial and Applied Mathematics 10(1), 196–210 (1962)
17. Itai, A., Rodeh, M.: Finding a minimum circuit in a graph. SIAM Journal on Computing 7(4), 413–423 (1978)
18. Kerr, L.R.: The effect of algebraic structure on the computational complexity of matrix multiplications. Ph.D. thesis, Cornell University (1970)
19. Knuth, D.E.: The Art of Computer Programming, 3rd edn. Seminumerical Algorithms, vol. 2. Addison–Wesley, Upper Saddle River (1998)
20. Koutis, I., Williams, R.: Limits and Applications of Group Algebras for Parameterized Problems. In: Albers, S., Marchetti-Spaccamela, A., Matias, Y., Nikoletseas, S., Thomas, W. (eds.) ICALP 2009, Part I. LNCS, vol. 5555, pp. 653–664. Springer, Heidelberg (2009)
21. Lokshtanov, D., Nederlof, J.: Saving space by algebraization. In: 2010 ACM International Symposium on Theory of Computing (STOC 2010), pp. 321–330. ACM, New York (2010)
22. Valiant, L.G.: Negation is powerless for boolean slice functions. SIAM Journal on Computing 15, 531–535 (1986)
23. Vassilevska, V., Williams, R.: Finding, minimizing, and counting weighted subgraphs. In: 2009 ACM International Symposium on Theory of Computing (STOC 2009), pp. 455–464. ACM, New York (2009)
24. Vassilevska, V., Williams, R., Yuster, R.: Finding heaviest $H$-subgraphs in real weighted graphs, with applications. ACM Transactions on Algorithms 6(3), Art. 44, 23 (2010)
25. Williams, R.: Finding paths of length $k$ in $O^*(2^k)$ time. Information Processing Letters 109(6), 315–318 (2009)
26. Yates, F.: The Design and Analysis of Factorial Experiments. Imperial Bureau of Soil Science, Harpenden, England (1937)

# Weighted Counting of $k$-Matchings Is #W[1]-Hard

Markus Bläser and Radu Curticapean

Saarland University, Dept. of Computer Science
{mblaeser,curticapean}@cs.uni-saarland.de

**Abstract.** In the seminal paper for parameterized counting complexity [1], the following problem is conjectured to be #W[1]-hard: Given a bipartite graph $G$ and a number $k \in \mathbb{N}$, which is considered as a parameter, count the number of matchings of size $k$ in $G$.

We prove hardness for a natural weighted generalization of this problem: Let $G = (V, E, w)$ be an edge-weighted graph and define the weight of a matching as the product of weights of all edges in the matching. We show that exact evaluation of the sum over all such weighted matchings of size $k$ is #W[1]-hard for bipartite graphs $G$.

As an intermediate step in our reduction, we also prove #W[1]-hardness of the problem of counting $k$-partial cycle covers, which are vertex-disjoint unions of cycles including $k$ edges in total. This hardness result even holds for unweighted graphs.

## 1 Introduction

Counting problems are an important class of problems in theoretical computer science and also appear in other areas, such as machine learning and statistical physics. Within computer science, they were formally introduced in [2] together with the complexity class #P. In the same paper, #P-hardness of the permanent evaluation problem for 0-1-matrices was proven, which is equivalent to the statement that counting perfect matchings on bipartite graphs is #P-hard.

Given this hardness result, the complexity of relaxed versions of the permanent evaluation and other hard counting problems was investigated. Typical relaxations consist of restricting the input to *tame* graphs, such as planar graphs or graphs of bounded degree. For instance, the number of perfect matchings is polynomial-time computable on planar graphs, as shown in [3,4], but remains hard on general graphs of maximum degree 3, cf. [5]. Other relaxations, such as *approximate* counting [6], are also studied.

### 1.1 Parameterized Counting Problems

Among the most recent relaxations for hard counting problems are *parameterized* counting problems, cf. [1]. Inputs $x$ to such problems come with an additional parameter $k$, and a parameterized counting problem is *fixed-parameter tractable* if it can be solved in time $f(k)|x|^{O(1)}$ on inputs $x$ with parameter $k$. In [1],

D.M. Thilikos and G.J. Woeginger (Eds.): IPEC 2012, LNCS 7535, pp. 171–181, 2012.

the class #W[1] is defined analogously to the well-known class W[1], and #W[1]-hardness under parameterized reductions is introduced.

In this paper, we consider only parameterized counting problems on *graphs*. Here, $k$ typically either measures some notion of intricacy of the *input graph* or the intricacy of the structures *to be counted*.

Typical parameters associated with the input graph are, for instance, its treewidth, cliquewidth or genus. A well-known result for parameterization by treewidth is a counting analogue [7] of Courcelle's Theorem [8]: If $\phi(X)$ is a formula in monadic second-order logic over graphs with a free set variable $X$, then computing the number of sets $X$ that satisfy $\phi$ is fpt in the treewidth of $G$. This implies, among others, that counting perfect matchings of a graph $G$ is fpt in the treewidth of $G$. Furthermore, by [9], counting the perfect matchings of a graph is fpt in its genus.

A commonly used parameter associated with the structures to be counted is simply given by their size. Results based on this kind of parameterization include that computing the number of $k$-vertex covers is fpt in $k$, while computing the number of $k$-paths, $k$-cliques or $k$-cycles is #W[1]-hard, all proven in [1]. If $G$ is a graph of bounded local treewidth, such as a planar graph or a graph of bounded degree, then many tractability results for counting structures of small size can be derived from the following meta-theorem [10]: If $\phi(x_1, \ldots, x_k)$ is a formula of first-order logic with free individual variables $x_1, \ldots, x_k$, then counting the number of tuples $(a_1, \ldots, a_k) \in V^k$ with $G \models \phi(a_1, \ldots, a_k)$ is fpt in $k$.

### 1.2 Counting $k$-Matchings

As an open problem in [1], it is conjectured that counting $k$-matchings (matchings with $k$ edges) in bipartite graphs is #W[1]-hard in the parameter $k$. This conjecture is backed up by the fact that the best known algorithms for this counting problem have time bounds of the type $O^*(n^{\Theta(k)})$. Among these is [11], which runs in time $O^*(2^{k+o(k)}\binom{n}{k/2})$ on general graphs. This was subsequently improved by a slightly faster algorithm [12] that requires only polynomial space.

In this paper, we show that *weighted* counting of $k$-matchings is #W[1]-hard in the parameter $k$, even for bipartite graphs. In our setting, we consider edge-weighted bipartite graphs $G = (V, E, w)$ and assign to every matching $M \subseteq E$ the weight $\prod_{e \in M} w(e)$. This kind of weighting also appears often in the study of graph polynomials, see [13,14] for examples.

While our hardness result holds only for a weighted, and as such, generalized version of the problem of counting $k$-matchings, we still consider this to be a valid step towards proving #W[1]-hardness of the unweighted problem. As future work, we consider modifying our proof in such a way that all intermediate reductions that require edge weights are replaced by unweighted reductions.

## 2 Definitions and Proof Outline

In this paper, several notions of graphs are distinguished: A *graph* is an undirected simple graph, i.e., it has neither self-loops nor multiple edges. A *digraph* is a directed graph with self-loops $(v, v)$ allowed, but multiple edges forbidden. An *edge-weighted (di)graph* is a triple $(V, E, w)$ with $(V, E)$ a (di)graph and $w : E \to \mathbb{Z}$. For all graphs, paths and cycles are defined to be simple.[1] By standard convention, we consider self-loops as cycles, but not as paths.

We denote the falling factorial by $(x)_n := x(x-1)\dots(x-n+1)$. If $M$ is a set, we define $\#M := |M|$ for convenience.

For an undirected graph $G$ and $k \in \mathbb{N}$, we write $\mathcal{M}_k[G]$ for the set of matchings of size $k$ in $G$. Our goal is to prove #W[1]-hardness of the parameterized matching problem p#Match: Given a bipartite graph $G$ and $k \in \mathbb{N}$, compute $\#\mathcal{M}_k[G]$. While we do not obtain a proof for this statement, we can show #W[1]-hardness for p#wMatch, a weighted version of this problem, where $G = (V, E, w)$ is edge-weighted and every matching $M$ is weighted by $\prod_{e \in M} w(e)$:

| p#Match | p#wMatch |
|---|---|
| **Input:** Digraph $G$, $k \in \mathbb{N}$ | **Input:** Weighted digraph $G$, $k \in \mathbb{N}$ |
| **Parameter:** $k$ | **Parameter:** $k$ |
| **Output:** $\#\mathcal{M}_k[G]$ | **Output:** $\sum_{M \in \mathcal{M}_k[G]} \prod_{e \in M} w(e)$ |

Our result is proven in a series of four reductions, starting from p#Clique, whose #W[1]-hardness was established in [1]. First, we observe in Section 3 that p#Match is equivalent to the problem of counting disjoint unions of paths and cycles on digraphs. We call these structures $k$-partial path-cycle covers:

**Definition 1.** *Let $G = (V, E)$ be a digraph and $k \in \mathbb{N}$. A $k$-partial path-cycle cover $C$ in $G$ is a set $C \subseteq E$ with $|C| = k$ that consists of a vertex-disjoint union of paths and cycles.*[2]

*The number of cycles in $C$ is denoted by $\sigma(C)$, that of paths by $\rho(C)$, and that of isolated vertices by $\iota(C)$. We call $C$ a $k$-partial* cycle *cover if $\rho(C) = 0$.*

*The set of $k$-partial path-cycle covers in $G$ is denoted by $\mathcal{PC}_k[G]$, and that of $k$-partial cycle covers by $\mathcal{C}_k[G]$.*

We denote the parameterized problem of counting $k$-partial path-cycle covers by p#PCC, and that of counting $k$-partial cycle covers by p#CC. As for matchings, we can define the weight of a path-cycle cover $C$ as the product of all weights of edges in $C$, and obtain the problem p#wPCC:

| p#CC | p#wPCC |
|---|---|
| **Input:** Digraph $G$, $k \in \mathbb{N}$ | **Input:** Weighted digraph $G$, $k \in \mathbb{N}$ |
| **Parameter:** $k$ | **Parameter:** $k$ |
| **Output:** $\#\mathcal{C}_k[G]$ | **Output:** $\sum_{C \in \mathcal{PC}_k[G]} \prod_{e \in C} w(C)$ |

In Section 4, we introduce a graph gadget that allows to reduce the problem of counting partial *cycle* covers to that of counting weighted partial

[1] "Non-simple cycles" *will* appear in this paper, but they will never be called cycles.

[2] If $k$ is not relevant in the context, we simply call $C$ a *partial* path-cycle cover.

path-cycle covers. The weights are essential in this reduction, as they are required to cancel out unwanted paths. We obtain:

$$\mathsf{p\#CC} \leq^{T}_{fpt} \mathsf{p\#wPCC}.$$

It remains to prove #W[1]-hardness of p#CC. For our reduction, we introduce a combinatorial structure that could be described as a "union of closed walks without distinguished start vertices". For notational simplicity, we call such a structure a UCW:

**Definition 2.** *Let $G = (V, E)$ be a digraph. Let $(v_1, \ldots, v_k) \in V^k$ such that $(v_i, v_{i+1}) \in E$ for all $i < k$ and $(v_k, v_1) \in E$. Write $[v_1, \ldots, v_k]$ for the set of all cyclic shifts of $(v_1, \ldots, v_k)$ and call $W = [v_1, \ldots, v_k]$ a* CW *of* length $l(W) := k$.
*A UCW is a multiset $U = \{W_1, \ldots, W_t\}$ of CWs. We set $l(U) := \sum_{i=1}^{t} l(W_i)$.*

To each UCW, we can associate a particular polynomial, its *type*. We define types analogously to the types of homomorphisms from [1]:

**Definition 3.** *Let $W$ be a CW and $v \in V$. We write $f_W(v)$ for the number of appearances of $v$ in (an arbitrary element of) $W$. For $U = \{W_1, \ldots, W_t\}$ a UCW, we set $f_U(v) := \sum_{i=1}^{t} f_{W_i}(v)$.* The type *$\theta_U$ of $U$ is defined as the polynomial*

$$\theta_U(x) := \prod_{v \in V} (x)_{f_U(v)}.$$

*We write $\mathcal{U}_k[G, \theta]$ for the set of UCWs of length $k$ and type $\theta$ in $G$.*

The type $\theta_U$ can be seen as an encoding of the multiset $\{f_U(v) \mid v \in V\}$. Note that the sum of this multiset is equal to $l(U)$ and $\deg(\theta_U)$.[3]

In analogy to the problem of counting typed directed cycles in a digraph, which was proven to be #W[1]-hard in [1], we define the problem of counting typed UCWs in digraphs:

| p#typUCW |
|---|
| **Input:** Digraph $G = (V, E)$, type $\theta$, and $k \in \mathbb{N}$ |
| **Parameter:** $k$ |
| **Output:** $\#\mathcal{U}_k[G, \theta]$ |

In Section 5, we use a graph construction from [1] to obtain

$$\mathsf{p\#typUCW} \leq^{T}_{fpt} \mathsf{p\#CC}.$$

Finally, in Section 6, we prove

$$\mathsf{p\#Clique} \leq^{T}_{fpt} \mathsf{p\#typUCW}.$$

In summary, our final reduction consists of the following intermediate reductions, starting from the #W[1]-hard problem p#Clique:

$$\mathsf{p\#Clique} \leq^{T}_{fpt} \mathsf{p\#typUCW} \leq^{T}_{fpt} \mathsf{p\#CC} \leq^{T}_{fpt} \mathsf{p\#wPCC} \leq_{fpt} \mathsf{p\#wMatch}.$$

In this chain, only the third reduction requires the introduction of weights.

[3] Thus, $\mathcal{U}_k[G, \theta] \neq \emptyset$ implies $k = \deg(\theta)$. We add the subscript $k$ only for clarity.

## 3 p#wPCC $\leq_{fpt}$ p#wMatch

This reduction follows from a well-known graph transformation, which has already been used for the study of the cover polynomial, see [15] and [13].

**Definition 4.** *Given a digraph $G = (V, E)$, replace each vertex $v \in V$ by vertices $v^{in}$ and $v^{out}$, and replace each $(u, v) \in E$ by the undirected edge $\{u^{out}, v^{in}\}$. We call the resulting graph the* split graph $S(G)$ *of* $G$.

Every split graph is bipartite with bipartition of its vertices into in- and out-vertices. Furthermore, the matchings of $S(G)$ are related to the path-cycle covers of $G$, as stated in the following lemma:

**Lemma 1.** *Let $G = (V, E)$ be a digraph and $S(G)$ be its split graph. For all $k \in \mathbb{N}$, there is a bijection $\mathcal{PC}_k[G] \simeq \mathcal{M}_k[S(G)]$.*

*Proof.* Every $k$-partial cycle cover $C \in \mathcal{PC}_k[G]$ is a subset $C \subseteq E$ of size $k$ such that $H = (V, C)$ has both indegree and outdegree upper-bounded by 1. Thus, the graph $S(H)$ is a matching of size $k$ in $S(G)$, and $S$ is injective if considered as $S : \mathcal{PC}_k[G] \to \mathcal{M}_k[S(G)]$.

$S$ is also surjective, since every $M \in \mathcal{M}_k[S(G)]$ can be transformed to some $C \in \mathcal{PC}_k[G]$ with $S(C) = M$ by identifying $v^{in}$ and $v^{out}$ for every $v \in V$ and orienting edges correspondingly. □

From this bijection, the reduction p#wPCC $\leq_{fpt}$ p#wMatch follows trivially.

## 4 p#CC $\leq_{fpt}^{T}$ p#wPCC

Let $G = (V, E)$ be a digraph and $k \in \mathbb{N}$. For $e, p \in \mathbb{N}$, denote by $c_G(e, p)$ the number of $e$-partial path-cycle covers of $G$ with $p$ paths. We wish to compute $c_G(k, 0)$, the number of $k$-partial cycle covers of $G$, using oracle access to p#wPCC.

By attaching to each vertex $v \in V$ a self-loop of weight $a$ and an edge of weight $b$ connected to a fresh vertex $u_v$, we obtain a new graph $G'$. The effect of this graph transformation on the partial path-cycle covers can be described using a graph polynomial $\gamma(G)$, which we define as follows:

**Definition 5.** *Let $G$ be a digraph. The edge-generating path-cycle polynomial $\gamma(G)$ is defined as*

$$\gamma(G; x) := \sum_{C \in \mathcal{PC}[G]} x^{|C|} w(C) = \sum_{k=0}^{n} x^k \left( \sum_{C \in \mathcal{PC}_k[G]} w(C) \right).$$

For any $k \in \mathbb{N}$, the weighted sum over $k$-partial path-cycle covers in $G$ is obviously equal to the $k$-th coefficient of $\gamma(G)$. We now show:

**Lemma 2.** *For any digraph $G = (V, E)$, it holds that*

$$\gamma(G'; x) = \sum_{C \in \mathcal{PC}[G]} x^{|C|} (1 + xb)^{\rho(C)} (1 + x(a + b))^{\iota(C)}.$$

*Proof.* For any path-cycle cover $D \in \mathcal{PC}[G']$, there is a unique $C \in \mathcal{PC}[G]$ with $C = D \cap E$. Defining $\mathcal{A}_C := \{D \in \mathcal{PC}[G'] \mid C = D \cap E\}$, we can partition $\mathcal{PC}[G']$ into classes $\{\mathcal{A}_C\}_{C \in \mathcal{PC}[G]}$ and obtain

$$\gamma(G'; x) = \sum_{C \in \mathcal{PC}[G]} \underbrace{\sum_{D \in \mathcal{A}_C} x^{|D|} w(D)}_{=:A_C}.$$

Consider $A_C$ for some $C \in \mathcal{PC}[G]$. Every $D \in \mathcal{A}_C$ can be decomposed as $D = C \dot{\cup} F$, where $F$ consists of gadget edges that can be appended independently to endpoints of paths or isolated vertices in $C$.

At each endpoint of a path, we can choose to extend the path using the edge of weight $b$, or not to extend. In total, this yields the factor $(1 + xb)^{\rho(C)}$. At each isolated vertex, we can choose to include the self-loop of weight $a$, the edge of weight $b$, or not to include a new edge. This yields the factor $(1 + x(a + b))^{\iota(C)}$. □

From now on, we will always choose $a = -b$. This ensures that extensions to *isolated* vertices cancel out in $\gamma(G'; x)$ and we thus obtain

$$\gamma(G'; x) = \sum_{C \in \mathcal{C}[G]} x^{|C|} (1 + xb)^{\rho(C)}.$$

This expression can be rewritten as

$$\begin{aligned} \gamma(G'; x) &= \sum_{e=0}^{n} x^e \sum_{p=0}^{n} c_G(e, p) \cdot (1 + xb)^p \\ &= \sum_{e,p=0}^{n} c_G(e, p) \cdot \left( x^e \sum_{i=0}^{p} \binom{p}{i} x^i b^i \right). \end{aligned}$$

With this identity, we are ready to prove the wanted reduction:

**Lemma 3.** *The problem* p#CC *admits an fpt Turing reduction to* p#wPCC.

*Proof.* Define $\alpha_j$ to be the $j$-th coefficient of $\gamma(G'; x)$. For sake of clarity, we decompose $j = i + (j - i)$, with $i$ denoting the number of "internal" edges, and $(j - i)$ denoting the number of "additional" edges, which come from gadgets. Using this decomposition, we can write

$$\alpha_j = \sum_{i=0}^{j} b^{j-i} \underbrace{\sum_{p=0}^{j} \binom{p}{j-i} c_G(i, p)}_{=:\alpha_{j,i}}.$$

Observe that $\alpha_j$ is in fact a polynomial $\alpha_j(b)$ of degree $\leq j$. Using oracle calls to p#wPCC, we can evaluate $\alpha_j(b)$ at $j$ different values while keeping $a = -b$. This allows us to interpolate $\alpha_j$ and to recover all its coefficients $\alpha_{j,i}$, for $0 \leq i \leq j$.

Using interpolation, compute $\alpha_{k,k}, \ldots, \alpha_{2k+1,k}$. This requires $\leq (2k+1)^2$ oracle calls and the parameter value of each oracle call is bounded by $2k+1$. By the definition of $\alpha_{j,i}$, we obtain the following system of linear equations:

$$\begin{pmatrix} \binom{0}{0} & \cdots & \binom{k}{0} \\ \vdots & \binom{i}{i} & \vdots \\ \binom{0}{k} & \cdots & \binom{k}{k} \end{pmatrix} \begin{pmatrix} c_G(k,0) \\ \vdots \\ c_G(k,k) \end{pmatrix} = \begin{pmatrix} \alpha_{k,k} \\ \vdots \\ \alpha_{2k+1,k} \end{pmatrix}.$$

This system is upper triangular, features only ones on its diagonal, and is thus invertible and can be solved for $c_G(k,0)$. □

*Remark 1.* In this reduction, the edge of weight $b$ at any vertex $v$ could safely be replaced by $b$ edges between $v$ and $b$ fresh vertices. However, it is not clear how to remove the weight $a$ from self-loops without introducing multiple self-loops.

## 5 p#typUCW $\leq^T_{fpt}$ p#CC

Let $G$ be a digraph, $k \in \mathbb{N}$, and $\theta$ be a type. We wish to count the UCWs of type $\theta$ and length $k$ in $G$, given oracle access to p#CC. Our reduction is based on a graph transformation from [1], which was used there to reduce the problem of counting typed cyclic walks to the problem of counting directed cycles.

**Definition 6.** *[1, Proof of Lemma 23] Let $G = (V, E)$ be a digraph and $l, m \in \mathbb{N}$. Define the graph $G_{l,m}$ as follows:*

1. *Replace each $v \in V$ by $L_v := \{(v, i, j) \mid 1 \leq i \leq l, 1 \leq j \leq m\}$ and add, for all* $1 \leq i < l$ and $1 \leq j, j' \leq m$, the edge $((v, i, j), (v, i+1, j'))$. *Call the resulting graph fragment "ladder at $v$".*
2. *Replace each $e = (u, v) \in E$ by $P_e := \{((u, l, j), (v, 1, j')) \mid 1 \leq j, j' \leq m\}$. Call the obtained edges "external edges at $e$".*

It follows from the construction that every cycle in $G_{l,m}$ must pass through ladders and external edges alternatingly. Thus, every cycle and every partial cycle cover in $G_{l,m}$ has length $il$ for some $i \in \mathbb{N}$. Given $k \in \mathbb{N}$, we partition the partial cycle covers $C \in \mathcal{C}_{kl}[G_{l,m}]$ into classes by associating with every such $C$ a particular UCW, its so-called *projection* $\pi(C)$. Here, $\pi$ is a function

$$\pi : \mathcal{C}_{kl}[G_{l,m}] \to \mathcal{U}_k[G].$$

**Definition 7.** *Let $C \in \mathcal{C}[G_{l,m}]$ with cycles $C_1 \cup \ldots \cup C_{\sigma(C)} = C$. For each $C_i$, define the CW $W_i$ by contracting, for each $v \in V$, the ladder at $v$ to the vertex $v$. Then define the* projection *of $C$ as $\pi(C) := \{W_1, \ldots, W_{\sigma(C)}\} \in \mathcal{U}_k[G]$.*

We observe that the number of partial cycle covers whose projection is $U$ depends only on the type $\theta_U$:

**Proposition 1.** *For any $l, m \in \mathbb{N}$, type $\theta$ and $U \in \mathcal{U}_k[G, \theta]$, the number of $C \in \mathcal{C}_{kl}[G_{l,m}]$ with $\pi(C) = U$ is equal to $\theta(m)^l$.*

*Proof.* Recall Def. 3 for the definition of $f_U(v)$. Since $\pi(C) = U$, for each $v \in V$, the cover $C$ contains $f_U(v)$ vertex-disjoint paths passing through the ladder at $v$. At each ladder, these paths can be chosen in $(m)^l_{f_U(v)}$ different ways, and paths can be chosen independently at different ladders. Once all ladder paths are fixed, the choice of external edges is also fixed. In total, this yields

$$\prod_{v \in V} (m)^l_{f_U(v)} = \theta(m)^l$$

different partial cycle covers $C \in \mathcal{C}_{kl}[G_{l,m}]$ with projection $\pi(C) = U$. □

This can be used to prove the wanted reduction:

**Lemma 4.** *The problem* p#typUCW *admits an fpt Turing reduction to* p#CC.

*Proof.* The proof follows [1, Lemma 23]. Let $\Theta_k$ be the set of types of degree $k$. Then, for $t := |\Theta_k|$, it clearly holds that $t \leq f(k)$ for some computable function $f$. Thus, there is some $m \leq g(k)$, with $g$ computable, such that $\theta_a(m) \neq \theta_b(m)$ holds for all types $\theta, \theta'$ of degree $k$ with $\theta \neq \theta'$. Compute such an $m$ and note that oracle calls to p#CC can be used to compute

$$\alpha_l := \#\mathcal{C}_{kl}[G_{l,m}].$$

Let $\beta_\theta$ be the number of UCWs of type $\theta$ in $G$. By Proposition 1, we can write

$$\alpha_l = \sum_{\theta \in \Theta_k} \beta_\theta \cdot \theta(m)^l.$$

We compute $\alpha_1, \ldots, \alpha_t$ with calls to p#CC and obtain the equation system

$$\begin{pmatrix} \theta_1(m)^1 & \ldots & \theta_t(m)^1 \\ \vdots & & \vdots \\ \theta_1(m)^t & \ldots & \theta_t(m)^t \end{pmatrix} \begin{pmatrix} \beta_{\theta_1} \\ \vdots \\ \beta_{\theta_t} \end{pmatrix} = \begin{pmatrix} \alpha_1 \\ \vdots \\ \alpha_t \end{pmatrix}. \tag{1}$$

This system has a Vandermonde matrix, allowing it to be solved for any $\beta_\theta$. □

## 6 p#Clique $\leq^T_{fpt}$ p#typUCW

Let $G = (V, E)$ be a graph and $k \in \mathbb{N}$ be fixed throughout this section. We describe how to compute the number of $k$-cliques in $G$ in fpt-time when given an oracle for p#typUCW, adapting the reduction in [1, Lemma 25] in large parts and reusing some of its notation where appropriate.

First, let $G'$ be the graph obtained from $G$ by replacing each edge by a pair of antiparallel edges, followed by adding a self-loop to each vertex. The number of $k$-cliques in $G$ is equal to the number of induced subgraphs in $G'$ which are isomorphic to the complete digraph $K = K_k := ([k], [k]^2)$.

Let $\mathcal{H} = \{H_1, \ldots, K\}$ be the set of graphs on $k$ vertices, where isomorphic graphs are identified, and the complete digraph $K$ is defined as above. Let $h :=$

$|\mathcal{H}|$ and $\mathcal{H}^- := \mathcal{H} \setminus \{K\}$. For $H \in \mathcal{H}$, let $x_H$ be the number of $U \subseteq V$ such that $G[U] \simeq H$. Our goal is to determine $x_K$, the number of $k$-cliques in $G$.

To this goal, let

$$\gamma^{(l)} := \#\mathcal{U}[G', (x)_l^k]$$

be the number of UCWs of type $(x)_l^k$ in $G'$. This value can be computed with an oracle call to p#typUCW, provided that $l \leq f(k)$. Writing

$$\beta_H^{(l)} := \#\mathcal{U}[H, (x)_l^k],$$

we observe that

$$\gamma^{(l)} = \sum_{H \in \mathcal{H}} x_H \beta_H^{(l)}.$$

We generate such equations for $\gamma^{(1)}, \ldots, \gamma^{(l)}$, where $l \leq f(k)$ will be determined later. This yields the system

$$\begin{pmatrix} \beta_A^{(1)} & \ldots & \beta_K^{(1)} \\ \vdots & \ddots & \vdots \\ \beta_A^{(l)} & \ldots & \beta_K^{(l)} \end{pmatrix} \begin{pmatrix} x_A \\ \vdots \\ x_K \end{pmatrix} = \begin{pmatrix} \gamma^{(1)} \\ \vdots \\ \gamma^{(l)} \end{pmatrix}. \tag{2}$$

A solution to (2) can be found in time polynomial in $l, h$ and $n$. While the system (2) does not feature full rank, we can show as in as in [1, Lemma 25] that there exists an $l \in \mathbb{N}$ such that the last column is linearly independent of all other columns. This implies that *all* solutions to (2) agree on their values for $x_K$.

To prove the existence of $l$, we first need a technical lemma. A similar statement was shown in [1] for cycles instead of UCWs, but using a fundamentally different proof approach:

**Lemma 5.** *For all graphs $H \in \mathcal{H}^-$, it holds that*

$$\lim_{\ell \to \infty} \frac{\beta_H^{(\ell)}}{\beta_K^{(\ell)}} = 0. \tag{3}$$

*Proof.* Let $\ell \in \mathbb{N}$. Recall Definition 6 and consider $K_{1,\ell}$, which is isomorphic to the complete graph $K_{k\ell}$.[4] Every $k\ell$-partial cycle cover in $K_{1,\ell}$ uses all vertices of the graph. Thus, its projection is some $U \in \mathcal{U}[K, (x)_\ell^k]$. By Proposition 1, every $U \in \mathcal{U}[K, (x)_\ell^k]$ is the projection of exactly $(\ell!)^k$ cycle covers in $K_{1,\ell}$.

Now let $(u, v) \notin E(H)$, let $\delta_{unres}^{(\ell)}$ denote the number of cycle covers in $K_{1,\ell}$, and let $\delta_{res}^{(\ell)}$ denote the number of cycle covers in $K_{1,\ell}$ that do not use any of the external edges at $(u, v)$. Since every $U \in \mathcal{U}[K, (x)_\ell^k]$ has the same number of cycle covers in $K_{1,\ell}$ projecting to it, we have

$$\frac{\beta_H^{(\ell)}}{\beta_K^{(\ell)}} \leq \frac{\delta_{res}^{(\ell)}}{\delta_{unres}^{(\ell)}}. \tag{4}$$

[4] Please note that, in this section, the notation $K_{1,\ell}$ will never refer to the complete bipartite graph with 1 and $\ell$ vertices on its sides.

Let $O_t$ be the $t \times t$ all-ones matrix. Then, $\delta_{res}^{(\ell)}$ is easily seen to be equal to $\mathrm{perm}(B \otimes O_\ell)$, where $\otimes$ denotes the Kronecker product and $B$ is defined to be $O_k$, but with one entry, say $B_{1,1}$, set to 0.

We number rows and columns of $B \otimes O_\ell$ from 1 to $k\ell$. Any permutation that contributes to $\mathrm{perm}(B \otimes O_\ell)$ maps every element from $[\ell]$ to an element from $[k\ell] \setminus [\ell]$, which gives $((k-1)\ell)_\ell$ choices for these elements, as opposed to $(k\ell)_\ell$ choices for an arbitrary permutation. Thus,

$$\frac{\delta_{res}^{(\ell)}}{\delta_{unres}^{(\ell)}} = \frac{((k-1)\ell)_\ell}{(k\ell)_\ell},$$

which can be upper-bounded by

$$\frac{((k-1)\ell)_\ell}{(k\ell)_\ell} = \prod_{i=0}^{\ell-1} \frac{k-1-i/\ell}{k-i/\ell} = \prod_{i=0}^{\ell-1} \left(1 - \frac{1}{k-i/\ell}\right) \leq \left(1 - \frac{1}{k}\right)^\ell.$$

Since $k$ is fixed, this value converges to 0 for $\ell \to \infty$. □

*Remark 2.* The correspondence between $\mathcal{U}[K, (x)_\ell^k]$ and $\mathcal{C}_{k\ell}[K_{k\ell}]$ established in the previous proof also shows that $\#\mathcal{U}[K, (x)_\ell^k] = \frac{(k\ell)!}{(\ell!)^k}$.

**Lemma 6.** *There exists a computable function $l = f(k)$ such that*

$$(\beta_K^{(1)}, \ldots, \beta_K^{(l)})^T \notin \mathrm{span}\{(\beta_H^{(1)}, \ldots, \beta_H^{(l)})^T \mid H \in \mathcal{H}^-\}.$$

*Proof.* Having proved Lemma 5 for UCWs, the proof continues exactly as in [1, Lemma 25]. □

**Acknowledgements.** We are grateful to the anonymous referees, whose comments improved the presentation of this paper.

## References

1. Flum, J., Grohe, M.: The parameterized complexity of counting problems. SIAM Journal on Computing, 538–547 (2002)
2. Valiant, L.G.: The complexity of computing the permanent. Theoretical Computer Science 8(2), 189–201 (1979)
3. Temperley, H.N.V., Fisher, M.E.: Dimer problem in statistical mechanics - an exact result. Philosophical Magazine 68(6), 1478–6435 (1961)
4. Kasteleyn, P.: The statistics of dimers on a lattice: I. The number of dimer arrangements on a quadratic lattice. Physica 27(12), 1209–1225 (1961)
5. Vadhan, S.P.: The complexity of counting in sparse, regular, and planar graphs. SIAM J. Comput. 31(2), 398–427 (2001)
6. Jerrum, M., Sinclair, A.: Approximating the permanent. SIAM J. Comput. 18(6), 1149–1178 (1989)
7. Makowsky, J.A.: Algorithmic uses of the Feferman-Vaught theorem. Annals of Pure and Applied Logic 126(1-3), 159–213 (2004); Provinces of logic determined. Essays in the memory of Alfred Tarski. Parts I, II and III

8. Courcelle, B.: The monadic second-order logic of graphs. I. Recognizable sets of finite graphs. Information and Computation 85(1), 12–75 (1990)
9. Galluccio, A., Loebl, M.: On the theory of Pfaffian orientations. I. Perfect matchings and permanents. Electronic Journal of Combinatorics 6 (1998)
10. Frick, M.: Generalized Model-Checking over Locally Tree-Decomposable Classes. In: Alt, H., Ferreira, A. (eds.) STACS 2002. LNCS, vol. 2285, pp. 632–644. Springer, Heidelberg (2002)
11. Vassilevska, V., Williams, R.: Finding, minimizing, and counting weighted subgraphs. In: Proceedings of the 41st Annual ACM Symposium on Theory of Computing, STOC 2009, pp. 455–464. ACM, New York (2009)
12. Björklund, A., Husfeldt, T., Kaski, P., Koivisto, M.: Counting Paths and Packings in Halves. In: Fiat, A., Sanders, P. (eds.) ESA 2009. LNCS, vol. 5757, pp. 578–586. Springer, Heidelberg (2009)
13. Bläser, M., Curticapean, R.: The Complexity of the Cover Polynomials for Planar Graphs of Bounded Degree. In: Murlak, F., Sankowski, P. (eds.) MFCS 2011. LNCS, vol. 6907, pp. 96–107. Springer, Heidelberg (2011)
14. Bläser, M., Dell, H.: Complexity of the Cover Polynomial. In: Arge, L., Cachin, C., Jurdziński, T., Tarlecki, A. (eds.) ICALP 2007. LNCS, vol. 4596, pp. 801–812. Springer, Heidelberg (2007)
15. Chung, F.R.K., Graham, R.L.: On the cover polynomial of a digraph. J. Combin. Theory Ser. B 65, 273–290 (1995)

# Computing Directed Pathwidth in $O(1.89^n)$ Time

Kenta Kitsunai, Yasuaki Kobayashi, Keita Komuro,
Hisao Tamaki, and Toshihiro Tano

Department of Computer Science, Meiji University
Kawasaki, Japan 214-8571

**Abstract.** We give an algorithm for computing the directed pathwidth of a digraph with $n$ vertices in $O(1.89^n)$ time. This is the first algorithm with running time better than the straightforward $O^*(2^n)$. As a special case, it computes the pathwidth of an undirected graph in the same amount of time, improving on the algorithm due to Suchan and Villanger which runs in $O(1.9657^n)$ time.

## 1 Introduction

The *pathwidth* [2,15] of an undirected graph $G$ is defined as follows. A *path-decomposition* $G$ is a sequence $\{X_i\}$, $1 \leq i \leq t$, of vertex sets of $G$ that satisfies the following three conditions:

1. $\bigcup_{1\leq i \leq t} X_i = V(G)$,
2. for each edge $\{u, v\}$ of $G$, there is some $i$, $1 \leq i \leq t$ such that $u, v \in X_i$, and
3. for each $v \in V(G)$, the set of indices $i$ such that $v \in X_i$ is contiguous, i.e., is of the form $\{i \mid a \leq i \leq b\}$.

The *width* of a path-decomposition $\{X_i\}$, $1 \leq i \leq t$, is $\max_{1\leq i\leq t} |X_i| - 1$ and the *pathwidth* of $G$ is the smallest integer $k$ such that there is a path-decomposition of $G$ whose width is $k$.

The *directed path-decomposition* of a digraph $G$ is defined analogously. A sequence $\{X_i\}$, $1 \leq i \leq t$, of vertex sets is a directed path-decomposition of $G$ if, together with conditions 1 and 3 above, the following condition 2' instead of condition 2 is satisfied:

2'. for each directed edge $(u, v)$ of $G$, there is a pair $i$, $j$ of indices such that $i \leq j$, $u \in X_i$, and $v \in X_j$.

The directed pathwidth of $G$ is defined similarly to the pathwidth of an undirected graph. According to Barát [1], the notion of directed pathwidth was introduced by Reed, Thomas, and Seymour around 1995.

For an undirected graph $G$, let $\hat{G}$ denote the digraph obtained from $G$ by replacing each edge $\{u, v\}$ by a pair of directed edges $(u, v)$ and $(v, u)$. Then, condition 2 for $G$ implies condition 2' for $\hat{G}$ and, conversely, condition 2' for $\hat{G}$ together with condition 3 implies condition 2 for $G$. Therefore, a directed path-decomposition

D.M. Thilikos and G.J. Woeginger (Eds.): IPEC 2012, LNCS 7535, pp. 182–193, 2012.

of $\hat{G}$ is a path-decomposition of $G$ and vice versa. Thus, the problem of computing the pathwidth of an undirected graph is a special case of the problem of computing the directed pathwidth of a digraph. On the other hand, directed pathwidth is of interest since some problems, such as Directed Hamiltonicity, are polynomial time solvable on digraphs with bounded directed pathwidth [10] although not necessarily on those whose underlying graphs have bounded pathwidth. Directed pathwidth is also studied in the context of search games [1,20].

Computing pathwidth is NP-hard [11] even for bounded degree planar graphs [14], chordal graphs [8], cocomparability graphs [9] and bipartite distance hereditary graphs [13] (although it is polynomial time solvable for permutation graphs [5], cographs [6], and circular-arc graphs [17]). Consequently, computing directed pathwidth is NP-hard even for digraphs whose underlying graphs lie in these classes. On the positive side, pathwidth is fixed parameter tractable [16] with running time linear in $n$ [3]. In contrast, it is open whether directed pathwidth is fixed parameter tractable. Recent work of one of the present authors [19] shows that directed pathwidth admits an XP algorithm, that is, an algorithm with running time $n^{O(k)}$, where $k$ is the directed pathwidth of the given digraph with $n$ vertices.

Without parameterization, both problems can be solved in $O^*(2^n)$ time, where $n$ is the number of vertices and the $O^*$ notation hides polynomial factors, using Bellman-Held-Karp style dynamic programming for vertex ordering problems [4]. Suchan and Villanger [18] improved the running time for pathwidth to $O(1.9657^n)$ and also gave an additive constant approximation of pathwidth in $O(1.89^n)$ time. On the other hand, no algorithm faster than $O^*(2^n)$ time was known for directed pathwidth before the present work.

Our result is as follows.

**Theorem 1.** *The direct pathwidth of a digraph with $n$ vertices can be computed in $O(1.89^n)$ time.*

Our algorithm can be viewed as one based on Bellman-Held-Karp style dynamic programming. For each $U \subseteq V(G)$, let $N^-(U)$ denote the set of in-neighbors of $U$. Given a positive integer $k$ and a digraph $G$, we build the collection of "feasible" subsets of $V(G)$, where $U \subseteq V(G)$ is considered feasible if $G[U \cup N^-(U)]$ has a directed path-decomposition of width $\leq k$ whose last subset $X_t$ contains $N^-(U)$ (for a more precise definition of feasibility see Section 2). To get a non-trivial bound on the number of subsets $U$, we adopt a strategy essentially due to Suchan and Villanger [18]. Since either $U$, $N^-(U)$, or $V(G) \setminus (U \cup N^-(U))$ has cardinality at most $n/3$, the hope is that we may be able to obtain a non-trivial upper bound on the number of relevant subsets $U$ using the well-known bound $2^{H(\alpha)n}$ on the number of subsets of $V(G)$ with cardinality at most $\alpha n$, where $H(x)$ is the binary entropy function, with $\alpha = \frac{1}{3}$. Note that $2^{H(1/3)} < 1.89$. The difficulty, as observed in [18], is that there can be an exponential number of subsets $U$ with $N^-(U) = S$ for a fixed $S$. The larger constant 1.9657 or the relaxation to additive constant approximation in [18] comes from the need to deal with this problem.

A key to overcoming this difficulty is the following observation whose undirected version is used in [18]. Suppose we are to decide whether the directed pathwidth of $G$ is at most $k$ and $S$ is a vertex set of $G$ with $|S| < k - d$. Let $\mathcal{C}$ be the set of strongly connected components of $G[V(G) \setminus S]$ with cardinality greater than $d$. For each subset $\mathcal{A}$ of $\mathcal{C}$, we are able to define at most one subset $U$ in a "canonical form" such that $N^-(U) = S$, $U$ contains all components in $\mathcal{A}$, but disjoint from all components in $\mathcal{C} \setminus \mathcal{A}$. Although there are variants of $U$ that satisfy these conditions, it can be shown that only the canonical one is needed in our dynamic programming computation. Thus, the number of relevant subsets $U$ with $N^-(U) = S$ is bounded by $2^{|\mathcal{C}|} \leq 2^{\frac{n}{d+1}}$. See Lemma 5 for a more formal treatment.

Our result is established by two algorithms, which we call LARGE-WIDTH and SMALL-WIDTH. Algorithm LARGE-WIDTH deals with the case where $k \geq (\frac{1}{3} + \delta)n$, $\delta$ being a small constant, while algorithm SMALL-WIDTH deals with the other case. In algorithm LARGE-WIDTH, the slack of $\delta n$ allows us to bound the number of subsets $U$ with $N^-(U) = S$ for each $S$ with $|S| \leq n/3$ by $2^{1/\delta}$. In algorithm SMALL-WIDTH, we force a slack of $d$ where $d$ is a large enough constant: we record only those feasible sets $U$ with $|N^-(U)| < k - d$. To process those subsets $U$ with $|N^-(U)| \geq k - d$, we use an algorithm based on the XP algorithm in [19], which runs in $n^{d+(1)}$ time and either decides that the directed pathwidth of $G$ is at most $k$, decides that $U$ is irrelevant, or produces some proper superset $W$ with $|N^-(W)| < k - d$, which can safely replace $U$ in the search. See Lemma 12 for details.

The rest of this paper is organized as follows. In Section 2 we define basic concepts and prove some lemmas needed by our algorithms. In Section 3, we state and analyze algorithm LARGE-WIDTH. In Section 4, we review the XP algorithm given in [19] to prepare for the next section. Then, in Section 5, we state and analyze algorithm SMALL-WIDTH. Finally, in Section 6, we combine the two algorithms to prove Theorem 1.

## 2 Preliminaries

Let $G$ be a digraph. We use the standard notation: $V(G)$ is the set of vertices of $G$, $E(G)$ is the set of edges of $G$, and $G[U]$, where $U \subseteq V(G)$, is the subgraph of $G$ induced by $U$. For each vertex $v \in V(G)$, we denote by $N_G^-(v)$ the set of in-neighbor of $v$, i.e., $N_G^-(v) = \{u \in V(G) \setminus \{v\} | (u, v) \in E(G)\}$. For each subset $U$ of $V(G)$, we denote by $N_G^-(U)$ the set of in-neighbors of $U$, i.e., $N_G^-(U) = \bigcup_{v \in U} N_G^-(v) \setminus U$. When $G$ is clear from the context, we drop $G$ from this notation. We also use the notation $\tilde{U} = V(G) \setminus (U \cup N^-(U))$ where $G$ is implicit.

We call a sequence $\sigma$ of vertices of $G$ *non-duplicating* if each vertex of $G$ occurs at most once in $\sigma$. We denote by $\Sigma(G)$ the set of all non-duplicating sequences of vertices of $G$. For each sequence $\sigma \in \Sigma(G)$, we denote by $V(\sigma)$ the set of vertices constituting $\sigma$ and by $|\sigma| = |V(\sigma)|$ the length of $\sigma$.

For each pair of sequences $\sigma, \tau \in \Sigma(G)$ such that $V(\sigma) \cap V(\tau) = \emptyset$, we denote by $\sigma\tau$ the sequence in $\Sigma(G)$ that is $\sigma$ followed by $\tau$. If $\sigma' = \sigma\tau$ for some $\tau$, then we say that $\sigma$ is a *prefix* of $\sigma'$ and that $\sigma'$ is an *extension* of $\sigma$; we say that $\sigma$ is a *proper* prefix of $\sigma'$ and that $\sigma'$ is a *proper* extension of $\sigma$ if $\tau$ is nonempty.

Let $G$ be a digraph and $k$ a positive integer. We say $\sigma \in \Sigma(G)$ is *k-feasible* for $G$ if $|N_G^-(\sigma')| \leq k$ for every prefix $\sigma'$ of $\sigma$. We say that $\sigma$ is *strongly k-feasible* for $G$ if moreover $\sigma$ is a prefix of a $k$-feasible sequence $\tau$ with $V(\tau) = V(G)$. We may drop the reference to $G$ and say $\sigma$ is $k$-feasible (or strongly $k$-feasible) when $G$ is clear from the context.

For each $U \subseteq V(G)$, we say that $U$ is $k$-feasible (strongly $k$-feasible) if there is a $k$-feasible (strongly $k$-feasible) sequence $\sigma$ with $V(\sigma) = U$.

The directed vertex separation number of digraph $G$, denoted by $\text{dvsn}(G)$, is the minimum integer $k$ such that $V(G)$ is $k$-feasible.

It is known that the directed pathwidth of $G$ equals $\text{dvsn}(G)$ for every digraph $G$ [20](see also [12] for the undirected case). Based on this fact, we work on the directed vertex separation number in the remaining of this paper.

The following lemma formulates a straightforward reasoning used twice in the sequel.

**Lemma 1.** *Let $U \subseteq V(G)$, let $X \subseteq V(G) \setminus U$ be such that $N^-(X) \subseteq U \cup N^-(U)$, and let $W = U \cup X$. Suppose that $W$ is k-feasible and $U$ is strongly k-feasible. Then, $W$ is also strongly k-feasible.*

*Proof.* Since $U$ is strongly $k$-feasible, there is a sequence $U = U_0, U_1, \ldots, U_h = V(G)$ of $k$-feasible sets of vertices such that $U_{i-1} \subseteq U_i$ and $|U_i| = |U_{i-1}| + 1$ for $1 \leq i \leq h$. Let $W_i = U_i \cup X$ for $0 \leq i \leq h$. Since $N^-(X) \subseteq U \cup N^-(U)$ by assumption, we have $N^-(W_i) \subseteq N^-(U_i)$ and hence $|N^-(W_i)| \leq k$, for $0 \leq i \leq h$. Since $W_{i-1} \subseteq W_i$ and either $|W_i| = |W_{i-1}|$ or $|W_i| = |W_{i-1}| + 1$ for $1 \leq i \leq h$, a straightforward induction shows that $W_i$ for each $0 \leq i \leq h$ is $k$-feasible. Therefore, $W_0 = W$ is strongly $k$-feasible. □

We call $U \subseteq V(G)$ a *full set* (with respect to $G$), if there is no $v \in N^-(U)$ with $N^-(v) \subseteq U \cup N^-(U)$. For each $U$, there is a unique superset of $U$ that is a full set, which we denote by $\text{fullset}(U)$. Indeed, $\text{fullset}(U)$ is defined by

$$\text{fullset}(U) = U \cup \{v \in N^-(U) \mid N^-(v) \subseteq U \cup N^-(U)\}.$$

Note that $N^-(\text{fullset}(U)) \subseteq N^-(U)$.

**Lemma 2.** *Let $U$ be an arbitrary subset of $V(G)$. If $U$ is k-feasible then so is* $\text{fullset}(U)$. *Moreover, if $U$ is strongly k-feasible then so is* $\text{fullset}(U)$.

*Proof.* Let $X = \text{fullset}(U) \setminus U$ and let $v_1, v_2, \ldots$, and $v_h$, where $h = |X|$, be the elements of $X$ listed in an arbitrary order. Let $U_0 = U$ and $U_i = U_{i-1} \cup \{v_i\}$, $1 \leq i \leq h$. Since $N^-(v_i) \subseteq U \cup N^-(U)$ for $1 \leq i \leq h$ by the definition of $\text{fullset}(U)$, we have $N^-(U_i) = N^-(U_{i-1}) \setminus \{v_i\}$ for $1 \leq i \leq h$. Therefore, the first claim of the lemma follows.

Suppose next that $U$ is strongly $k$-feasible. By the first claim, $\text{fullset}(U)$ is $k$-feasible. Since $N^-(X) \subseteq U \cup N^-(U)$, by Lemma 1 it follows that $\text{fullset}(U) = U \cup X$ is strongly $k$-feasible. □

Let $U \subseteq V(G)$, $H = G[V(G) \backslash N^-(U)]$. Observe that, for each strongly connected component $C$ of $H$, either $C \subseteq U$ or $C \subseteq \tilde{U}$, as otherwise $N^-(U)$ would contain a vertex in $C$.

An undirected counterpart of the following lemma is called the *component push rule* in [18].

**Lemma 3.** *Let $U$ and $H$ be as above and let $C$ be a strongly connected component of $H$ such that $C \subseteq \tilde{U}$, $N^-(C) \subseteq U \cup N^-(U)$, and $|N^-(U)| + |C| \leq k+1$. If $U$ is $k$-feasible then $U \cup C$ is $k$-feasible. Moreover, if $U$ is strongly $k$-feasible then $U \cup C$ is strongly $k$-feasible.*

*Proof.* Let $W = U \cup C$. First observe that $N^-(W) = N^-(U)$: $N^-(W) \subseteq N^-(U)$ since $N^-(C) \subseteq U \cup N^-(U)$ by assumption; $N^-(U) \subseteq N^-(W)$ since $C \cap N^-(U) = \emptyset$. The first claim of the lemma is trivial since $|N^-(U \cup A)| \leq |N^-(U)| + |C| - 1 \leq k$ for every nonempty subset $A$ of $C$.

Suppose next that $U$ is strongly $k$-feasible. Since $N^-(C) \subseteq U \cup N^-(U)$ by assumption, by Lemma 1 it follows that $W$ is strongly $k$-feasible. □

For each digraph $H$ let $\mathcal{C}(H)$ denote the set of all strongly connected components of $H$. Consider the natural partial ordering $\prec$ on $\mathcal{C}(H)$: $C \prec D$ if and only if $H$ contains a directed path from a vertex in $C$ to a vertex in $D$. For each $U \subseteq V(G)$ with $|N^-(U)| \leq k$, we denote by $\text{push}_k(U)$ the superset of $U$ defined as follows. Let $H = G[V(G) \setminus N^-(U)]$, $s = k - |N^-(U)| + 1$, and define

$$\mathcal{P} = \{C \in \mathcal{C}(H) \mid C \subseteq \tilde{U}, |C| \leq s, \text{and there is no } D \in \mathcal{C}(H) \text{ with } D \subseteq \tilde{U}, |D| > s, \text{and } D \prec C\}.$$

Then, we let $\text{push}_k(U) = U \cup \bigcup_{C \in \mathcal{P}} C$.

By a repeated application of Lemma 3, we obtain the following lemma.

**Lemma 4.** *Let $U \subseteq V(G)$. If $U$ is $k$-feasible then so is $\text{push}_k(U)$. Moreover, if $U$ is strongly $k$-feasible then so is $\text{push}_k(U)$.*

Following [18] we use component push rules not only as an algorithmic technique but also as a tool for analysis. The following lemma formalizes this latter aspect.

**Lemma 5.** *Let $S \subseteq V(G)$ with $|S| \leq k$. Then, the number of vertex sets $U \subseteq V(G)$ with $N^-(U) = S$ and $\text{push}_k(U) = U$ is at most $2^{\frac{n}{s+1}}$ where $s = k - |S| + 1$.*

*Proof.* Fix $S \subseteq V(G)$ with $|S| \leq k$ and let $s = k - |S| + 1$. Let $H = G[V(G) \setminus S]$ and $\mathcal{C}_s$ the set of strongly connected components of $H$ with cardinality strictly larger than $s$. Note that, for each $U \subseteq V(G)$ with $N^-(U) = S$ and a strongly connected component $C$ of $H$, either $C \subseteq U$ or $C \subseteq \tilde{U}$. Thus, each such $U$ induces a bipartition $(\mathcal{A}_U, \mathcal{B}_U)$ of $\mathcal{C}_s$: $C \subseteq U$ for each $C \in \mathcal{A}_U$ and $C \subseteq \tilde{U}$ for each $C \in \mathcal{B}_U$.

We call a bipartition $(\mathcal{A}, \mathcal{B})$ of $\mathcal{C}_s$ *valid* if there is no pair $C \in \mathcal{B}$ and $D \in \mathcal{A}$ such that $C \prec D$ where $\prec$ is the partial ordering defined before on the set of strongly connected components of $H$. For some $U$ to exist such that $\mathcal{A}_U = \mathcal{A}$ and $\mathcal{B}_U = \mathcal{B}$, $(\mathcal{A}, \mathcal{B})$ must be a valid partition of $\mathcal{C}_s$ because, for each $U$ with $N^-(U) = S$, we cannot have two strongly connected components $C$ and $D$ of $H$ with $C \prec D$, $C \subseteq \tilde{U}$, and $D \subseteq U$.

For each valid bipartition $(\mathcal{A}, \mathcal{B})$ of $\mathcal{C}_s$, there is exactly one $U \subseteq V(G)$ such that $N^-(U) = S$, $\text{push}_k(U) = U$, and $(\mathcal{A}_U, \mathcal{B}_U) = (\mathcal{A}, \mathcal{B})$. Indeed, such $U$ is defined by $\bigcup_{C \in \mathcal{A} \cup \mathcal{D}} C$, where $\mathcal{D}$ is the set of strongly connected components $C$ of $H$ such that $|C| \leq s$ and there is no $D \in \mathcal{B}$ with $D \prec C$.

Since $|\mathcal{C}_s| \leq \frac{n}{s+1}$, the number of bipartitions of $\mathcal{C}_s$ is at most $2^{\frac{n}{s+1}}$ and this bound applies to the number of vertex sets $U$ with $N^-(U) = S$ and $\text{push}_k(U) = U$. □

Let $H(x) = -x \log x - (1-x) \log(1-x)$, $0 < x < 1$, denote the binary entropy function. We freely use the following well-known bound on the number of subsets of bounded cardinality.

**Proposition 1.** (see [7], for example) *Let $S$ be a set of $n$ elements and let $0 < \alpha \leq \frac{1}{2}$. Then the number of subsets of $S$ with cardinality at most $\alpha n$ is at most $2^{H(\alpha)n}$.*

## 3 Algorithm LARGE-WIDTH

Given an integer $k > 0$ and a digraph $G$ with $n$ vertices, Algorithm LARGE-WIDTH decides whether $\text{dvsn}(G) \leq k$ in the following steps.

The algorithm uses function $f^*$ defined as follows. Define $f : 2^{V(G)} \to 2^{V(G)}$ by $f(U) = \text{fullset}(\text{push}_k(U))$. Since $U \subseteq f(U)$, there is some $h$ for each $U$ such that $f^h(U) = f^{h+1}(U)$. We denote this $f^h(U)$ by $f^*(U)$. Note that if $W = f^*(U)$ for some $U$ then $W = \text{fullset}(W) = \text{push}_k(W)$.

1. Set $\mathcal{U}_1 := \{\{v\} \mid v \in V(G), |N^-(v)| \leq k\}$ and $\mathcal{U}_i := \emptyset$ for $2 \leq i \leq n$.
2. Repeat the following for $i = 1, 2, \ldots, n-1$.
   (a) For each $U \in \mathcal{U}_i$ and for each $v \in V(G) \setminus U$ with $|N^-(U \cup \{v\})| \leq k$, let $W = f^*(U \cup \{v\})$ and reset $\mathcal{U}_j := \mathcal{U}_j \cup \{W\}$ where $j = |W|$.
3. If $\mathcal{U}_n$ is not empty then answer "YES"; otherwise answer "NO".

**Lemma 6.** *Algorithm LARGE-WIDTH is correct.*

*Proof.* In the following proof, we abuse the notation and let $\mathcal{U}_i$, $1 \leq i \leq n$, denote the final value of the program variable $\mathcal{U}_i$. To justify the answer at step 3, we show that $V(G)$ is $k$-feasible if and only if $\mathcal{U}_n$ is nonempty.

First note that, by Lemmas 2 and 4, every member of $\mathcal{U}_i$ is $k$-feasible for $1 \leq i \leq n$. Therefore if $\mathcal{U}_n \neq \emptyset$ then $V(G)$ is $k$-feasible.

To prove the other direction, suppose $V(G)$ is $k$-feasible. We show that $\mathcal{U}_n \neq \emptyset$. Let $t$ be the largest $i$ such that $\mathcal{U}_i$ contains a strongly $k$-feasible vertex set. If $t = n$ then we are done. Suppose not and let $U \in \mathcal{U}_t$ be strongly $k$-feasible. Then, by definition, there must be some $v \notin U$ such that $U \cup \{v\}$ is strongly $k$-feasible. Then, by Lemmas 2 and 4, $W = f^*(U \cup \{v\})$ is strongly $k$-feasible. Since the algorithm puts $W$ in $\mathcal{U}_j$ where $j = |W| > t$, this is a contradiction and therefore it must be that $t = n$. □

We analyze the complexity of this algorithm for particular values of $k$ for which this algorithm is intended.

**Lemma 7.** *Let $\delta > 0$ be fixed. For $k > (\frac{1}{3} + \delta)n$, algorithm LARGE-WIDTH runs in $O^*(2^{H(\frac{1}{3})n})$ time.*

*Proof.* Let $\mathcal{W}$ denote the set $\bigcup_{2\leq i\leq n} \mathcal{U}_i$ after the algorithm execution. We consider the following subsets of $\mathcal{W}$:

$$\mathcal{W}_1 = \{U \in \mathcal{W} \mid |U| \leq n/3\},$$
$$\mathcal{W}_2 = \{U \in \mathcal{W} \mid |U \cup N^-(U)| \geq 2n/3\}, \text{and}$$
$$\mathcal{W}_3 = \mathcal{W} \setminus (\mathcal{W}_1 \cup \mathcal{W}_2).$$

By Proposition 1, we have $|\mathcal{W}_1| \leq 2^{H(\frac{1}{3})n}$. To bound $|\mathcal{W}_2|$, observe that $U =$ fullset$(U)$ for each member $U$ of $\mathcal{W}$ and hence, for each $S \subseteq V(G)$, there is at most one $U \in \mathcal{W}$ such that $U \cup N^-(U) = S$. Since the number of $S$ with $|S| \geq 2n/3$ is at most $2^{H(\frac{1}{3})n}$, we have $|\mathcal{W}_2| \leq 2^{H(\frac{1}{3})n}$.

For each $U \in \mathcal{W}_3$, we have $|N^-(U)| < n/3$. For each $S \subseteq V(G)$ with $|S| < n/3$, let $\mathcal{W}_S = \{U \in \mathcal{W}_3 \mid N^-(U) = S\}$. Since $U = \text{push}_k(U)$ for each $U \in \mathcal{W}$, we have, by Lemma 5, $|\mathcal{W}_S| \leq 2^{\frac{n}{k-|S|+2}} \leq 2^{\frac{1}{\delta}}$ for each $S$. Since the number of $S \subseteq V(G)$ with $|S| < n/3$ is at most $2^{H(\frac{1}{3})n}$, we have $|\mathcal{W}_3| = O(2^{H(\frac{1}{3})n})$. Since each member of $\mathcal{W}$ is involved in a computation of $n^{O(1)}$ time, the total running time of the algorithm is $O^*(2^{H(\frac{1}{3})n})$. □

## 4 XP Algorithm

We review the XP algorithm for directed pathwidth due to Tamaki [19] which is an essential ingredient in algorithm SMALL-WIDTH.

**Theorem 2.** [19] *Given a positive integer $k$ and a digraph $G$ with $n$ vertices and $m$ edges, it can be decided in $O(mn^{k+1})$ time whether $V(G)$ is $k$-feasible.*

The algorithm claimed in this theorem is based on the natural search tree consisting of all $k$-feasible sequences in $\Sigma(G)$. The running time is achieved by pruning this search tree of potentially factorial size into one with $O(n^{k+1})$ search nodes. The following lemma is used to enable this pruning. We say that a proper extension $\tau$ of $\sigma \in \Sigma(G)$ is *non-expanding* if $|N^-(\tau)| \leq |N^-(\sigma)|$.

**Lemma 8.** (Commitment Lemma [19]) *Let $\sigma$ be a strongly $k$-feasible sequence in $\Sigma(G)$ and let $\tau$ be a shortest non-expanding $k$-feasible extension of $\sigma$, that is,*

1. *$|N^-(V(\tau))| \leq |N^-(V(\sigma))|$, and*
2. *$|N^-(V(\tau'))| > |N^-(V(\sigma))|$ for every $k$-feasible proper extension $\tau'$ of $\sigma$ with $|\tau'| < |\tau|$.*

*Then, $\tau$ is strongly $k$-feasible.*

Suppose sequence $\sigma$ is in the search tree and has a non-expanding $k$-feasible extension $\tau$. Then the commitment lemma allows $\sigma$ to "commit to" this descendant $\tau$: we may remove from the search tree all the descendants of $\sigma$ with length $|\tau|$ but $\tau$. It is shown in [19] that the resulting search tree contains $O(n^{k+1})$ sequences.

To adapt this result for our purposes, we need some details of the pruned search tree. Let $\sigma$ and $\tau$ be two $k$-feasible sequences of the same length. We say that $\sigma$ is *preferable to* $\tau$ if either $|N^-(V(\sigma))| < |N^-(V(\tau))|$ or $|N^-(V(\sigma))| = |N^-(V(\tau))|$ and $\sigma < \tau$ in the lexicographic ordering on $\Sigma(G)$ based on some fixed total order on $V(G)$. We say $\sigma$ *suppresses* $\tau$, if $\sigma$ is preferable to $\tau$ and there is some common prefix $\sigma'$ of $\sigma$ and $\tau$ such that $\sigma$ is a shortest non-expanding $k$-feasible extension of $\sigma'$.

Let $S_i$, $1 \leq i \leq n$, denote a set of $k$-feasible sequence with length $i$ defined inductively as follows. Each member of $S_i$ will represent a node in our search tree at level $i$.

1. $S_1 = \{v \mid |N^-(v)| \leq k\}$.
2. For $1 \leq i < n$, let $T_{i+1} = \{\sigma v \mid \sigma \in S_i, v \in V(G) \setminus V(\sigma), \text{and } |N^-(V(\sigma) \cup \{v\})| \leq k\}$. We let $S_{i+1}$ be the set of elements of $T_{i+1}$ not suppressed by any elements of $T_{i+1}$.

To analyze the size of each set $S_i$, [19] assigns a sequence $\text{sgn}(\sigma)$, called the *signature* of $\sigma$, to each $k$-feasible sequence $\sigma$ as follows.

Call a non-expanding $k$-feasible extension $\tau$ of $\sigma$ *locally shortest*, if no proper prefix of $\tau$ is a non-expanding extension of $\sigma$. We define $\text{sgn}(\sigma)$ inductively as follows.

1. If $\sigma$ is empty then $\text{sgn}(\sigma)$ is empty.
2. If $\sigma$ is nonempty and is a locally shortest non-expanding extension of some prefix of $\sigma$, then $\text{sgn}(\sigma) = \text{sgn}(\tau)$, where $\tau$ is the shortest prefix of $\sigma$ with the property that $\sigma$ is a locally shortest non-expanding $k$-feasible extension of $\tau$.
3. Otherwise $\text{sgn}(\sigma) = \text{sgn}(\sigma')v$, where $v$ is the last vertex of $\sigma$ and $\sigma = \sigma' v$.

**Lemma 9.** [19] *For each $i$, $1 \leq i \leq n$, if $\sigma$ and $\tau$ are two distinct elements of $S_i$ then neither* $\text{sgn}(\sigma)$ *nor* $\text{sgn}(\tau)$ *is the prefix of the other.*

The following properties of the pruned search tree follow from this lemma.

**Lemma 10.** *Suppose $\sigma \in S_{|\sigma|}$ is a non-expanding extension of a singleton sequence $v$. Then, $\sigma$ is the only extension of $v$ in $S_{|\sigma|}$.*

*Proof.* Let $\sigma_0 = v$ and, for $i = 1, \ldots$, let $\sigma_i$ be the shortest prefix of $\sigma$ that is a non-expanding extension of $\sigma_{i-1}$. We see that $\sigma_1$ is well-defined since $\sigma$ is a non-expanding extension of $\sigma_0$ and that there is some $j > 0$ such that $\sigma_i$ for $1 \leq i \leq j$ is well-defined and $\sigma_j = \sigma$. Observe that, for $1 \leq i \leq j$, $\sigma_i$ is the locally shortest non-expanding extension of $\sigma_{i-1}$ and $\sigma_{i-1}$ is the shortest prefix $\alpha$ of $\sigma_i$

such that $\sigma_i$ is the locally shortest non-expanding extension of $\alpha$. Therefore, by the definition of the signature, we have $\mathrm{sgn}(\sigma_i) = v$ for $0 \leq i \leq j$.

Suppose now that $S_{|\sigma|}$ contains some extension $\tau$ of $v$ that is distinct from $\sigma$. By Lemma 9, $\mathrm{sgn}(\tau)$ must start with some $u \neq v$. This is possible only if $\tau$ has a prefix $\tau'$ that is a non-expanding extension of the empty sequence, which implies that $N^-(V(\tau')) = \emptyset$. This is impossible since if such $\tau'$ exists, then some prefix of $\sigma$ would be suppressed contradicting the presence of $\sigma$ in $S_{|\sigma|}$. □

**Lemma 11.** *Let $1 \leq j \leq n$ and let $h$ be the minimum value of $|N^-(V(\sigma))|$ over all sequences $\sigma$ in $\bigcup_{1 \leq i \leq j} S_i$. Then, we have $|S_i| \leq n^{k-h}$ for $1 \leq i \leq j$.*

*Proof.* In [19], it is shown that $|\mathrm{sgn}(\sigma)| \leq |N^-(V(\sigma))|$ for each $k$-feasible sequence $\sigma$. The proof therein in fact shows that $|\mathrm{sgn}(\sigma)| \leq |N^-(V(\sigma))| - h$ for $\sigma \in \bigcup_{1 \leq i \leq j} S_i$. By Lemma 9, we have $|S_i| \leq n^{k-h}$ for $1 \leq i \leq j$. □

## 5 Algorithm SMALL-WIDTH

Fix $\epsilon > 0$ and fix an integer $d > 1/\epsilon$. The following description of our algorithm depends on $d$. We assume $k$, an input to the algorithm, satisfies $k > d$; otherwise the algorithm in Theorem 2 runs in $n^{O(1)}$ time.

Our strategy is to record only those sets $U$ with $|N^-(U)| < k - d$ and $U = \mathrm{push}_k(U)$ in our computation. For each $S$ with $|S| < k - d$, by Lemma 5, the number of $U$ such that $N^-(U) = S$ and $U = \mathrm{push}_k(U)$ is at most $2^{\epsilon n}$.

To process $U$ with $|N^-(U)| \geq k - d$, we use the XP algorithm in [19]. The following lemma is at the heart of our algorithm.

**Lemma 12.** *There is an algorithm that, given a $k$-feasible vertex set $U \subseteq V(G)$ with $k - d \leq |N^-(U)| \leq k$, runs in $n^{d+O(1)}$ time and either*

1. *proves that $U$ is strongly $k$-feasible,*
2. *proves that $U$ is not strongly $k$-feasible, or*
3. *produces some proper superset $W$ of $U$ with $|N^-(W)| < k - d$ such that $U$ is strongly $k$-feasible if and only if $W$ is strongly $k$-feasible.*

*Proof.* Let $G/U$ denote the graph obtained from $G$ by shrinking $U$ into one vertex $v_U$: $V(G/U) = (V(G) \setminus U) \cup \{v_U\}$ and $E(G/U) = E_1 \cup E_2 \cup E_3$ where

$$\begin{aligned} E_1 &= \{(u, v) \in E(G) \mid u, v \in V(G) \setminus U\}, \\ E_2 &= \{(v_U, u) \mid u \notin U, \exists v \in U : (v, u) \in E(G)\}, \text{ and} \\ E_3 &= \{(u, v_U) \mid u \notin U, \exists v \in U : (u, v) \in E(G)\}. \end{aligned}$$

We run the algorithm in Theorem 2 on $H = G/U$, setting the root of the search tree to be the singleton sequence consisting of $v_U$. Suppose first that the search tree does not contain any sequence $\sigma$ with $|N_H^-(V(\sigma))| < k - d$. In this case, the size of the search tree is $O(n^{d+1})$ by Lemma 11 and therefore the running time of this algorithm execution is $n^{d+O(1)}$. If the search tree contains a sequence $\sigma$ with $V(\sigma) = V(H)$, we know that $\{v_U\}$ is strongly $k$-feasible in $H$ and hence $U$ is strongly $k$-feasible in $G$. Otherwise we conclude that $U$ is not strongly $k$-feasible in $G$.

Suppose next that the search tree contains a sequence $\sigma$ with $|N_H^-(V(\sigma))| < k-d$: let $\sigma$ be the shortest among such. We expand the search tree in the breadth-first manner and stop the search as soon as $\sigma$ is encountered. By Lemma 10, $\sigma$ is the only sequence with length $|\sigma|$ in the search tree. Since $\sigma$ is obtained from the singleton sequence $v_U$ via a series of commitments (see the proof of Lemma 10), $\sigma$ is strongly $k$-feasible in $H$ if and only if $\{v_U\}$ is strongly $k$-feasible in $H$. In terms of the original graph $G$, we let $W = U \cup (V(\sigma) \setminus \{v_U\})$ and see that $U$ is strongly $k$-feasible if and only if $W$ is strongly $k$-feasible. We return $W$ for the third case of the algorithm. The size of the search tree is $O(n^{d+1})$, and hence the running time is $n^{d+O(1)}$, in this case as well. □

Algorithm SMALL-WIDTH, given $G$ and $k$, decides if $\mathrm{dvsn}(G) \leq k$ in the following steps.

1. Set $\mathcal{U}_1 := \{\{v\} \mid v \in V(G), |N^-(v)| \leq k\}$ and $\mathcal{U}_i := \emptyset$ for $2 \leq i \leq n$.
2. Repeat the following for $i = 1, 2, \ldots n-1$.
   For each $U \in \mathcal{U}_i$ and for each $v \in V(G) \setminus U$ with $|N^-(U \cup \{v\})| \leq k$, let $U' = U \cup \{v\}$ and do the following.
   (a) If $|N^-(U')| < k-d$ then let $W = \mathrm{push}_k(U')$ and reset $\mathcal{U}_j := \mathcal{U}_j \cup \{W\}$, where $j = |W|$.
   (b) If $|N^-(U')| \geq k-d$ then apply the algorithm of Lemma 12 to $G$ and $U'$.
      i. If $U'$ is found strongly $k$-feasible, then stop the entire algorithm answering "YES".
      ii. If $U'$ is found not strongly $k$-feasible, then do nothing.
      iii. If a proper superset $W$ of $U'$ with $|N^-(W)| < k-d$ is returned, then reset $\mathcal{U}_j := \mathcal{U}_j \cup \{\mathrm{push}_k(W)\}$, where $j = |\mathrm{push}_k(W)|$.
3. If $\mathcal{U}_n$ is nonempty then answer "YES"; otherwise answer "NO".

**Lemma 13.** *Algorithm SMALL-WIDTH is correct.*

*Proof.* As a consequences of Lemma 12, the answer "YES" at step (2-b-i) is correct. So suppose that the algorithm terminates at step 3. In the following proof, we abuse the notation and let $\mathcal{U}_i$, $1 \leq i \leq n$, denote the final value of the program variable $\mathcal{U}_i$. To justify the answers at step 3, we prove that $V(G)$ is $k$-feasible if and only if $\mathcal{U}_n$ is nonempty.

First observe that every member of $\bigcup_{1 \leq i \leq n} \mathcal{U}_i$ is $k$-feasible. This is certainly true for every member of $\mathcal{U}_1$ and the algorithm ensures only $k$-feasible sets are put into $\mathcal{U}_j$, for any $j$, at subsequent steps. Note that we rely on Lemma 4 when we put $\mathrm{push}_k(X)$ for some $k$-feasible set $X$ into $\mathcal{U}_j$ at steps (2-a) and (2-b-iii). Therefore, if $\mathcal{U}_n$ is nonempty, then $V(G)$ is $k$-feasible.

To prove the other direction, suppose $V(G)$ is $k$-feasible. We show that $\mathcal{U}_n$ is nonempty.

Let $t$ be the maximum $i$ such that $\mathcal{U}_i$ contains a strongly $k$-feasible vertex set. If $t = n$ we are done. Suppose not and let $U \in \mathcal{U}_t$ be strongly $k$-feasible. Since $U$ is strongly $k$-feasible, there must be some $v \in V(G) \setminus U$ such that $U' = U \cup \{v\}$

is strongly $k$-feasible. If $|N^-(U')| < k - d$ then the algorithm puts $\text{push}_k(U')$ in $\mathcal{U}_j$ for some $j > t$, a contradiction since $\text{push}_k(U')$ is strongly $k$-feasible by Lemma 4. Otherwise, the algorithm of Lemma 12 is applied to $U'$. Under our assumptions that case (2-b-i) does not happen and that $U'$ is strongly $k$-feasible, the algorithm must return some proper superset $W$ of $U'$ with $|N^-(W)| < k-d$ that is strongly $k$-feasible. A contradiction since the algorithm puts $\text{push}_k(W)$ in $\mathcal{U}_j$ for some $j > t$.

In either case, we have a contradiction and conclude that $t = n$. □

**Lemma 14.** *For $k \le n/2$, algorithm SMALL-WIDTH runs in $O(2^{(H(k/n)+\epsilon)n})$ time.*

*Proof.* It is clear that the algorithm spends $n^{d+O(1)} = n^{O(1)}$ time to process each element in $\mathcal{U} = \bigcup_{1 \le i \le n} \mathcal{U}_i$. Therefore, it is enough to show that $|\mathcal{U}| \le O(2^{(H(k/n)+\epsilon')n})$ for some $\epsilon' < \epsilon$.

Obviously, we have $|\mathcal{U}_1| \le n$. Let $U \in \mathcal{U} \setminus \mathcal{U}_1$. Since $U = \text{push}_k(X)$ for some $X$, we have $U = \text{push}_k(U)$. By Lemma 5, for each $S \subseteq V(G)$ with $|S| < k - d$, the number of $U$ with $N^-(U) = S$ and $\text{push}_k(U) = U$ is at most $2^{\epsilon' n}$ where $\epsilon' = \frac{1}{d+3} < \epsilon$. The desired upper bound on $|\mathcal{U}|$ follows since the number of vertex sets $S$ with $|S| < k - d$ is smaller than $2^{H(k/n)n}$. □

## 6 Combining the Two Algorithms

Theorem 1 immediately follows from a combination of the two algorithms given in previous sections. Observe $2^{H(1/3)} < 1.89$ and choose positive $\delta$ and $\epsilon$ so that $2^{H(1/3+\delta)+\epsilon} < 1.89$. If $k > (\frac{1}{3} + \delta)n$, we apply algorithm LARGE-WIDTH; otherwise, we apply algorithm SMALL-WIDTH. From Lemmas 7 and 14, we see that the running time of the algorithm is $O(1.89^n)$ in both cases.

## References


1. Barát, J.: Directed path-width and monotonicity in digraph searching. Graphs and Combinatorics 22(2), 161–172 (2006)
2. Bodlaender, H.L.: A Tourist Guide Through Treewidth. Acta Cybernetica 11, 1–23 (1993)
3. Bodlaender, H.L.: A linear-time algorithm for finding tree-decompositions of small treewidth. SIAM Journal on Computing 25, 1305–1317 (1996)
4. Bodlaender, H.L., Fomin, F.V., Kratsch, D., Thilikos, D.: A Note on Exact Algorithms for Vertex Ordering Problems on Graphs. Theory of Computing Systems 50(3), 420–432 (2012)
5. Bodlaender, H.L., Kloks, T., Kratsch, D.: Treewidth and Pathwidth of Permutation Graphs. In: Lingas, A., Carlsson, S., Karlsson, R. (eds.) ICALP 1993. LNCS, vol. 700, pp. 114–125. Springer, Heidelberg (1993)
6. Bodlaender, H.L., Möhring, R.H.: The Pathwidth and Treewidth of Cographs. SIAM Journal on Discrete Mathematics 6(2), 181–188 (1992)
7. Fomin, F.V., Kratsch, D.: Exact Exponential Algorithms. Springer (2010)

8. Gusted, J.: On the pathwidth of chordal graphs. Discrete Applied Mathematics 45(3), 233–248 (1993)
9. Habib, M., Möhring, R.H.: Treewidth of cocomparability graphs and a new order-theoretic parameter. Order 11(1), 44–60 (1994)
10. Johnson, T., Robertson, N., Seymour, P.D., Thomas, R.: Directed tree-width. Journal of Combinatorial Theory, Series B 82(1), 138–154 (2001)
11. Kashiwabara, T., Fujisawa, T.: NP-completeness of the problem of finding a minimum-clique-number interval graph containing a given graph as a subgraph. In: Proceedings of International Symposium on Circuits and Systems, pp. 657–660 (1979)
12. Kinnersley, G.N.: The vertex separation number of a graph equals its path-width. Information Processing Letters 42(6), 345–350 (1992)
13. Kloks, T., Bodlaender, H.L., Müller, H., Kratsch, D.: Computing treewidth and minimum fill-in: All you need are the minimal separators. In: Proceedings of the 1st Annual European Symposium on Algorithms, pp. 260–271 (1993)
14. Monien, B., Sudborough, I.H.: Min cur is NP-complete for edge weighted trees. Theoretical Computer Science 58(1-3), 209–229 (1988)
15. Robertson, N., Seymour, P.D.: Graph minors. I. Excluding a forest. Journal of Combinatorial Theory, Series B 35(1), 39–61 (1983)
16. Robertson, N., Seymour, P.D.: Graph minors VIII The disjoint paths peoblem. Journal of Combinatorial Theory, Series B 63, 65–110 (1995)
17. Suchan, K., Todinca, I.: Pathwidth of Circular-Arc Graphs. In: Brandstädt, A., Kratsch, D., Müller, H. (eds.) WG 2007. LNCS, vol. 4769, pp. 258–269. Springer, Heidelberg (2007)
18. Suchan, K., Villanger, Y.: Computing Pathwidth Faster Than $2^n$. In: Chen, J., Fomin, F.V. (eds.) IWPEC 2009. LNCS, vol. 5917, pp. 324–335. Springer, Heidelberg (2009)
19. Tamaki, H.: A Polynomial Time Algorithm for Bounded Directed Pathwidth. In: Kolman, P., Kratochvíl, J. (eds.) WG 2011. LNCS, vol. 6986, pp. 331–342. Springer, Heidelberg (2011)
20. Yang, B., Cao, Y.: Digraph searching, directed vertex separation and directed pathwidth. Discrete Applied Mathematics 156(10), 1822–1837 (2008)

# MSOL Restricted Contractibility to Planar Graphs*

James Abello[1], Pavel Klavík[2], Jan Kratochvíl[2], and Tomáš Vyskočil[2]

[1] DIMACS Center for Discrete Mathematics and Theorethical Computer Science, Rutgers University, Piscataway, NJ
abello@dimacs.rutgers.edu

[2] Department of Applied Mathematics, Faculty of Mathematics and Physics, Charles University, Prague
{klavik,honza,whisky}@kam.mff.cuni.cz

**Abstract.** We study the computational complexity of graph planarization via edge contraction. The problem CONTRACT asks whether there exists a set $S$ of at most $k$ edges that when contracted produces a planar graph. We give an FPT algorithm in time $\mathcal{O}(n^2 f(k))$ which solves a more general problem $P$-RESTRICTEDCONTRACT in which $S$ has to satisfy in addition a fixed inclusion-closed MSOL formula $P$.

For different formulas $P$ we get different problems. As a specific example, we study the $\ell$-subgraph contractability problem in which edges of a set $S$ are required to form disjoint connected subgraphs of size at most $\ell$. This problem can be solved in time $\mathcal{O}(n^2 f'(k,l))$ using the general algorithm. We also show that for $\ell \geq 2$ the problem is NP-complete. And it remains NP-complete when generalized for a fixed genus (instead of planar graphs).

## 1 Introduction

Graph visualization techniques are thoroughly studied. In many applications visual understanding of the graph under consideration is important or required. It is commonly accepted that edge crossings make a plane drawing of a graph less clear, and thus the goal is to avoid them, or reduce their number. It is now well-known that one can decide fast whether crossings can be avoided at all, as planarity testing is linear time decidable [10], while determining the minimum number of crossings needed to draw a non-planar graph is NP-hard [8]. Several variants of planar visualization of graphs have been then considered and explored, including simultaneous embeddings, book-embeddings, embeddings on surfaces of higher genus, etc.

Marx et al. [12] considered planarization of a graph by removing its vertices. Another possible way to planarize a graph is by contracting some of its edges.

* The first author acknowledges support of Special focus on Algorithmic Foundations of the Internet, NSF grant #CNS-0721113 and mgvis.com http://mgvis.com. The latter three authors acknowledge support of ESF Eurogiga project GraDR as GAČR GIG/11/E023.

D.M. Thilikos and G.J. Woeginger (Eds.): IPEC 2012, LNCS 7535, pp. 194–205, 2012.

If the number of contracted edges is not limited, every connected graph can trivially be contracted into a single vertex, and thus becomes planar. A graph is *k-contractible* if the number of contracted edges is limited by a number $k$. If $k$ is a part of the input, testing $k$-contractibility is NP-complete [1]. Polynomial-time algorithms are known if one asks about contraction to a particular fixed planar graph (so called $H$-contractibility); for nice overviews see [4,11]. In this paper, we present a fixed-parameter tractable algorithm for contractability to planar graphs.

**Restricted Contractibility.** We address the following more general problem. A subset of edges $S$ is called a *planarizing set* if by contracting $S$ in $G$ the graph becomes planar. We want to find a planarizing set $S$ of size at most $k$ that satisfies an additional restriction: a monadic second-order logic (MSOL) formula $P$ fixed for the problem.[1]

| | |
|---|---|
| **Problem:** | $P$-RESTRICTEDCONTRACT |
| **Input:** | An undirected graph $G$ and an integer $k$. |
| **Output:** | Is there a planarizing set $S \subseteq E(G)$ of size at most $k$ satisfying $P$ that when contracted produces a planar graph? |

An MSOL formula $P$ is *inclusion-closed* if for every $S$ satisfying $P$ also every $S' \subseteq S$ satisfies $P$. This is a natural condition which, for instance, allows to look for an inclusion minimal matching in $G$. The main result of this paper is:

**Theorem 1.** *For every inclusion-closed MSOL formula P, the problem P-*RESTRICTEDCONTRACT *is solvable in time $\mathcal{O}(n^2 f(k))$ where n is the number of vertices of G.*

Our algorithm is a modification of a quadratic-time FPT algorithm for crossing number of Grohe [9]. The most significant change is a different proof of Lemma 1. We cannot use the same approach as in [9] because $k$-contractible graphs do not have bounded genus [5] which is essential in [9].

For a trivial formula $P$ that is true for every set of edges, we get the $k$-contractibility problem considered above. We note that $k$-contractibility was independently proved to be solvable in time $\mathcal{O}(n^{2+\varepsilon}\bar{f}(k))$ for every $\varepsilon > 0$ in a recent paper of Golovach et al. [5]. We obtained independently our results at a similar date [6]. The algorithm described here uses similar techniques but improves the time complexity and in fact solves a more general problem.

**$\ell$-subgraph Contractibility.** For different formulas $P$, we get problems different from $k$-contractibility having new specific properties. As one example, we describe a problem called $\ell$-subgraph contractibility. A graph is called *$\ell$-subgraph contractible* if there exists a planarizing set $S$ such that its edges form disjoint connected subgraphs of size at most $\ell$. We show that this property of the planarizing set $S$ can be described in MSOL.

[1] More precisely, we have different formulas $P_k(e_1, \ldots, e_k)$ for each $k$ where $S = \{e_1, \ldots, e_k\}$. So the length of the formula may depend on $k$.

| | |
|---|---|
| **Problem:** | $\ell$-SUBCONTRACT |
| **Input:** | An undirected graph $G$ and an integer $k$. |
| **Output:** | Is $G$ $\ell$-subgraph contractible by a set $S$ having at most $k$ edges? |

This problem is closely related to graph clustering: The mentioned subgraphs of size at most $\ell$ are non-trivial clusters of the graph and are such that the resulting cluster graph is planar. In comparison to $k$-contractibility, the contracted edges have to be more equally distributed in $G$, and thus the contractions do not change the graph too much.

From a graph drawing perspective this approach offers a drawing such that all crossings happen in disjoint areas nearby the clusters and the rest of the meta-drawing is crossing-free. Such a meta-drawing resembles well the original graph and can be well grasped by a glance from the distance. The local crossings get inspected by taking a magnifying glass for particular clusters.

**Proposition 1.** *For every fixed $\ell$, the problem $\ell$-SUBCONTRACT can be solved in time $\mathcal{O}(n^2 f'(k))$ where $n$ is the number of vertices.*

If $\ell = 1$, the problem is solvable in linear time as it becomes just planarity testing. For $\ell \geq 2$, we prove:

**Theorem 2.** *For $\ell \geq 2$, the problem $\ell$-SUBCONTRACT is NP-complete.*

We note that in the case of $\ell = 2$, the planarizing set $S$ is a matching in $G$. In the Conclusions Section, we show how the hardness result generalizes to surfaces of higher genus.

### Definitions and Notation

For a graph $G$, we denote by $V(G)$ its vertices and by $E(G)$ its edges (or simply $V$ and $E$, when the graph is clear from the context). We denote by $G \circ e$ the (multi)graph obtained by contracting an edge $e$ in $G$ (in contractions, we keep parallel edges and loops; they do not influence planarity of $G \circ e$ anyway).

For a set of edges $S$, we denote a graph created from $G$ by contracting all edges of $S$ by $G \circ S$. We call $S \subseteq E$ *a planarazing set* of $G$ if $G \circ S$ is a planar graph. We say that $G$ is *$k$-contractible* if there exists a planarizing set $S$ of size at most $k$.

## 2 Restricted Contractibility Is Fixed-Parameter Tractable

In this section, we show that the problem $P$-RESTRICTEDCONTRACT for a fixed inclusion-closed MSOL formula $P$ is fixed-parameter tractable with parameter $k$. Namely, we exhibit an algorithm, to solve this parameterized version, that runs in time $\mathcal{O}(n^2 \cdot f(k))$.

A basic structure of our algorithm is based on the following well-known idea invented by Grohe [9]. If the graph has small tree-width, we can solve the problem in a "brute-force way" using MSOL. If the tree-width is large, we find an embedded large hexagonal grid and produce a smaller graph to which we can apply the procedure recursively.

## 2.1 Definitions

We first introduce notation similar to that of Grohe's in [9].

**Topological Embeddings.** A topological embedding $h : G \hookrightarrow H$ of $G$ into $H$ consists of two mappings: $h_V : V(G) \to V(H)$ and $h_E : E(G) \to P(H)$, where $P(H)$ denotes paths in $H$. These mappings must satisfy the following properties:

- The mapping $h_V$ is injective, distinct vertices of $G$ are mapped to distinct vertices of $H$.
- For distinct edges $e$ and $f$ of $G$, the paths $h_E(e)$ and $h_E(f)$ are distinct, do not share internal vertices and share possibly at most one endpoint.
- If $e = uv$ is an edge of $G$ then $h_V(u)$ and $h_V(v)$ are the endpoints of the path $h_E(e)$. If $w$ is vertex of $G$ different from $u$ and $v$ then path $h_E(e)$ does not contain the vertex $h_V(w)$.

It is useful to notice that there exists a topological embedding $h : G \hookrightarrow H$, if there exists a subdivision of $G$ which is a subgraph of $H$. For a subgraph $G' \subseteq G$, denote by $h \restriction G'$ the restriction of $h$ to $G'$. For simplicity, we use the term *embeddings* instead of topological embeddings.

**Hexagonal Grid.** We define recursively the hexagonal grid $H_r$ of radius $r$ (see Figure 1). The graph $H_1$ is a hexagon (the cycle of length six) The graph $H_{r+1}$ is obtained from $H_r$ by adding $6r$ hexagonal faces around $H_r$ as indicated in Figure 1.

The nested *principal cycles* $C^1, \dots, C^r$ are called the boundary cycles of $H_1, \dots, H_r$ . From the inductive construction of $H_r$, $H_k$ is obtained from $H_{k-1}$ by adding $C^k$ and connecting it to $C^{k-1}$. A *principal subgrid* $H_r^s$ where $s \leq r$ denotes the subgraph of $H_r$ isomorphic to $H_s$ and bounded by the principal cycle $C^s$ of $H_r$.

**Flat Topological Embeddings.** Let $H$ be a subgraph of a graph $G$. An *$H$-component* $C$ of $G$ is

- either a connected component of $G \setminus H$ together with the edges connecting $C$ to $H$, or
- an edge $e$ of $G \setminus H$ with both endpoints in $H$.

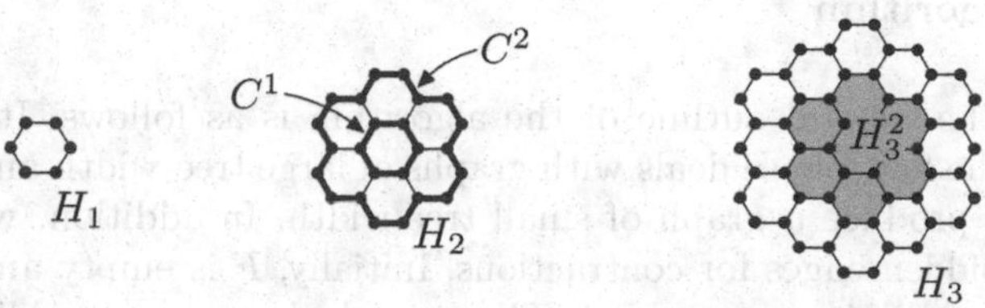


**Fig. 1.** Hexagonal grids $H_1$, $H_2$ and $H_3$. Inside $H_2$, the principal cycles $C^1$ and $C^2$ are depicted in bold. Inside $H_3$, the principal subgrid $H_3^2$ is shown.

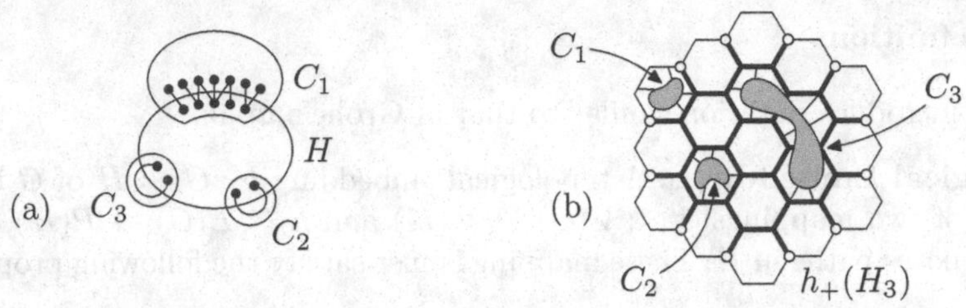


**Fig. 2.** (a) $H$-components $C_1$, $C_2$ and $C_3$ of $G$. (b) A subgraph $h_+(H_3)$ consisting of $h(H_3)$ and three proper components $C_1$, $C_2$ and $C_3$ having attachments to inner vertices of $h(H_3)$ (highlighted in bold). The embedding $h$ is not flat since $C_3$ obstructs planarity.

The endpoints of edges of $C$ contained in $H$ are called the *vertices of attachment* of $C$. Figure 2a illustrates the notion of *H-components*.

Let $G$ be a graph and let $h : H_r \hookrightarrow G$ be an embedding of a hexagonal grid in $G$. An $h(H_r)$-component $C$ is called *proper* if $C$ has at least one vertex of attachment in $h(H_r) \setminus h(C^r)$, namely, the component is attached to an inner vertex of the grid. Let $h_+(H_r)$ denote the union of $h(H_r)$ with all proper $h(H_r)$-components. Notice that the proper $h(H_r)$-components may be obstructions to the planarity of $h_+(H_r)$. The embedding $h$ is called a *flat embedding* if $h_+(H_r)$ is a planar graph. For an example, see Figure 2b.

**Tree-width.** For a graph $G$, its tree-width is the smallest integer $k$ such that $G$ is subgraph of a $k$-tree. This parameter describes how much is $G$ "similar" to a tree. For our purposes, we use tree-width as a black box in our algorithm. The following two properties of tree-width are crucial.

**Theorem 3 (Robertson and Seymour [15], Boadlaender [2], Perkovič and Reed [14]).** *For every $s \geq 1$, there is an integer $w \geq 1$ and a linear-time algorithm that, given a graph $G$, either (correctly) recognizes that the tree-width of $G$ is at most $w$ or returns an embedding $h : H_s \hookrightarrow G$.*

**Theorem 4 (Courcelle [3]).** *For every graph $G$ of tree-width at most $t$ and every MSOL formula $\varphi$, there exists an algorithm that in time $\mathcal{O}(n \cdot g(t))$, where $n$ number of vertices of $G$, decides the formula $\varphi$ on $G$.*

## 2.2 The Algorithm

**Overview.** The general outline of the algorithm is as follows. It proceeds in two phases. The first phase deals with graphs of large tree-width and repeatedly modifies $G$ to produce a graph of small tree-width. In addition, we keep a set $F \subset E$ of forbidden edges for contractions. Initially, $F$ is empty and during the modification some edges are added. The second phase uses an MSOL formula and Courcelle's Theorem 4 to solve $P$-RESTRICTEDCONTRACT on this graph.

**Phase I.** We first prove the following lemma, which states that in an embedded large hexagonal grid $H_s$ into $G$, we either find a smaller flat hexagonal grid $H_r$, or else $G$ is not $k$-contractible. This lemma represents the most significant difference from the paper of Grohe [9].

**Lemma 1.** *Let $G$ be a $k$-contractible graph. For every $r \geq 1$, there exists $s \geq 1$ such that for every embedding $h : H_s \hookrightarrow G$ there is some subgrid $H_r \subseteq H_s$ such that $h \restriction H_r$ is a flat embedding.*

*Proof.* For given $r$ and $k$, we fix $s$ and $t$ large enough as follows: We choose $s \approx 2kt$ so that $H_s$ contains $2k+1$ *disjoint subgrids* $H_{t_1}, \ldots, H_{t_{2k+1}}$ of radius $t$ and let $H'_{t_i}$, formerly denoted by $H^{t-2}_{t_i}$, be a subgrid of $H_{t_i}$ without two outermost layers. We choose $t \approx 7rk$ so that each $H'_{t_i}$ contains $7k+1$ *disjoint subgrids* $H_{t_i,r_1}, \ldots, H_{t_i,r_{7k+1}}$ of radius $r$.[2] In this way, we get a hierarchy of nested subgrids:

$$H_s \supset H_{t_i} \supset H'_{t_i} \supset H_{t_i,r_j}, \qquad 1 \leq i \leq 2k+1, \quad 1 \leq j \leq 7k+1.$$

We argue next that in this hierarchy, some $H_{t_i,r_j}$ is $h \restriction H_{t_i,r_j}$ a flat embedding.

Since we are assuming that $G$ is $k$-contractible, we can fix a matching planarizing set $S$ and consider one subgrid $H_{t_i}$. Let a *cell* be an $h$-image of a hexagon of $H'_{t_i}$. We call an $h(H'_{t_i})$-component *bad* if it contains an edge from $S$. A cell is considered *bad* if it contains an edge of $S$ or if there is at least one bad $h(H'_{t_i})$-component attached to the cell. Since bad cells have some obstructions to planarity attached to them we will exhibit some grid $H_{t_i,r_j}$ such that its embeddiing $h(H_{t_i,r_j})$ avoids all bad cells.

To proceed, call an $h(H'_{t_i})$-component $C$ *large* if the vertices of attachment of $C$ are not contained in one cell (as an example, in Figure 2b, components $C_1$ and $C_2$ are not large but $C_3$ is large). Large $h(H'_{t_i})$-components posses the following useful properties which we prove afterwards in a series of claims.

1. If an $h(H'_{t_i})$-component is large, then we can embed $K_{3,3}$ into $h_+(H_{t_i})$. This means that there must be some $H_{t_i}$ having no large $h(H'_{t_i})$-component, otherwise the graph would not be $k$-contractible.
2. On the other hand, if a bad $h(H'_{t_i})$-component is not large, it can produce at most seven bad cells.This implies that $h(H'_{t_i})$ must have a number of bad cells bounded by $7k$ and therefore for some $j$, $h(H_{t_i,r_j})$ is a flat embedding.

*Claim.* Let $C$ be a large $h(H'_{t_i})$-component. Then we can embed $K_{3,3}$ into $h_+(H_{t_i})$ such that $K^-_{3,3} := K_{3,3} - e$ is embedded directly into the grid $h(H_{t_i})$.

*Proof.* Instead of a tedious formal proof, we illustrate the main idea in Figure 3. If $C$ is large, it has two vertices $u$ and $v$ not contained in one cell. Thus there exists a path $P$ going across the grid "between" $u$ and $v$. Using $C$ as a "bridge" from $u$ to $v$, we can cross $P$ by another path across the grid. These two paths together with two outer layers of $h(H_{t_i})$ allow us to embed $K_{3,3}$ into $h_+(H_{t_i})$ such that $K^-_{3,3}$ is embedded into $h(H_{t_i})$. ◇

[2] Actually just $s \approx \sqrt{2kt}$ and $t \approx \sqrt{7kr}$ would be sufficient.

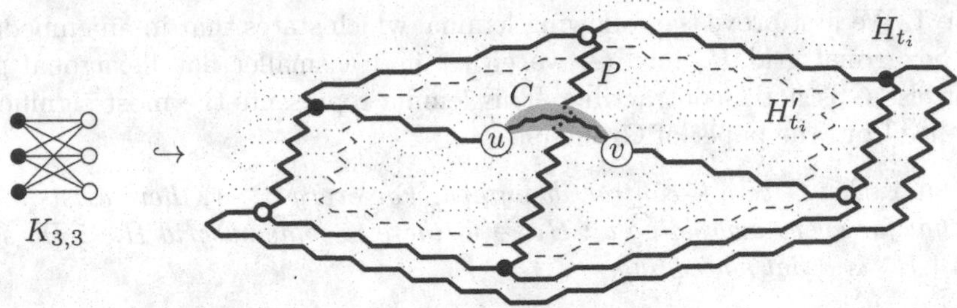


**Fig. 3.** The component $C$ acts as a bridge allowing two non-crossing paths across the grid. Thus we can embed $K_{3,3}$ into $h(H_{t_i})$.

*Claim.* There is some $H_{t_i}$ such that there is no large $h(H'_{t_i})$-component.

*Proof.* According to the above claim, if $H_{t_\ell}$ contains a large $h(H'_{t_\ell})$-component, we can embed $K_{3,3}$ into $h_+(H_{t_\ell})$. Since $G \circ S$ is a planar graph, this embedding of $K_{3,3}$ has to be contracted by $S$. To contract this $K_{3,3}$, there has to be an edge $e \in S$ incident with $h(K^-_{3,3})$ and therefore incident with $h(H_{t_\ell})$ directly. But since $|S| \leq k$ and we have $2k+1$ disjoint grids, there is some $H_{t_i}$ not having any edge $S$ incident with $h(H_{t_i})$. Thus there is no large $h(H'_{t_i})$-component. ◇

*Claim.* For this $H_{t_i}$, $h(H'_{t_i})$ contains at most $7k$ bad cells.

*Proof.* Let $e \in S$. Since no $h(H'_{t_i})$-component $C$ is large, it is attached to at most seven cells. Therefore, if $e \in C$, we get at most seven bad cells. If $e$ belongs to a cell directly, we get two bad cells. Since $|S| \leq k$, we get at most $7k$ bad cells. ◇

To conclude the proof of Lemma 1, by the pigeon-hole principle one of $h(H_{t_i,r_1}), \ldots, h(H_{t_i,r_{7k+1}})$ containing no bad cells forces the existence of one $H_{t_i,r_j}$ such that $h_+(H_{t_i,r_j})$ has no edges of $S$ and it is planar; in other words $h(H_{t_i,r_j})$ is flat. □

The next lemma shows that a small part of $h_+(H_r)$ is never contracted by a minimal set $S$. Let a *kernel* $K$ of $h_+(H_r)$ denote the $h$-image of the central principal cycle $h(C^1)$ together with the $h(H_r)$-components attached only to $h(C^1)$.

**Lemma 2.** *Let $G$ be a $k$-contractible graph and let $S$ be a minimal planarizing set of $G$. Let $h : H_r \hookrightarrow G$, $r \geq 2k+3$ be a flat embedding and let $K$ be a kernel of $h_+(H_r)$. Then the edges of $G$ incident with the vertices of $K$ do not belong to $S$.*

Details of the proof are omited due to space limitations but it is very similar to the one in Grohe [9] (see full version for details).

Recall that $F$ is the set of forbidden edges to contract. If the graph $G$ is $k$-contractible by a planarizing set $S$ such that $S \cap F = \emptyset$, we say that $G$ is $(k, F)$-contractible. Proceeding with our algorithm, we use Lemma 2 to modify the input $G$ and $F$ to a smaller graph $G'$ and an extended set $F'$ as follows: $G' := G \circ K$ (the kernel $K$ is contracted into a single vertex), $F'$ is equal to $F$ together with edges connecting $K$ to $G \setminus K$.

**Lemma 3.** *The graph $G$ is $(k, F)$-contractible if and only if the graph $G'$ is $(k, F')$-contractible.*

*Proof.* According to Lemma 2, a minimal matching planarizing set $S$ for $G$ avoiding $F$ does not contain any edges of $K$ and $F'$. Therefore, it also works for $G'$.

On the other hand, if $G'$ has a matching contractible set $S$ disjoint from $F$, consider an embedding of $G' \circ S$ and replace the contracted vertex of the kernel by an embedding of $K$ in a manner completely analog to the one described in the proof of Lemma 2. □

**Phase II.** We show next that when the tree-width of $G$ is small, we can solve $(k, F)$-contractibility with respect to $P$ using Courcelle's theorem [3]. To this effect all we need to show is that is possible to express $(k, F)$-contractibility in monadic second-order logic (MSOL).

**Lemma 4.** *For a fixed graph $H$, there exists an MSOL formula $\mu_H(e_1, \ldots, e_k)$ which is satisfied if and only if $G' := G \circ e_1 \circ \cdots \circ e_k$ contains $H$ as a minor.*

Due to space limitations, the proof is moved to the full version.

An MSOL formula $\varphi_{(k,F)}$ for testing $(k, F)$-contractibility can be defined as follows. A formula $\sigma$ tests whether edges $e_1, \ldots, e_k$ avoid $F$. Then solvability of $P$-RestrictedContract can be checked using MSOL formula:

$$\sigma(F, e_1, \ldots, e_k) := \bigwedge_i (e_i \notin F),$$

$$\varphi_{(k,F)} := \sigma(F, e_1, \ldots, e_k) \wedge P(e_1, \ldots, e_k) \wedge \wedge \neg\mu_{K_5}(e_1, \ldots, e_k) \wedge \neg\mu_{K_{3,3}}(e_1, \ldots, e_k).$$

The formula $\varphi_{(k,F)}$ is satisfiable if and only if $G$ is $(k, F)$-contractible with respect to the MSOL formula $P$.

**Putting All the Pieces Together.** We finish this section with a proof of the announced Theorem 1 stating that $P$-RestrictedContract for a inclusion-closed MSOL formula $P$ is solvable in time $\mathcal{O}(n^2 \cdot f(k))$ disjoint from $F$.

*Proof (Theorem 1).* In Phase I, we repeat the following steps until the tree-width of $G$ is small enough. We find an embedding $h$ of a large hexagonal grid $H_s$, by Theorem 3 in linear time. Using Lemma 1, we can find a subgrid $H_r$ such that $h(H_r)$ is flat. Moreover, we can find such $H_r$ in linear time $\mathcal{O}(k^2 n)$ by testing planarity for all $h_+(H_{t_i,r_j})$.

Using Lemma 2, we transform the graph $G$ and the forbidden set $F$ to a smaller graph $G'$ and an extended forbidden set $F'$ without changing the solution to our problem (see Lemma 3). After each modification, we get a smaller graph $G$. Therefore we need to repeat the steps at most $\mathcal{O}(n)$ times for a total time for Phase I of $\mathcal{O}(n^2 \cdot p(k))$ for some function $p$.

Phase II uses Theorem 4 to solve the problem when $G$ has small tree-width using an MSOL formula $\varphi_{(k,F)}$ in time $\mathcal{O}(n \cdot q(k))$. Phase I modifies the graphs and removes edges which do not appear in any minimal planarizing set. Since the formule $P$ is inclusion-closed, this does not pose a problem; there exists a minimal planarizing set satisfying $P$ if and only if there exists any planarizing set satisfying $P$.

Therefore, the overall complexity of the algorithm is $\mathcal{O}(n^2 \cdot f(k))$ for some function $f$. □

## 3 $\ell$-Subgraph Contractibility

Proposition 1 follows from the fact that $\ell$-subgraph contractibility is expressible using MSOL (details are contained in the full version).

**Matching Contractibility.** Concerning NP-completeness, we first introduce a new problem called *matching contractibility.* The graph $G$ is *$F$-matching contractible* if there exists a planarizing set $S$ which forms a matching in $G$ and $S \cap F = \emptyset$.

| | |
|---|---|
| **Problem:** | MATCHINGCONTRACT |
| **Input:** | An undirected graph $G$ and a set of forbidden edges $F \subseteq E$. |
| **Output:** | Is $G$ an $F$-matching contractible graph? |

First, we show that $\ell$-subgraph contractibility can solve matching contractibility.

**Lemma 5.** *Matching contractibility is reducible from $\ell$-subgraph contractibility.*

*Proof.* For an input $G$ and $F$, we produce a graph $G'$ which is $\ell$-subgraph contractible if and only if $G$ is $F$-matching contractible. We replace edges of $G$ by paths; if $e \in F$, we replace it by a path of length $\ell$, if $e \notin F$, we replace it by a path of length $\ell - 1$. Also put $k = |E(G')|$ so only $\ell$-subgraphs limit $S$.

If a planarizing set $S$ is a matching in $G$ avoiding $F$, then we can contract the corresponding paths in $G'$ by $\ell$-subgraphs. On the other hand, if a path in $G'$ corresponding to $e \in E(G)$ is contracted, it has to be contracted by a single $\ell$-subgraphs. In such a case, $e \notin F$ (otherwise the path is too long) and the contracted paths have to form a matching (the $\ell$-subgraphs cannot share the end-vertices belonging to $G$). So the planarazing set $S'$ of $G'$ gives a planarizing set $S$ of $G$ which is a matching and which avoids $F$ (we ignore the $\ell$-subgraphs not contracting entire paths). □

**Overview of the Reduction.** To show the NP-hardness of MATCHINGCONTRACT, we present a reduction from CLAUSE-LINKED PLANAR 3-SAT. The NP-completeness of this problem was shown by Fellows et al. [7]. An instance

$I$ of CLAUSE-LINKED PLANAR 3-SAT is a Boolean formula in CNF such that each variable occurs in exactly three clauses, once negated and twice positive, each clause contains two or three literals and the incidence graph of $I$ is planar.

Given a formula $I$, we construct a graph $G_I$ with a set $F_I$ of forbidden edges such that $G_I$ is $F_I$-matching contractible if and only if $I$ is satisfiable. The construction is performed by replacing each variable $x$ by a variable gadget $G_x$ (all of them are isomorphic), and replacing each clause $c$ by a clause gadget $H_c$. Both the variable gadget and the clause gadget are graphs $K_5$, containing most of the edges in $F_I$. In Figure 4, the edges not contained in $F_I$ are represented by thick lines. Each variable gadget contains three pendant edges that are identified with certain edges of the clause gadgets, thus connecting the variable and clause gadgets.

**Variable Gadget.** Let a variable $x$ be positive in clauses $c_1$ and $c_2$, and negative in clause $c_3$. The corresponding variable gadget $G_x$ is depicted in Figure 4a. It consists of 4 copies of $K_5$, each having all but two edges in $F_I$. Three of the copies of $K_5$ have pendant edges attached, denoted by

$$e(x, c_i) = v(x, c_i)w(c_i), \qquad i \in \{1, 2, 3\}$$

(refer to Figure 4a). These edges also belong to the clause gadgets. All other vertices and edges are private to the variable gadget $G_x$.

The main idea behind the variable gadget is that exactly one of the edges $t_x$ and $f_x$ will be contracted, encoding in this way the assignment of the variable $x$: $t_x$ for true and $f_x$ for false. The edges shared with the clause gadgets, $e(x, c_1)$ and $e(x, c_2)$, can be contracted if and only if $t_x$ is contracted, and $e(x, c_3)$ can be contracted if and only if $f_x$ is contracted.

**Clause Gadget.** Let $c$ be a clause containing variables $x$, $y$ and possibly $z$. The clause gadget $H_c$ is a copy of $K_5$ with all but 2 or 3 edges in $F_I$. The edges that are not forbidden to contract share a common vertex $w(c)$ and they are the edges $e(x, c)$, $e(y, c)$ and possibly $e(z, c)$ shared with the variable gadgets $G_x$, $G_y$ and possibly $G_z$ (See Figure 4b).

To make the clause gadget planar, we need to contract exactly one of the edges $e(x, c)$, $e(y, c)$ and possibly $e(z, c)$. This is possible only if the clause is satisfied by some variable evaluated as true in the clause.

**Lemma 6.** *The graph $G_I$ is $F_I$-matching contractible if and only if $I$ is satisfiable.*

*Proof.* $\Longrightarrow$: Suppose first that $G_I$ is $F_I$-matching contractible, and let $S \subseteq E(G_I)$ be a matching planarizing set. Using $S$, we construct a satisfying assignment of $I$. Consider a variable, say $x$. Each copy of $K_5$ needs to have at least one edge contracted by $S$.

Exactly one of $t_x$ and $f_x$ is in $S$. If $t_x \in S$, then $t'_x$ cannot be in $S$ (note that $S$ is a matching), hence $e'(x, c_3) \in S$, and $e(x, c_3)$ cannot be in $S$. On the other hand, if $t_x \notin S$, necessarily $f_x \in S$, and by a similar sequence of arguments, none of $e(x, c_1)$ and $e(x, c_2)$ is in $S$.

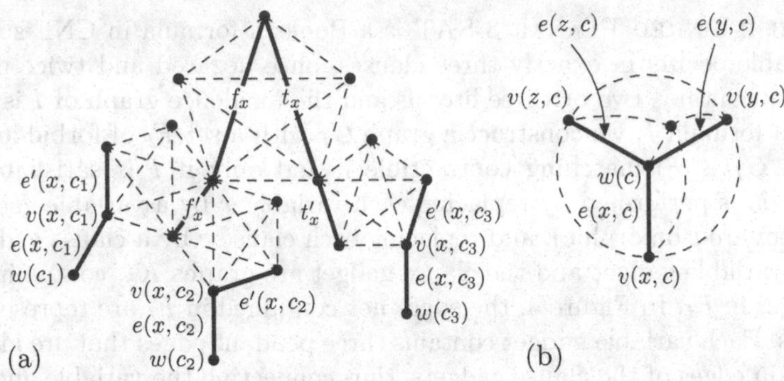


**Fig. 4.** The bold edges can be contracted, dashed edges are forbidden edges from $F_I$. (a) The variable gadget where the three outgoing edges are shared with clause gadgets. (b) The clause gadget; the edge $e(z,c)$ may also be in $F_I$ if the clause contains only two variables. The two or three contractible edges are shared with variable gadgets.

Define a truth assignment for the variables of $I$ so that $x$ is true if and only if $t_x \in S$. It follows that if $x$ is evaluated as false in a clause $c$, then the edge $e(x,c)$ is not in $S$. Since $S$ has to contain at least one edge of $H_c$, in each clause gadget at least one variable must be evaluated to true. Thus $I$ is satisfiable.

$\Longleftarrow$: Suppose that $I$ is satisfiable and fix a satisfying truth assignment $\phi$. We set

$$S = \{t_x, f'_x, e(x,c_1), e(x,c_2), e'(x,c_3) \mid \phi(x) = \text{true}\} \cup \\ \{f_x, t'_x, e(x,c_3), e'(x,c_1), e'(x,c_2) \mid \phi(x) = \text{false}\}.$$

For the variable gadgets, $S$ is a matching. Each clause gadget contains at least one edge of $S$. But if a clause, say $c$, contains more than one variable evaluated as true, its clause gadget contains more edges with a common vertex. In such a case perform a *pruning operation* on $c$, i.e, remove all edges from $S \cap E(H_c)$ but one. The resulting set $S'$ is a matching such that each $K_5$ in $G_I$ contains exactly one edge of $S'$.

It only remains to argue that this set $S'$ is a planarizing set. The graph $G' = G \circ S'$ consists of copies of $K_4$ glued together by vertices or edges. Each copy is attached to other copies by at most three vertices. Since $K_4$ itself has a non-crossing drawing in the plane such that three of its vertices lie on the boundary of the outer face, and these vertices can be chosen arbitrarily as well as their cyclic order on the outer face, the drawings of the variable and clause gadgets can be combined together along a planar drawing of the incidence graph of $I$. Therefore, $G'$ is planar. □

This shows that $\ell$-SUBCONTRACT is NP-complete for every $\ell \geq 2$:

*Proof (Theorem 2).* The problem $\ell$-SUBCONTRACT is reducible from MATCHINGCONTRACT. The above reduction shows that MATCHINGCONTRACT is NP-hard. Clearly, $\ell$-SUBCONTRACT belongs to NP. □

## 4 Conclusions

The problem $\ell$-SubContract remains NP-complete when generalized for surfaces of a fixed genus $g$ (instead of planar graphs). Consider a graph $H_g$ such that for every embedding of $H_g$ into the surface, each face is homeomorphic to the disk. We modify our reduction by taking $G_I \cup H_g$ as the graph and by adding all the edges of $H_g$ into $F$ (see Section 3). For each surface, there exists such a graph $H_g$ (see triangulated surfaces in Mohar and Thomassen [13]).

Generalization of the FPT algorithm is an open problem, with the main difficulty being a generalization of Lemma 1:

*Problem 1.* Is a generalization of $P$-RestrictedContract on graphs on surfaces of genus $g$ fixed-parameter tractable with parameter $k$?

## References

1. Asano, T., Hirata, T.: Edge-Contraction Problems. Journal Comput. System Sci. 26, 197–208 (1983)
2. Bodlaender, H.L.: A linear-time algorithm for finding tree-decompositions of small tree-width. SIAM Journal on Computing 25, 1305–1317 (1996)
3. Courcelle, B.: The monadic second-order logic of graphs III: tree-decompositions, minor and complexity issues. ITA 26, 257–286 (1992)
4. van't Hof, P., Kamiński, M., Paulusma, D., Szeider, S., Thilikos, D.M.: On graph contractions and induced minors. Discrete Applied Mathematics 160, 799–809 (2012)
5. Golovach, P.A., van't Hof, P., Paulusma, P.: Obtaining Planarity by Contracting Few Edges. CoRR, abs/1204.5113 (2012), `http://arxiv.org/abs/1204.5113`
6. Abello, J., Klavík, P., Kratochvíl, J., Vyskočil, T.: Matching and $\ell$-Subgraph Contractibility to Planar Graphs. CoRR, abs/1204.6070 (2012), `http://arxiv.org/abs/1204.6070`
7. Fellows, M.R., Kratochvíl, J., Middendorf, M., Pfeiffer, F.: The Complexity of Induced Minors and Related Problems. Algorithmica 13(3), 266–282 (1995)
8. Garey, M.R., Johnson, D.S.: Crossing number is NP-complete. SIAM J. Alg. Discr. Meth. 4(3), 312–316 (1983)
9. Grohe, M.: Computing crossing numbers in quadratic time. J. Comput. Syst. Sci. 68(2), 285–302 (2004)
10. Hopcroft, J.E., Tarjan, R.E.: Efficient Planarity Testing. J. ACM (JACM) 21(4), 549–568 (1974)
11. Heggernes, P., van't Hof, P., Lokshtanov, D., Paul, C.: Obtaining a bipartite graph by contracting few edges. In: Proc. FSTTCS 2011, pp. 217–228 (2011)
12. Marx, D., Schlotter, I.: Obtaining a Planar Graph by Vertex Deletion. Algorithmica 62, 807–822 (2012)
13. Bojan, M., Thomassen, C.: Graphs on Surfaces. Johns Hopkins University Press
14. Perkovič, L., Reed, B.: An improved algorithm for finding tree decompositions of small width. International Journal of Foundations of Computer Science 11(3), 365–371 (2000)
15. Robertson, N., Seymour, P.D.: Graph minors XIII. The disjoint paths problem. Journal of Combinatorial Theory, Series B 63, 65–110 (1995)

# On the Space Complexity of Parameterized Problems

Michael Elberfeld, Christoph Stockhusen, and Till Tantau

Institute for Theoretical Computer Science
Universität zu Lübeck
D-23562 Lübeck, Germany
{elberfeld,stockhus,tantau}@tcs.uni-luebeck.de

**Abstract.** Parameterized complexity theory measures the complexity of computational problems predominantly in terms of their parameterized *time* complexity. The purpose of the present paper is to demonstrate that the study of parameterized *space* complexity can give new insights into the complexity of well-studied parameterized problems like the feedback vertex set problem. We show that the undirected and the directed feedback vertex set problems have different parameterized space complexities, unless L = NL; which explains why the two problem variants seem to necessitate different algorithmic approaches even though their parameterized time complexity is the same. For a number of further natural parameterized problems, including the longest common subsequence problem and the acceptance problem for multi-head automata, we show that they lie in or are complete for different parameterized space classes; which explains why previous attempts at proving completeness of these problems for parameterized time classes have failed.

**Keywords:** parameterized complexity theory, logarithmic space, fixed-parameter tractable problems, feedback vertex set, reachability problems.

## 1 Introduction

When designing classical or parameterized algorithms, the focus often lies on the *time* complexity of a problem rather than its *space* complexity. Nevertheless, the study of space classes like logarithmic space is an integral part of classical complexity theory since many natural problems (like reachability and distance problems in graphs or satisfiability problems for powerful logics) do not appear to be complete for time classes like P or NP, rather they turn out to be complete for space classes like L, NL, or PSPACE.

It stands to reason that we may also expect some parameterized problems to be complete not for standard parameterized time classes like FPT or W[1] or XP, but rather for parameterized space classes. Furthermore, by analogy to findings from classical complexity theory, we may expect that parameterized problems with low space complexity can be solved quickly. A typical result supporting

D.M. Thilikos and G.J. Woeginger (Eds.): IPEC 2012, LNCS 7535, pp. 206–217, 2012.

this reasoning is that the parameterized vertex cover problem has low space complexity—it lies in the class para-L [5,11]—and, indeed, this problem can be solved *very* quickly in the parameterized setting.

In the present paper we investigate the space complexity of a number of natural parameterized problems whose space complexity has not yet been studied in this context. We will argue that the differences in their space complexities are the reasons why the problems could not be shown to be complete for standard parameterized time classes. Moreover, our results suggest a possible explanation of why the problems vary with respect to how quickly they can be solved in practice even though they lie in the same parameterized time class. As prominent examples, we compare the directed and the undirected version of the feedback vertex set problem as well as the treewidth problem. All of these problems are indistinguishable with respect to the parameterized time classes they belong to (namely FPT), but with regard to their space complexity they can be classified using new natural complexity classes that reflect their different complexities. As a corollary of these efforts we obtain that the directed and the undirected feedback vertex set problem have different parameterized space complexity, unless L = NL.

Previous research on parameterized space complexity [5,7,11] was directed at the logspace analogues para-L and para-NL of para-P, which is commonly known as FPT, as well as at the logspace analogues XL and XNL of XP. When one tries to determine the parameterized space complexity of natural parameterized problems, it quickly turns out, however, that these space classes do not suffice to paint a complete picture of the complexity landscape of the parameterized world. For this reason, we introduce new classes that are motivated by parameterized versions of natural reachability problems. An overview of the classes and our containment and completeness results is given in Figure 1.

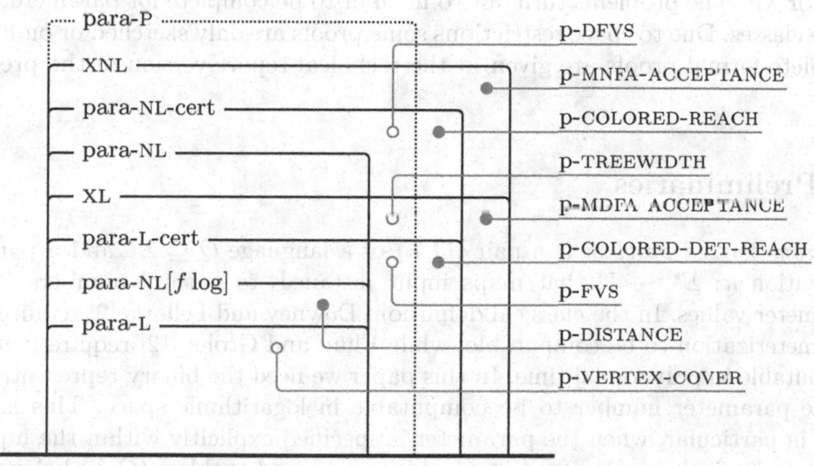


**Fig. 1.** Overview of the parameterized space classes studied in this paper and parameterized problems that lie in them. A filled circle indicates that the problem does not only lie in the class, but is even complete for it under para-L-reductions.

*Related Work.* In 1997, Cai et al. [5] introduced the class UNIFORM-LOGSPACE+ADVICE, which is the same as para-L, and they also considered its nondeterministic version. They showed that several problems in para-P also belong to para-L or para-NL, including the naturally parameterized vertex cover problem, which lies in para-L, and the parameterized $k$-leaf spanning tree problem, which lies in para-NL. In 2003, Flum and Grohe [11] continued the study of para-L and para-NL by showing that the parameterized model-checking problem of first-order formulas on graphs of bounded degree lies in para-L. This implies that many standard parameterized graph problems belong to para-L when we restrict our attention to bounded-degree graphs. In this context they asked, which other problems in para-P also belong to para-NL or para-L, and whether there are problems that are contained in para-P but are not contained in para-L or para-NL under the assumption that para-P $\neq$ para-L. In this paper, we answer this question affirmatively by presenting natural problems that belong to para-P, but are not contained in para-L under standard assumptions.

*Organization of This Paper.* In Section 2 on preliminaries we fix the terminology and review the definitions of parameterized space classes from the literature as well as of parameterized reductions. In Section 3 we investigate parameterized problems that are known to be *tractable* (that is, they are known to lie in para-P = FPT), but whose space complexity has not yet been investigated. The objective of this section is to get a deeper understanding of why some problems inside para-P appear more difficult than others even though they are all tractable. In Section 4 we turn our attention to parameterized problems that presumably lie outside of para-P. This time, our findings on the space complexity of natural problems provide explanations why researchers have failed to prove completeness of these problems for natural parameterized time classes like W[1] or XP: The problems turn out to lie in or to be complete for parameterized space classes. Due to space restrictions some proofs are only sketched or omitted; complete formal proofs are given in the technical report version of the present paper.

## 2 Preliminaries

A *parameterized problem* is a pair $(Q, \kappa)$ of a language $Q \subseteq \Sigma^*$ and a parameterization $\kappa \colon \Sigma^* \to \mathbb{N}$ that maps input instances to natural numbers, their parameter values. In the classical definition, Downey and Fellows [9] require the parameterization to be computable, while Flum and Grohe [12] require it to be computable in polynomial time. In this paper we need the binary representation of the parameter number to be computable in logarithmic space. This is the case, in particular, when the parameter is specified explicitly within the input.

For a classical complexity class $C$, a parameterized problem $(Q, \kappa)$ belongs to the *para-class* para-$C$ if there is an alphabet $\Pi$, a computable function $\pi \colon \mathbb{N} \to \Pi^*$, and a language $A \subseteq \Sigma^* \times \Pi^*$ with $A \in C$ such that for all $x \in \Sigma^*$ we have $x \in Q \iff (x, \pi(\kappa(x))) \in A$. Flum and Grohe [12] phrase this as "$(Q, \kappa)$ *is* *in C after a precomputation on the parameter"*. In particular, FPT is the same as para-P and all fixed-parameter tractable problems are in P after a precomputation on the parameter. Analogously, we can define the classes para-L and para-NL as the family of problems that are in L and in NL after a precomputation on the parameter, respectively. In terms of the $O$-notation, a parameterized problem $(Q, \kappa)$ over $\Sigma$ is in para-L if there is a function $f \colon \mathbb{N} \to \mathbb{N}$ such that the question $x \in Q$ can be decided within space $f(\kappa(x)) + O(\log |x|)$.

A different way of defining a parameterized version of a classical complexity class $C$ is to consider its so-called X-class [9]. A problem $(Q, \kappa)$ is in X$C$ if for every number $w \in \mathbb{N}$ the slice $Q_w = \{ x \mid x \in Q \text{ and } \kappa(x) = w\}$ lies in $C$. It is immediate from the definition that para-$C \subseteq$ X$C$ holds. The class XP is in wide use in parameterized complexity theory; the logarithmic space classes XL and XNL have previously been studied Chen, Flum, and Grohe [7,11].

In order to compare the complexity of parameterized problems, we use parameterized logspace reductions, called para-L-reductions in the following. Following Flum and Grohe, we define them similarly to FPT-reductions: Let $(Q_1, \kappa_1)$ and $(Q_2, \kappa_2)$ be parameterized problems over some alphabets $\Sigma_1$ and $\Sigma_2$. A para-L-*reduction* is a mapping $R \colon \Sigma_1^* \to \Sigma_2^*$ such that

1. for all $x \in \Sigma_1^*$ we have $x \in Q_1 \iff R(x) \in Q_2$,
2. $\kappa_2(R(x)) \le g(\kappa_1(x))$ for some computable function $g$,
3. $R$ is para-L-computable with respect to $\kappa_1$.

By standard arguments one can show that all classes in this paper are closed with respect to para-L-reductions.

## 3 Space Complexity of Tractable Problems

From the perspective of classical fpt-theory, the complexity of a parameterized problem is "settled in principle" once it has been shown to be solvable in time $f(\kappa(x)) \cdot n^c$; further research then focusses on making $f$ as slowly growing as possible. In the following, instead of trying to differentiate between problems in para-P through their $f$-functions, we try to determine differences caused by their different space complexities.

We start our exploration with a deceptively simple problem, namely the distance problem with the distance as the parameter:

*Problem 3.1* (p-DISTANCE).
*Instance.* A directed graph $G$, two vertices $s$ and $t$ of $G$, and a natural number $l$.
*Parameter.* $l$.
*Question.* Is there a path from $s$ to $t$ in $G$ of length at most $l$?

This problem is clearly in para-P and, since the distance problem lies in NL, also in para-NL. It might also seem like a good candidate for a para-NL-complete problem. Indeed, Flum and Grohe [11] showed that classes like para-P, para-NL, and para-L are "uninteresting" from the parameterized point of view in the sense that complete problems for the underlying classical complexity classes are always complete for the parameterized versions when considering the trivial parameterization. In particular, the standard distance problem for directed graphs is complete for para-NL with the trivial parameterization $\kappa(x) = 1$. However, this argument does not carry over to the above version of the distance problem with its more natural parameterization.

In order to describe the complexity of p-DISTANCE precisely, it turns out that a new class is needed that is derived from transferring the definition of the class W[P] to the space setting in a certain way. A classical definition of W[P] is as the para-P-reduction closure of the weighted circuit satisfiability problem [1]. Alternatively, W[P] is also the class of all problems $(Q, \kappa)$ that are solvable by NTMs deciding $Q$ in para-P-time, making at most $f(\kappa(x)) \cdot \log |x|$ nondeterministic steps on any input $x$. Finally, W[P] also contains exactly the problems decidable via DTMs that are provided with a proof certificate of length $f(\kappa(x)) \cdot \log |x|$ (the certificate is on a special read-only tape and the machine must accept all $x \in Q$ for some certificate and must reject $x \notin Q$ for all certificates).

In the context of parameterized time complexity all of the three characterizations of W[P] yield the same classes. When we look at the space analogues of these classes, the situation is somewhat different: First, the para-L-reduction closure (rather than the para-P-reduction closure) of the parameterized weighted circuit satisfiability problem is still W[P], see [7], and this class clearly does not capture the complexity of p-DISTANCE. Second, when we replace the para-P time bound in the second characterization by a para-L space bound, a subclass of para-NL arises. Third, when we make the same replacement in the third definition, a presumably different class arises that is a subclass of XNL.

**Definition 3.2.** *A parameterized problem $(Q, \kappa)$ is in the class* para-NL[$f$ log] *if it can be decided by a* para-NL *Turing machine which makes at most $f(\kappa(x)) \cdot O(\log |x|)$ many nondeterministic steps.*

**Definition 3.3.** *A parameterized problem $(Q, \kappa)$ is in the class* para-L-cert *if it can be decided by a* para-L *Turing machine which is provided proof certificates of length $f(\kappa(x)) \cdot O(\log |x|)$.*

It turns out that para-NL[$f$ log] is precisely what we need to characterize the complexity of the distance problem with its natural parameterization:

**Theorem 3.4.** p-DISTANCE *is complete for* para-NL[$f$ log] *with respect to* para-L *reductions.*

*Sketch of Proof.* The containment p-DISTANCE $\in$ para-NL[$f$ log] is shown by an algorithm that guesses a path of length $l$ from $s$ to $t$. For hardness, consider a problem $(Q, \kappa)$ with a Turing machine $M$ that witnesses $(Q, \kappa) \in$ para-NL[$f$ log] and an input $x$. Compute the configuration graph of $M$ on input $x$. Unfortunately we cannot pass this graph to the p-DISTANCE problem because the length of the path from the initial configuration $s$ to the accepting configuration $t$ is not bounded exclusively by the parameter. However, the number of configurations that are followed by nondeterministic steps on any path from $s$ to $t$ is bounded by $f(\kappa(x)) \cdot O(\log |x|)$. We use this fact by first contracting the deterministic parts of the configuration graph and then shortening the paths in the resulting graph by a logarithmic factor such that the length of the paths from $s$ to $t$ is exclusively bound by the parameter. □

In general, para-NL[$f$ log] seems to contain many parameterized versions of problems that are contained in NL in the classical sense and that follow the guess-and-forget approach, i.e. guessing an element from the input, processing this element, and guessing the next element while forgetting the previous one. With this observation, many NL-complete problems can be shown to be complete for para-NL[$f$ log] with natural parameterizations, e.g. many problems from [15]. On the other hand, para-NL[$f$ log] and para-NL do not coincide under standard assumptions. This can be shown with the observation that every NL-complete problem is complete for para-NL with respect to the trivial parameterization.

**Theorem 3.5.** para-NL[$f$ log] = para-NL *if, and only if,* L = NL.

*The Feedback Vertex Set Problems.* While it is useful to know that para-NL[$f$ log] has many natural complete problems, our original goal for looking at parameterized space classes was to study the parameterized space complexity of well-studied problems in para-P that do not come from the logspace world in the way p-DISTANCE does. We now look at the problem of identifying a set $S$ of vertices of a given graph $G$ such that $G - S$ has treewidth at most $w$ for a fixed natural number $w$. For $w = 0$ this is the vertex cover problem. For $w = 1$ we obtain the (undirected) feedback vertex set problem p-FVS. We also look at the problem of deciding whether a given graph has treewidth $w$ for a given $w$.

The first problem of this "treewidth hierarchy" has been studied in the context of parameterized space complexity theory by Cai et al. [5], who showed that p-VERTEX-COVER ∈ para-L. We look at the next problem of this hierarchy, the feedback vertex set problem p-FVS. It does not seem to lie in para-L nor even in para-NL. However, it turns out that the class para-L-cert contains it:

**Theorem 3.6.** p-FVS ∈ para-L-cert.

*Proof.* On input of an undirected graph $G$, our para-L-cert Turing machine $M$ interprets its certificate of length $\kappa(x) \cdot \lceil \log |V| \rceil$ as the description of the $k$ vertices of the feedback-vertex set. Then $M$ removes these vertices and its incident edges from $G$ and checks whether the resulting graph $G'$ is acyclic. Both, removal and test for acyclicity can be done by para-L Turing machines [8]. With the fact that para-L-cert is closed with respect to para-L-reductions we obtain the result. □

While the classes para-NL[$f$ log] and para-NL are contained in para-P, this is not the case for para-L-cert under the standard assumption para-P $\subsetneq$ W[SAT] since the weighted satisfiability problem for propositional formulas belongs to para-L-cert.

**Theorem 3.7.** *If* para-L-cert ⊆ para-P, *then* W[SAT] = para-P.

In particular, p-FVS is not complete for para-L-cert, unless para-P = W[SAT], which is considered to be unlikely (para-L-cert does, however, have natural complete problems as shown in Section 4). Nevertheless, para-L-cert turns out to be useful for contrasting the complexity of the feedback vertex set problem for undirected graph to the version for directed graphs.

The undirected version was shown to be fixed-parameter tractable in 1993 by Bodlaender [2]. The fixed-parameter tractability of the directed version remained an open problem until Chen et al. [6] presented a para-P-algorithm in 2008. So far, these apparently different complexities could only be felt by looking at the complexity of the proofs. In the setting of parameterized logarithmic space, the different complexities are not only mirrored, we are even able to prove that p-FVS and p-DFVS are of different complexity (under standard assumptions):

**Theorem 3.8.** p-DFVS $\in$ para-L-cert *if, and only if,* L = NL.

*Proof.* To prove the forward direction, we first show that the parameterized problem p-UNREACH, the complement of the reachability problem where we ask whether there does *not* exist a path from a given vertex $s$ to another vertex $t$ in a directed graph parameterized by the trivial parameterization, para-L-reduces to p-DFVS. The reduction works in two steps: The first step takes the $n$-vertex input graph and makes it acyclic by producing $n$ copies if it, conceptually arranging them as $n$ layers from left to right, and drawing edges between consecutive layers from left to right in the same way as they are present in the original graph. The second step inserts an edge from $t$'s copy in the last layer to $s$'s copy in the first layer. The resulting directed graph remains acyclic (has a feedback vertex set of size 0) exactly if there is no path from $s$ to $t$ in the input graph.

Next, assume p-DFVS $\in$ para-L-cert. This implies the existence of a para-L-TM that, provided with a certificate of length $f\big(\kappa(x)\big) \cdot \log|x|$, decides p-UNREACH. But because p-UNREACH is parameterized in the trivial way, this Turing machine is in fact a deterministic logarithmic-space Turing machine that decides the NL-complete reachability problem in directed graphs. This implies L = NL.

For the backward direction assume L = NL. Then the algorithm used above for p-FVS also works for p-DFVS because cycle detection in directed graphs is then possible in deterministic logarithmic space. □

This raises the question, what class captures p-DFVS. The central difference between p-FVS and p-DFVS is the complexity of the underlying cycle detection problem: In the undirected case this can be done in deterministic logarithmic space, in the directed case nondeterministic logarithmic space is required. Therefore, the natural approach is to consider a nondeterministic version of para-L-cert, namely para-NL-cert: This class is defined in a similar way as para-L-cert, but here we allow para-NL Turing machines. We immediately obtain the following result:

**Theorem 3.9.** p-DFVS $\in$ para-NL-cert.

For the same reasons as in the undirected case, we will not be able to prove that p-DFVS is complete for para-NL-cert.

Many more problems from para-P are contained in para-L-cert and its nondeterministic version. In fact, every problem that can be decided via a bounded search tree algorithm [9], where the aim is to find a root-to-leaf path within a search tree whose depth depends on the parameter, is a promising candidate to be contained in para-L-cert, because the certificate of a para-L-cert Turing machine can describe the desired root-to-leaf path.

*The Treewidth Problem and Slicewise Logarithmic Space.* We conclude this section with a study of the treewidth problem: Given a graph $G$ and a natural number $w$ as the parameter, the question is whether $G$ has treewidth at most $w$. Bodlaender showed that this problem is fixed-parameter tractable in 1996 [3], and a recent result [10] from 2010 showed that for every fixed $w$ this problem is decidable in deterministic logarithmic space. These theorems immediately give us the following result, but we will not be able to show that p-TREEWIDTH is complete for XL unless para-P = W[SAT]:

**Theorem 3.10.** p-TREEWIDTH $\in$ XL.

The following theorem relates XL and XNL to the surrounding complexity classes.

**Theorem 3.11**

*1.* para-L-cert $\subseteq$ XL *and* para-NL-cert $\subseteq$ XNL*,*
*2.* para-NL-cert $\subseteq$ XL *implies* L = NL*.*

## 4 Space Complexity of Intractable Problems

Most of the classes discussed in the previous section are not known to be contained in para-P. In fact, these classes contain problems that are far above of para-P inside the W-hierarchy, for example the clique problem or the weighted satisfiability problem for propositional formulas, both of which are contained in para-L-cert. Therefore, a natural question is to ask for complete problems for these classes. In this section, we will give natural complete problems for the classes that are not fully contained in para-P, that is, para-L-cert, para-NL-cert, XL, and XNL.

*Slicewise Logarithmic Space.* So far, only few problems are known to be complete for XL or XNL. Chen, Flum, and Grohe gave the first complete problems for these classes based on Turing machine simulations. They studied the compact Turing machine computation problem for deterministic and nondeterministic Turing machines, p-CTMC and p-CNTMC, respectively [7]. The p-CTMC problem is defined as follows: On input of an encoding of a deterministic Turing machine $M$, a string $x$ over $M$'s alphabet, and a natural number $k$ in binary, the question is if there is an accepting computation of $M$ on $x$ that uses at most $k$ work tape cells, where $k$ is the parameter. The problem p-CNTMC is defined in the same way, but with respect to nondeterministic Turing machines.

We add the following natural automata evaluation tasks to the list of problems complete for XL and XNL.

*Problem 4.1* (p-MDFA-ACCEPTANCE)
*Instance.* The code of a deterministic finite two-way automaton $A$ with $k$ heads and an input string $x$.
*Parameter.* $k$.
*Question.* Does $A$ accept $x$?

The problem p-MNFA-ACCEPTANCE is defined in the same way, but with respect to nondeterministic automata.

**Theorem 4.2**
1. p-MDFA-ACCEPTANCE *is complete for* XL *with respect to* para-L *reductions.*
2. p-MNFA-ACCEPTANCE *is complete for* XNL *with respect to* para-L *reductions.*

*Proof.* Hartmanis showed [13] that $A \in \mathrm{L}$ holds if, and only if, there is a deterministic multi-head automaton that decides $A$. The basic idea is that the position of an automata's head on the input tape can be used to store a number between 0 and $n$, where $n$ is the input tape length, and, thus, the position of a head can store up to $\log_2 n$ bits of information. A fixed number of heads hence suffice to store the information of the work tape of a machine using $O(\log n)$ space and modifications of this work tape correspond to appropriate sequences of auxiliary heads computing the right number of steps to be made by one of the heads. The details of the construction are not important for proving the theorem. It suffices to note that p-MDFA-ACCEPTANCE lies in XL because the construction of Hartmanis is uniform. To show hardness, given a parameterized problem $(Q, \kappa) \in \mathrm{XL}$ that is solved by a machine in space $f\big(\kappa(x)\big) \cdot O\big(\log |x|\big) + f\big(\kappa(x)\big)$, by Hartmanis's result there is a multi-head two-way automaton deciding $Q$ whose number of heads depends only on $f(\kappa(x))$. A para-L-reduction must simply map the input $x$ to this automaton together with $x$. The proof for the nondeterministic version is analogous. □

The next results show that, outside of para-P, the classes XL and XNL enable us to make statements about problems that could not be classified within the W-hierarchy. Consider the following well-known problem of finding a longest common subsequence of $k$ strings, parameterized by $k$:

*Problem 4.3* (p-LCS).
*Instance.* A set $\{s_1, \dots, s_k\}$ of strings over an alphabet encoded in the input and a natural number $l$.
*Parameter.* $k$.
*Question.* Is there a subsequence of the given strings with length at least $l$?

Bodlaender et al. [4] showed that p-LCS is hard for W[$t$] for every level $t$. Pietrzak [16] showed that the variant p-FLCS, where the alphabet is fixed, is hard for W[1]. Pietrzak also conjectured that an exact classification of this problem within the parameterized hierarchy is not possible, because it seems not to be contained in W[P], the highest class of the W-hierarchy. While classical dynamic-programming approaches show that the longest common subsequence problem is contained in XP, completeness for XP could not be shown. We give a strong argument that this is not possible using the class XNL:

**Theorem 4.4.** p-LCS $\in$ XNL.

*Proof.* To decide on a given input of $k$ strings whether there exists a longest common subsequence of length $l$, the XNL Turing machine scans the first string from left to right and guesses the subsequence, using a pointer to a symbol in the first string. Using $k-1$ additional pointers on the remaining strings, the machine verifies that the guessed subsequence is also contained within the $k-1$ remaining strings. □

By definition, X-classes inherit their inclusion structure from their underlying complexity classes. From this, we immediately get the following theorem:

**Theorem 4.5.** p-LCS *is not complete for* XP, *unless* NL $=$ P.

Because p-LCS is hard for every level of the W-hierarchy it is now tempting to conclude W[$t$] $\subseteq$ XNL for every $t$, but this is not the case. The error in this argument is that the hardness of p-LCS has been shown with respect to para-P reductions, but XNL is only closed with respect to para-L reductions. On the other hand, this shows that under standard assumptions it will not be possible to prove that p-LCS is hard for the levels of the W-hierarchy using only para-L reductions, because this would imply NL $=$ P. However, it is an open question whether one can show that p-LCS is complete for XL or XNL.

*Problems for Parameterized Logarithmic Space with Certificates.* In the previous section we introduced the classes para-L-cert and para-NL-cert to upper bound the parameterized space requirements of the fixed-parameter tractable problems p-FVS and p-DFVS, respectively. These problems are not complete for the corresponding classes under standard assumptions; in the present section we identify natural complete problems for the classes para-L-cert and para-NL-cert.

The reachability problem, which asks whether there exists a path from a start to a target vertex in a given directed graph, is among the most prominent problems in the study of logarithmic space-bounded computations. Its general version is NL-complete [14] and becomes L-complete if the path we ask for is *deterministic* [8]; that means, every vertex on the path has only a single outgoing edge in the graph. In the same way as one can understand the transition from the classical complexity class P to its parameterized version W[P] by considering the complete circuit evaluation problem for the former and the complete parameterized weighted circuit satisfiability problem for the later class, we use the following variant of the reachability problem to better understand the transition from the classical classes L and NL to their certificate-based parameterized counterparts para-L-cert and para-NL-cert, respectively.

*Problem 4.6* (p-COLORED-REACH).

*Instance.* A vertex set $V$, two vertices $s, t \in V$, a set of multi-colored edges $E \subseteq V \times V$, and a natural number $k$.

*Parameter.* $k$.

*Question.* Is there a set of $k$ colors, such that there is a path from $s$ to $t$ that uses only edges that have at least one of the chosen colors.

Let p-COLORED-DET-REACH denote the variant where the path must be deterministic.

**Theorem 4.7.** p-COLORED-REACH *is complete for* para-NL-cert *with respect to* para-L *reductions.*

*Sketch of Proof.* For proving p-COLORED-REACH $\in$ para-NL-cert, we use a Turing machine that guesses a path through the graph that is made up by edges of the $k$ colors chosen by the certificate. To prove hardness, we consider the configuration graph of a para-NL-cert machine that accesses a certificate that has to be given. Consequently, configurations are only partially known; they lack the information which symbol is read on the certificate tape, but contain the position of the head on it. Moreover, the present transitions are not determined completely, but each one is triggered by a particular position in the certificate that stores a particular symbol. The main idea of the reduction is now to partition the certificate into $k$ blocks of logarithmic length, establish a set of colors for each block and each possible string in the block, and color all transitions that are due to the corresponding string in the corresponding block. After making sure that at least one color is chosen for each block, we get the desired reduction. □

The following theorem immediately follows from the proof of the previous one and the fact that we consider deterministic instead of nondeterministic Turing machines for para-L-cert.

**Theorem 4.8.** p-COLORED-DET-REACH *is complete for* para-L-cert *with respect to* para-L *reductions.*

Similar to the transition from the classical reachability problem to the deterministic version p-COLORED-DET-REACH, many results on logarithmic space-bounded computations can be adjusted to work in the context of graphs that are constructed by choosing $k$ sets of edges.

## 5 Conclusion

We have introduced new parameterized space classes that capture the complexity of parameterized problems and help in understanding the complexity of fixed-parameter tractable problems. We have seen that, under the assumption L $\neq$ NL, well-known problems like p-FVS and p-DFVS are separated. Moreover, classes like para-L-cert and para-NL-cert are not fully contained in para-P, but capture natural reachability problems that could not be classified in an exact way before.

The next step is to reconsider other parameterized problems (both fixed-parameter tractable and intractable under standard assumptions like W[1] $\neq$ para-P) with the presented framework in mind: How can these problems be classified within this framework? What techniques from classical space complexity help in the parameterized setting and vice versa? In particular, it might be interesting to use the framework to better understand the complexity of problems whose unparameterized versions are known to lie in P and, thus, are normally not studied from the parameterized point of view. An interesting problem in this direction is the associative generability problem parameterized by the size of the generator set: On input of the description of an associative operator $\circ\colon U \times U \to U$ and a parameter $k$, decide whether there is a set of *generators* $G \subseteq U$ of size $k$ such that all of $U$ can be generated using only elements from $G$. We believe that this problem is complete for para-NL-cert and currently prepare a proof of this statement for the technical report version of this paper.

As also suggested by our reviewers, one might even better understand the complexity of parameterized problems whose slices are not P-complete by considering parameterized versions of classical complexity classes based on parallel and circuit computations like NC, $\mathrm{NC}^1$, $\mathrm{TC}^0$, and $\mathrm{AC}^0$.

## References

1. Abrahamson, K.A., Downey, R.G., Fellows, M.R.: Fixed-parameter tractability and completeness IV: On completeness for W[P] and PSPACE analogues. Annals of Pure and Applied Logic 73(3), 235–276 (1995)
2. Bodlaender, H.L.: On linear time minor tests with depth-first search. Journal of Algorithms 14(1), 1–23 (1993)
3. Bodlaender, H.L.: A linear time algorithm for finding tree-decompositions of small treewidth. Siam Journal on Computing 25(6), 1305–1317 (1996)
4. Bodlaender, H.L., Downey, R.D., Fellows, M.R., Wareham, H.T.: The parameterized complexity of sequence alignment and consensus. Theoretical Computer Science 147(1), 31–54 (1995)
5. Cai, L., Chen, J., Downey, R.G., Fellows, M.R.: Advice classes of parameterized tractability. Annals of Pure and Applied Logic 84(1), 119–138 (1997); Asian Logic Conference
6. Chen, J., Liu, Y., Lu, S., O'Sullivan, B., Razgon, I.: A fixed-parameter algorithm for the directed feedback vertex set problem. Journal of the ACM 55(5), 21:2–21:19 (2008)
7. Chen, Y., Flum, J., Grohe, M.: Bounded nondeterminism and alternation in parameterized complexity theory. In: CCC 2003, pp. 13–29 (2003)
8. Cook, S.A., McKenzie, P.: Problems complete for deterministic logarithmic space. Journal of Algorithms 8(3), 385–394 (1987)
9. Downey, R.G., Fellows, M.R.: Parameterized Complexity. Springer (1999)
10. Elberfeld, M., Jakoby, A., Tantau, T.: Logspace versions of the theorems of Bodlaender and Courcelle. In: FOCS 2012, pp. 143–152 (2010)
11. Flum, J., Grohe, M.: Describing parameterized complexity classes. Information and Computation 187, 291–319 (2003)
12. Flum, J., Grohe, M.: Parameterized Complexity Theory. Springer (2006)
13. Hartmanis, J.: On non-determinancy in simple computing devices. Acta Informatica 1, 336–344 (1972)
14. Jones, N.D.: Space-bounded reducibility among combinatorial problems. Journal of Computer and System Sciences 11(1), 68–85 (1975)
15. Jones, N.D., Edmund Lien, Y., Laaser, W.T.: New problems complete for nondeterministic log space. Mathematical Systems Theory 10(1), 1–17 (1976)
16. Pietrzak, K.: On the parameterized complexity of the fixed alphabet shortest common supersequence and longest common subsequence problems. Journal of Computer and System Science 67(4), 757–771 (2003)

# On Tractable Parameterizations of Graph Isomorphism

Adam Bouland[1], Anuj Dawar[2], and Eryk Kopczyński[3]

[1] Massachusetts Institute of Technology, Cambridge, MA, USA
[2] University of Cambridge, Cambridge, UK
[3] University of Warsaw, Warsaw, Poland

**Abstract.** The fixed-parameter tractability of graph isomorphism is an open problem with respect to a number of natural parameters, such as tree-width, genus and maximum degree. We show that graph isomorphism is fixed-parameter tractable when parameterized by the *tree-depth* of the graph. We also extend this result to a parameter generalizing both tree-depth and max-leaf-number by deploying new variants of cops-and-robbers games.

## 1 Introduction

The fixed-parameter complexity of the graph isomorphism problem (GI) remains open with respect to a number of interesting graph parameters. Several parameterizations of graph isomorphism are known to yield tractable algorithms. For instance, graph isomorphism is known to be fixed-parameter tractable in the following parameters: size of the smallest feedback vertex set [17], tree-distance width [28], largest multiplicity of an eigenvalue of the adjacency matrix [8], size of the largest color class (in the case of colored graph isomorphism) [2][10][1], and maximum size of a simplicial component (in the case of chordal graph isomorphism) [27].

On the other hand, many natural parameterizations of the problem are known to produce algorithms which run in time $O(n^{f(k)})$, which places them in XP, but for which fixed-parameter tractability remains open. For instance, graph isomorphism is in XP when parameterized by the size of the smallest excluded minor [25][14] or topological minor [13] of a graph . This generalizes a long line of previous results that GI is in XP with respect to a number of other parameters, including genus [22], maximum degree [20], and tree-width [4]. It should be pointed out that in none of these cases do we know of any hardness result that indicates the problem is not fixed-parameter tractable.

One particular open question is whether or not GI is fixed-parameter tractable when parameterized by tree-width or path-width. In the present paper, we show that the problem is fixed-parameter tractable when parameterized by the *tree-depth* of a graph. The tree-depth of a graph measures how close a graph is to a star, in much the same way that tree-width measures how close a graph is to a tree. This parameter is natural in the context of sparse matrix factorization

D.M. Thilikos and G.J. Woeginger (Eds.): IPEC 2012, LNCS 7535, pp. 218–230, 2012.

[15][21] and descriptive complexity [7]. Our proof yields a natural generalization to a parameter we introduce and call *generalized tree-depth*, which generalizes the parameter *max-leaf-number* as well.

The key idea in our proof is to use a characterization of tree-depth in terms of cops and robbers games in order to show that in any graph $G$ of tree-depth at most $d$, the number of vertices that can serve as "roots" of a minimum height tree-depth decomposition is bounded by a function of $d$. This allows us to create an automorphism-invariant tree-depth decomposition algorithm based on Lindell's algorithm for logspace tree canonization [18].

## 2 Preliminaries

A language $Q$ over an alphabet $\Sigma$ is said to be *fixed-parameter tractable* with respect to parameterization $\kappa : \Sigma^* \to \mathbb{N}$, if it can be decided on input $x$ in time $O(f(k)n^c)$ where $n = |x|$, $c$ is a fixed constant, $k = \kappa(x)$ is the value of the parameter and $f$ is an arbitrary function.

The *max-leaf-number* of a graph $G$ is the maximum number of leaves in a spanning tree of $G$.

The *tree-depth* of a graph is defined recursively as follows:

**Definition 1.** *Let $G$ be a graph with connected components $G_1$, ..., $G_p$. Then the tree-depth of $G$, denoted* $\mathrm{td}(G)$*, is given by*

$$\mathrm{td}(G) = \begin{cases} 1 & \text{if } |V(G)| = 1 \\ 1 + \min\limits_{v \in V(G)} \mathrm{td}(G - v) & \text{if } p = 1 \text{ and } |V(G)| > 1 \\ \max\limits_{i=1\ldots p} \mathrm{td}(G_i) & \text{otherwise} \end{cases}$$

For example, the tree-depth of a star is 2, and the tree-depth of the complete graph $K_n$ is $n$. In some sense tree-depth measures how close a graph is to a star. Tree-depth occurs naturally in descriptive complexity, in which it was recently shown that monadic second order logic and first order logic coincide on a class of graphs $C$ iff $C$ has bounded tree-depth [7].

An alternative definition of tree-depth is also helpful for our results. The *height* of a rooted tree $T$ is the length of the longest path from the root to a leaf. The *closure* of a rooted tree $T$, denoted $\mathrm{clos}(T)$, is the graph obtained by adding edges from each vertex $v$ to all vertices $w$ which lie on a path from the root to $v$. Then the tree-depth of a connected graph $G$ is the minimum height of a tree such that $G$ is a subgraph of $\mathrm{clos}(T)$ [24].

Consider a connected graph $G$ and a tree $T$ over $V(G)$ such that $G$ is a subgraph of $\mathrm{clos}(T)$. We will call such a tree $T$ a *tree-depth decomposition* of $G$ if it obeys the following property: for every rooted subtree of $T$, the subgraph of $G$ induced by the subtree is connected. It can be easily shown that a graph has tree-depth $\leq d$ iff it has a tree-depth decomposition of depth $d$. The *root* of the decomposition is the root of the tree $T$. Note that if $G$ is disconnected, we define the tree-depth decomposition as a rooted forest consisting of the tree-depth decompositions of its connected components.

Tree-depth is well-behaved with respect to the operation of taking minors. Also, we can test if a graph has tree-depth $d$ efficiently:

**Claim 2.** *If $H$ is a minor of $G$, then $\text{td}(H) \leq \text{td}(G)$ [24].*

**Claim 3.** *Given a graph $G$, we can find the tree-depth of $G$ in time $O(f(d)n^2)$ for some computable function $f$, where $d = \text{td}(G)$ [23].*

We note that the tree-depth of a graph gives a lower bound on the vertex cover number of $G$. To see this, note that any graph of vertex cover number $k$ has a depth $k+1$ decomposition tree $T$ taken as follows: Order the vertices of a minimal vertex cover arbitrarily and place them in a path. At the bottom of the path, attach all remaining nodes of the graph as leaves. This obeys $E(G) \subseteq \text{clos}(T)$ by the definition of vertex cover, so for any graph $G$, $\text{td}(G) \leq \text{vcn}(G) + 1$.

Likewise, it can be easily shown that the path-width of a graph is a lower-bound on its tree-depth [3]. So if GI were FPT when parameterized by path-width, then it would trivially also be FPT parameterized by tree-depth. However GI is not known to be fixed-parameter tractable when parameterized by path-width.

## 3 Games

Suppose that $G$ is a connected graph of tree-depth $d$. We will show that the number of vertices in $G$ which can serve as a root of a minimal tree-depth decomposition is bounded as a function of $d$. In order to prove this result, we will descibe yet another equivalent definition of tree-depth in terms of cops and robbers games. Such games are frequently used in the context of logic and graph isomorphism, e.g. [5]. The bound we obtain on the number of roots will play a crucial role in our isomorphism algorithm.

### 3.1 A Characterization of Tree-Depth in Terms of Cops-and-Robbers

We define a cops-and-robbers game in which the cops do not move once they land on the graph. Thus, the number of moves in the game is limited by the number of cops. We make this precise below.

Consider the following game played on a connected graph $G = (V, E)$. The game is played by two players, called Cop and Robber. The Cop player controls $d$ cops, and the Robber player controls one robber.

First, Robber places the robber on any vertex in $G$, and announces his position. Cop announces a position where he will place his next cop. In response, the robber can move along a path in the graph to another position, including the one announced by Cop, but he cannot move through positions previously occupied by cops.

The Cop player wins if, at the end of the move, the robber is on the vertex just occupied by a cop, and the Robber player wins if all $d$ cops are on the graph and the robber is still not caught. This game is known to capture tree-depth [11][12] in the following way:

**Claim 4.** *For a connected graph $G$, $\mathrm{td}(G)$ is equal to the least $d$ for which the Cop player has a winning strategy.*

*Proof.* If $\mathrm{td}(G) = d$, consider the following strategy for the Cop player: place the first cop on the root $r$ of the decomposition. Place the next cop on the root of the depth $d-1$ decomposition of the connected component of $G-r$ containing the robber. Repeat. Clearly this is a winning strategy for the Cop player with $d$ cops. On the other hand, given a winning strategy $\gamma$ with $d$ cops, construct a decomposition by taking the root of each (sub)decomposition to be the position played by $\gamma$ if the robber is in that connected component. Since $\gamma$ uses $d$ cops, the depth of the resulting tree $T$ is at most $d$, and we will have $E \subseteq \mathrm{clos}(T)$ because $\gamma$ is a winning strategy. □

Hence a vertex $v$ is a *root* of a tree-depth decomposition of minimal depth iff the Cop player has a winning strategy using $\mathrm{td}(G)$ cops which places the first cop at $v$. Let $\mathrm{root}(G)$ be the set of all roots. In the rest of this section we will provide a self-contained proof that $|root(G)|$ is bounded by a function of $d = \mathrm{td}(G)$. As pointed out by an anonymous referee, this fact also follows easily from a recent paper by Dvořák, Giannopoulou and Thilikos [6]. These authors showed that the class $C_d = \{G : \mathrm{td}(G) \leq d\}$ is characterized by a finite set of forbidden subgraphs, each with at most $2^{2^{d-1}}$ vertices. Now consider a graph $G$ of tree-depth $d$. Since $G \notin C_{d-1}$, there exists a subgraph $H$ of $G$ containing at most $2^{2^{d-2}}$ vertices with $\mathrm{td}(H) = d$. If $\gamma$ is a winning strategy for Cop on $G$ using at most $d$ cops, then $\gamma$ must make its first move in $H$. Indeed, if $\gamma$ makes its first move outside of $H$, then the robber player can move into $H$, and subsequently play an optimal Robber strategy for $H$, forcing $\gamma$ to use $d+1$ cops to win. Therefore $\mathrm{root}(G) \subseteq H$, so $|\mathrm{root}(G)| \leq 2^{2^{d-2}}$. We conjecture that $|\mathrm{root}(G)| = 2^{O(d)}$, but for our purposes it is enough to show that it is bounded.

### 3.2 Components and Isomorphisms

Consider the state of the game after $k$ rounds of play. Let $B$ be the set of $k$ vertices occupied by cops so far.

We say that $C \subseteq V$ is a *component* of $V - B$ if there are no edges between $C$ and $V - C - B$, i.e., the Cops have blocked all exit routes from $C$. We say that two components $C_1$ and $C_2$ are *isomorphic* iff there is a bijection $\phi : C_1 \cup B \to C_2 \cup B$ such that $\phi(b) = b$ for $b \in B$, and $E(v_1, v_2)$ iff $E(\phi(v_1), \phi(v_2))$.

### 3.3 Counting Components

We will show that for a connected graph $G$, $\mathrm{td}(G)$ and $\mathrm{root}(G)$ are unaffected by removing "extra" copies of isomorphic components which arise in the course of the game. This will be the key fact which allows us to bound the size of $\mathrm{root}(G)$ as a function of the tree-depth.

**Lemma 5.** *Let $G$ be a connected graph with* $\mathrm{td}(G) = d$*. Let $B \subseteq V$ be a set of $k$ vertices, and let $C_1, C_2, \ldots, C_u$ be isomorphic components of $V - B$, where $u \geq d+1$. Let $G'$ be the graph obtained from $G$ by removing all the components $C_i$ for $i > d+1$. Then* $\mathrm{td}(G) = \mathrm{td}(G')$ *and* $\mathrm{root}(G) = \mathrm{root}(G')$.

*Proof.* Without loss of generality we can assume that for each $b \in B$ there is an edge between $b$ and $C_i$.

Let $\rho$ be a Robber strategy which forces Cop to use $d$ cops. We will construct a Robber strategy $\rho'$ based on $\rho$ which forces the Cop player to use at least $d$ cops on $G'$. This will show $\mathrm{td}(G') \geq d$. Since $G'$ is a subgraph of $G$, $\mathrm{td}(G') \leq d$, so this will show the tree-depth is unaffected by removing the extra copies of isomorphic components.

Let $\gamma''$ be a Cop strategy on $G'$. We will play $\gamma''$ against $\rho'$, and construct $\rho'$ to force $\gamma''$ to use $d$ cops.

We start by having $\rho'$ place the robber on an arbitrary vertex of the graph (it does not matter since the graph is connected). Then we construct $\rho'$ in two stages. The basic idea is to mirror the strategy $\rho$ as closely as possible. We will only have to change the strategy if $\gamma''$ plays in a $C_i$ for $i \geq d+1$ (because we deleted these vertices), or once $B$ has been filled with cops.

The first stage begins, and ends iff cops have been placed on all vertices of $B$, and the robber has moved to a $C_i$. This ensures that throughout this stage, the robber can move between all copies of $C_i$ in $G'$ which do not contain cops. We now have $\rho'$ play the same move that $\rho$ would play in $G$, unless $\rho$ plays in a $C_i$ for $i \geq d+1$. If this occurs, we mimic the response of $\rho$ via the isomorphism in one of the copies of $C_i$ in $G'$ which currently does not contain any cops. Since any Cop player on $G$ can place cops in at most $d$ copies of $C_i$, and we have kept $d$ copies of the $C_i$ in $G'$, we will never run out of copies to mirror this strategy.

If $\gamma''$ never exits the first stage, it must use at least $d$ cops to win. Indeed if the robber is connected to the $C_i$'s which contain no cops, the robber can always move between these $C_i$'s via $B$, so $\gamma''$ loses. If the robber is confined outside the $C_i$'s, $\gamma''$ must use at least $d$ cops because $\rho'$ is identical to $\rho$ once it is confined outside the $C_i$'s.

If $\gamma''$ does place cops on all of $B$, with the robber confined to a $C_i$, we proceed to the second stage. In this stage we simply directly copy the behavior of $\rho$ on an isomorphic copy of $C_i$ as before. We can easily see that $\rho'$ forces the Cop player to use $d$ cops, and hence $\mathrm{td}(G) = \mathrm{td}(G')$.

Now to show $\mathrm{root}(G) = \mathrm{root}(G')$, consider any winning Cop strategy $\gamma'$ induced by a winning strategy $\gamma$ on $G$. If $\gamma$ makes its first move on a $C_i$, then we could remove this $C_i$ in constructing $G'$ to create a winning strategy for the Cop player on $G'$ using $d-1$ cops, which is a contradiction. Hence $\mathrm{root}(G)$ must be disjoint with $C_i$ for all $i$. Therefore $\gamma$ must place its first cop outside all $C_i$, and so does $\gamma'$. Therefore $root(G) \subseteq root(G')$. We can likewise reverse this entire argument by considering adding copies of isomorphic components to $G'$ to obtain a larger graph $G$, assuming $G'$ has at least $d$ copies of the component already. By constructing the Cop and Robber strategies for $G$ based on the strategies for $G'$, we can see that $\mathrm{root}(G') \subseteq \mathrm{root}(G)$. Thus $\mathrm{root}(G) = \mathrm{root}(G')$. □

### 3.4 Measuring Components

Let $G$ be an arbitrary graph. As long as we can find a set $B \subseteq V(G)$ such that the graph $G - B$ contains more than $d+1$ isomorphic components, we remove the extra components by Lemma 5. Ultimately we obtain a *minimal* graph $G'$ where each component appears at most $d+1$ times. From Lemma 5 we know that $\text{root}(G') = \text{root}(G)$. Thus, we have only to show that $|\text{root}(G)|$ is bounded for minimal graphs.

Let $\gamma$ be winning a strategy for the Cop player which uses at most $d$ cops, and let $\rho$ be any robber strategy. The the following holds.

**Lemma 6.** *Let $G$ be a connected graph which is minimal as described in Lemma 5. Let $B$ be the set of vertices blocked by cops after $i$ rounds of play between any such $\gamma$ and $\rho$. Then there exists a function $f$ such that the component of $G - B$ containting the robber consists of at most $f(d, i)$ vertices.*

*Proof.* The proof follows by reverse induction on $i$. For $i = d$ we know that Robber has been caught, so $f(d, d) = 0$.

For $i < d$, let $v$ be the vertex where $\gamma$ puts its next cop. Let $B' = B \cup \{v\}$. From the inductive assumption we know that each component of $V - B'$ has size at most $s = f(d, i+1)$. Up to isomorphism there are at most $S = 2^{\binom{s}{2}(i+1)}$ possible components of size $s$. Since the graph $G$ is minimal, each of them appears at most $d+1$ times. Thus, $f(d, i) < 1 + (d+1)S$. □

Thus, a minimal graph of tree-depth $d$ has at most $f(d, 0)$ vertices, and we have proven the following lemma:

**Lemma 7.** *If a connected graph $G$ has tree-depth $d$, then the number of roots of tree-depth decompositions of $G$ of minimal depth is at most $f(d)$ for somc function of $d$.*

## 4 Isomorphism Algorithm

We will now create an algorithm which shows that graph isomorphism parameterized by tree-depth is in FPT. The basic idea is to extend the logspace algorithm for tree isomorphism developed by Lindell [18] to test for isomorphism over tree-depth decompositions.

Lindell's algorithm works by establishing an ordering $<$ on the set of connected trees [18]. In his algorithm, two trees $S$ and $T$ obey $S < T$ if

1. $|S| < |T|$, where $|S|$ denotes the number of nodes in $S$.
2. $|S| = |T|$ and $\#s < \#t$, where $\#s$ is the number of children of the root of $S$
3. $|S| = |T|$, $\#s = \#t = k$ and $(S_1, S_2, ..., S_k) < (T_1, T_2, ..., T_k)$ lexicographically, where we inductively assume that $S_1 \leq S_2 \leq ... \leq S_k$ and $T_1 \leq T_2 \leq ... \leq T_k$ are the ordered subtrees of $S$ and $T$ obtained by removing the roots of $S$ and $T$.

Clearly $S \cong T$ iff neither $S < T$ nor $T < S$ [18].

We will extend this ordering on trees to an ordering of the tree-depth decompositions. To test for isomorphism, we will find a minimal, canonical decomposition of each graph and compare the decompositions.

Recall that a tree-depth decomposition of a connected graph $G$ consists of a rooted tree $T$ over $V(G)$ such that $E(G) \subseteq E(\text{clos}(T))$. A tree-depth decomposition also has the property that any induced subgraph of $G$ obtained by the vertices of a rooted subtree of $T$ is connected.

We say that two decompositions $T_1$ of $G_1$ and $T_2$ of $G_2$ are equivalent, denoted $T_1 \simeq T_2$, if there is an isomorphism $\phi$ between $T_1$ and $T_2$ which preserves the edges of the underlying graphs as well, i.e. both $(u,v) \in E(T_1) \Leftrightarrow (\phi(u), \phi(v)) \in E(T_2)$ and $(u,v) \in E(G_1) \Leftrightarrow (\phi(u), \phi(v)) \in E(G_2)$. In particular, $T_1 \simeq T_2$ implies $G_1 \cong G_2$.

Suppose that we are given a connected graph $G$ with $\text{td}(G) = d$. Given a tree-depth decomposition $T$ of $G$, define a sub tree-depth decomposition of a graph $G'$ induced by a subtree of $T'$ of $T$ to consist of the following: the tree $T'$ over $V(G')$ with root $t'$, as well as the path $P$ from the parent of $t'$ to the root of $T$. If $td(G') = d'$, then $P$ consists of vertices $r = r_1, r_2, ... r_{d-d'}$, where $r$ is the root of $T$ and $r_{d-d'}$ is the parent of $t'$ in $T$. See Figure 1 for clarification.

We inductively define an ordering of sub tree-depth decompositions of $G$ as follows. Let $S$ and $T$ be two depth $d'$ subdecompositions of $G_S$ and $G_T$, respectively, with roots $s$ and $t$, respectively, and which share the same path $P$ defined above. Note $S$ and $T$ must share the same path $P$ to be comparable. Also note that when considering the entire graph, $P$ is empty so this defines an ordering on all tree-depth decompositions.

We say that the subdecomposition $S$ of $G_S$ is less than the subdecomposition $T$ of $G_T$, denoted $S < T$, if one of the following conditions is satisfied:

1. $|G_S| < |G_T|$
2. $|G_S| = |G_T|$ and $\#s < \#t$, where $\#x$ is the number of connected components in $G_X - x$.
3. $|G_S| = |G_T|$, $\#s = \#t$, and $(E(r_1, s), E(r_2, s), ..., E(r_{d-d'}, s)) < (E(r_1, t), E(r_2, t), ..., E(r_{d-d'}, t))$ lexicographically, where $E(x, y) = 1$ if there is an edge from $x$ to $y$ and 0 otherwise. If $d' = d$ this condition is trivially satisfied.
4. $|G_S| = |G_T|$, $\#s = \#t$, $E(r_i, s) = E(r_i, t)$ $\forall i = 1...(d-d')$ and

$$(S_1, S_2, ..., S_k) < (T_1, T_2, ..., T_k)$$

   lexicographically, where we inductively assume $S_1 \leq S_2 \leq ... \leq S_k$ and $T_1 \leq T_2 \leq ... \leq T_k$ are the connected components of $G_S - s$ and $G_T - t$, ordered by their subdecompositions induced by $S$ and $T$. (Here $S \leq T$ means $S < T$ or $S \simeq T$).

This ordering has several nice properties. The following can be shown by simple induction on the tree-depth:

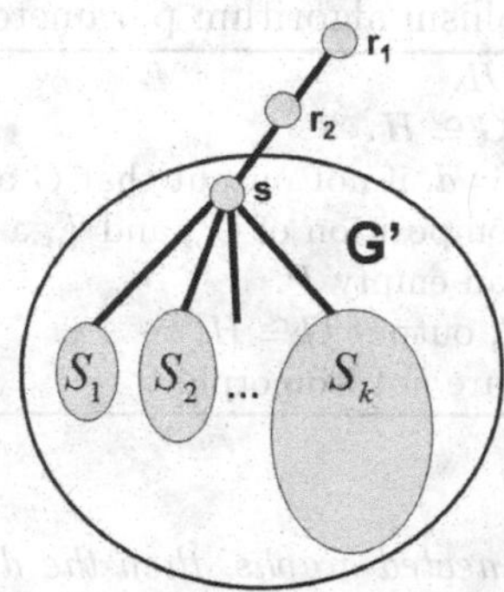


**Fig. 1.** A sub-decomposition of $G'$ with root $s$ and components $S_1...S_k$ of $G' - s$

**Claim 8.** *Suppose G and H are connected graphs, both of tree-depth d and the same size. Let S be a minimal tree-depth decomposition of G and T a minimal trec-depth decomposition of H according to the above ordering. Then if neither* $S < T$ *nor* $T < S$, *then* $S \simeq T$ *and* $G \cong H$.

By condition (4) of the ordering, we know that to find the minimal decomposition of $G$ rooted at $s$, we simply need to find the minimal decompositions of each of the connected components of $G - s$. This forms the basis of a recursive algorithm to compute the minimum depth-$d$ decomposition of a graph $G$, Algorithm 1. With this in hand, we can show that Algorithm 2 correctly tests for isomorphism.

---

**Algorithm 1.** Recursive construction of a minimal tree-decomposition

---

**Input**: A connected graph $G'$ of tree-depth $d'$ along with a specified path $P = r_1...r_l$.
**Output**: A sub tree-depth decomposition $S$ of $G'$ of depth $d'$ which is minimal with respect to $<$ for $P$.
**if** $\text{td}(G) = 1$ **then**
  Output the trivial decomposition of the graph.
**else**
  Find $R = \{v \in V(G') : \text{td}(G' - v) = d - 1\}$.
  Remove those elements $r \in R$ which do not have minimal values of $(E(r_1, r), E(r_2, r), ..., E(r_l, r))$ or $\#r$.
  **foreach** $r \in R$ **do**
    Compute minimal decompositions of $S_1...S_k$, the connected components of $G' - r$, using this algorithm and appending $r$ path $P$.
    Order $S_1...S_k$ by $<$ using $k \log k$ comparisons.
  **end**
  Find which $r \in R$ produces the decompositions $(S_1, ...S_k)$ which are minimal in lexicographic order, and output the decomposition obtained by making this the root of the decomposition.
**end**

---

**Algorithm 2.** An isomorphism algorithm parameterized by tree-depth

**Input**: Two graphs $G$ and $H$.
**Output**: Whether or not $G \cong H$.
Check that $\mathrm{td}(G) = \mathrm{td}(H) = d$, if not output that $G$ and $H$ are not isomorphic.
Compute $S$, a minimal decomposition of $G$ , and $T$, a minimal decomposition of $H$ using Algorithm 1 with an empty $P$.
If neither $S < T$ nor $T < S$, output $G \cong H$.
Else output that $G$ and $H$ are not isomorphic.

**Claim 9.** *If $G \cong H$ are connected graphs, then the decompositions produced by Algorithm 1 on G and H are isomorphic.*

*Proof.* Follows because all steps in Algorithm 1 are isomorphism invariant. □

**Corollary 10.** *Algorithm 2 correctly tests for isomorphism over connected graphs.*

**Theorem 11.** *Graph isomorphism is fixed-parameter tractable in tree-depth.*

*Proof.* We will upper bound $T(n, d)$, the runtime of Algorithm 1 on a connected graph $G$ with $n$ vertices and tree-depth $d$.

By Claim 3 we can check if $\mathrm{td}(G) = d - 1$ in time $f(d-1)n^2$, so finding $R = \{v \in V(G') : \mathrm{td}(G' - v) = d - 1\}$ can be done in time $f(d)n^3$.

Next we reduce the size of $R$ to only those vertices with minimal $\#r$. For each $r$, computing $\#r$ can be done in time $\Sigma_{i=1\ldots\#r} O(|S_i|^2) \leq O(n^2)$. Hence this step takes time $O(g(d)n^2)$, since by Lemma 7 $|R| \leq g(d)$.

Now the algorithm recurses. By Lemma 7 it will recurse on at most $g(d)$ different roots. For each of these roots $r$, if $k = \#r$ then it will compute a decomposition of each of the connected components $S_1 \ldots S_k$ of $G - r$, which will take time $\leq \Sigma_{i=1}^{k} T(|S_i|, d-1)$. To order these decompositions by $<$, it will then make $k \log k$ comparisons of the decompositions using $<$, each of which takes time $O(n^2)$, and subsequently make $g(d)k$ comparisons in time $g(d)kn^2$ between these sorted decompositions to find which of the $g(d)$ roots is minimal.

This yields a recursion relation for the run time given by

$$T(n, d) \leq f(d)n^3 + g(d)n^2 + g(d) \left\{ \left( \sum_{i=1}^{k} T(|S_i|, d-1) \right) + O(k \log kn^2) \right\} \quad (1)$$

One can easily check that a run time of $T(n, d) = h(d)n^3 \log(n)$ suffices. Plugging this ansatz into the recursion relation, and simplifying using the convexity of $n^3 \log(n)$ and the fact that $k \leq n$, one can see that

$$T(n, d) \leq (f(d) + g(d))n^3 \log(n) + g(d) \left\{ h(d-1)n^3 \log(n) + O(n^3 \log(n)) \right\} \quad (2)$$

Taking $h(d) = f(d) + g(d)\,(h(d-1) + O(1))$, we have $T(n, d) \leq h(d)n^3 \log(n)$. By setting $h(0) = 1$, this provides an inductive definition of $h$ as a function.

Therefore Algorithm 1 runs in time $O(h(d)n^3 \log(n))$. Since checking if two decompositions are equivalent takes $O(n^2)$ time (assuming the orderings of the sub-decompositions of each level are recorded), Algorithm 2 correctly decides isomorphism over connected graphs in time $O(h(d)n^3 \log(n))$. This algorithm extends easily to disconnected graphs by the convexity of $n^3$. □

## 5 Generalized Tree-Depth

We will now define a new parameter which generalizes both tree-depth and max-leaf-number. Recall that the max-leaf-number of a graph $G$, denoted $\mathrm{mln}(G)$, is the maximum number of leaves in a spanning tree of $G$. A crucial fact is that if $\mathrm{mln}(G) = k$, then the number of vertices of degree $\neq 2$ in $G$ is bounded by a function of $k$. This can be easily used to show that graph isomorphism is fixed-parameter tractable in max-leaf-number, by simply trying all bijections between vertices of degree $\neq 2$ and noting that all other vertices lie on simple paths.

**Claim 12 (From Kleitman and West [16] via [9]).** *If $G$ is a graph with $\mathrm{mln}(G) \leq k$, then $G$ is a subdivision of a graph $H$ with at most $4k - 2$ vertices.*

Low max-leaf-number means that the graph becomes a collection of paths after removing a small number of vertices. Low tree-depth, on the other hand, means that the graph quickly degenerates to an empty set after alternately removing vertices and considering disjoint components separately. Following the example of max-leaf number, we can generalize tree-depth by allowing a broader class of graphs as leftovers at the end of $k$-cops and robbers game. Previously, we said that the Cop player wins the game if a cop lands on the robber. Consider a modified version of the game, in which Cop wins if the robber is confined to either a simple path with cops at both endpoints or a simple cycle. The endpoints of this path (which must be occupied by cops) may be connected to other vertices of the graph, but the other points of the path may not have any edges to the rest of the graph. To win by confining the robber to a cycle, the cycle must be disconnected from the rest of the graph.

**Definition 13.** *A graph $G$ has generalized tree-depth $d$, denoted $\mathrm{gtd}(G) = d$, iff $d$ is the least $k$ such the Cop player has a winning strategy in the modified $k$ cops and robber game described above.*

It is clear that the generalized tree-depth of a graph is a lower bound on its tree-depth, since a single vertex is a simple path of length zero. Furthermore, this parameter bounds from below the max-leaf-number, because the Cop player has a winning strategy using $4\mathrm{mln}(G)-2$ cops by placing places cops on all vertices of degree $\neq 2$. Therefore for any graph $G$, $\mathrm{gtd}(G) \leq 4\mathrm{mln}(G)$ and $\mathrm{gtd}(G) \leq \mathrm{td}(G)$, so this parameter generalizes both tree-depth and max-leaf-number.

We can now extend our arguments from the previous sections to show that graph isomorphism is fixed-parameter tractable in the generalized tree-depth. First, note that having generalized tree-depth $d$ is equivalent to having a tree-depth decomposition of depth $d$ as before, except that the leaves of the tree can now consist of simple paths. The simple cycles, being disconnected from the graph, are omitted from the decomposition and are handled separately later. The endpoints of the path in each leaf are specified in the decomposition. By the same arguments as in [24], $C_d = \{G : \text{gtd}(G) \leq d\}$ is closed under taking minors for $d \geq 2$, while $C_0$ and $C_1$ have trivial poly-time membership tests, yielding a (non-constructive) FPT algorithm to compute generalized tree-depth in time $O(f(d)n^3)$ by the Robertson-Seymour Theorem [26][19].

Next, note that our arguments bounding the number of roots of a decomposition also carry through. When counting the number of vertices in each component, we count only the endpoints of the paths in the leaves, since only these vertices can connect to the rest of the graph. These specified endpoints are considered the roots of the leaf's decomposition. This ensures that the base case of Lemma 6 is still bounded above by two. We again obtain that the number of roots of a graph of generalized tree-depth $d$ is bounded by a function $g(d)$. The fact that we handle simple cycles separately is crucial to keeping this bound.

We can likewise extend Lindell's tree isomorphism algorithm to an FPT algorithm for generalized tree-depth exactly as before. To do so, we simply modify our ordering on decompositions to take in to account the number of nodes in the path of each leaf, and handle the simple cycles separately. This yields an algorithm to test for isomorphism in $O(h(d)n^4)$ time, where $d$ is the generalized tree-depth and $h(d)$ is a function which is not necessarily computable, as we have used the Robertson-Seymour theorem. We have thus shown:

**Theorem 14.** GI *is fixed-parameter tractable in generalized tree-depth.*

## 6 Conclusion

We have shown that graph isomorphism is fixed-parameter tractable when parameterized by generalized tree-depth. An open question is whether or not GI is FPT in the path-width of the graph. Unlike in the case of tree-depth, the number of valid path-width decompositions of a graph is exponential in the number of vertices, so our approach does not immediately generalize.

**Acknowledgments.** Research by Adam Bouland was partially supported by the NSF Graduate Research Fellowship under grant no. 1122374 and by a Marshall Scholarship. Research by Anuj Dawar was partially supported by EPSRC grant EP/H026835. Research by Eryk Kopczyński was partially supported by ESF Research Networking Programme GAMES and by the Polish National Science Centre (grant N N206 567140).

## References

1. Arvind, V., Das, B., Johannes, K., Toda, S.: Colored Hypergraph Isomorphism is Fixed Parameter Tractable. In: ECCC 93 (2009)
2. Babai, L.: Monte-Carlo algorithms in graph isomorphism testing. Tech. Rep. DMS 79-10, Université de Montréal, pp. 1–33 (1979)

3. Bodlaender, H., Hafsteinsson, H., Gilbert, J.R., Kloks, T.: Approximating treewidth, pathwidth, frontsize, and shortest elimination tree. J. Algorithms 18, 238–255 (1995)
4. Bodlaender, H.L.: Polynomial Algorithms for Graph lsomorphism and Chromatic Index on Partial k-Trees. Journal of Algorithms 11(4), 631–643 (1990)
5. Cai, J.-Y., Fürer, M., Immerman, N.: An Optimal Lower Bound on the Number of Variables for Graph Identification. Combinatorica 12(4), 389–410 (1992)
6. Dvořák, Z., Giannopoulou, A., Thilikos, D.M.: Forbidden graphs for tree-depth. European Journal of Combinatorics 33(5), 969–979 (2012)
7. Elberfeld, M., Grohe, M.: Where First-Order and Monadic Second-Order Logic Coincide. Arxiv preprint arXiv:1204.6291, pp. 1–15 (2012)
8. Evdokimov, S., Ponomarenko, I.: Isomorphism of Coloured Graphs with Slowly Increasing Multiplicity of Jordan Blocks. Combinatorica 19(3), 321–333 (1999)
9. Fellows, M., Lokshtanov, D., Misra, N., Mnich, M., Rosamond, F., Saurabh, S.: The Complexity Ecology of Parameters: An Illustration Using Bounded Max Leaf Number. Theory of Computing Systems 45(4), 822–848 (2009)
10. Furst, M., Hopcroft, J.: Luks: Polynomial-time algorithms for permutation groups. In: Proc. FOCS 1980, pp. 36–41 (1980)
11. Ganian, R., Hliněný, P., Kneis, J., Langer, A., Obdržálek, J., Rossmanith, P.: On Digraph Width Measures in Parameterized Algorithmics. In: Chen, J., Fomin, F.V. (eds.) IWPEC 2009. LNCS, vol. 5917, pp. 185–197. Springer, Heidelberg (2009)
12. Giannopoulou, A., Hunter, P., Thilikos, D.: LIFO-search: A min-max theorem and a searching game for cycle-rank and tree-depth. Submitted to J. Discrete Math. (2011)
13. Grohe, M., Marx, D.: Structure Theorem and Isomorphism Test for Graphs with Excluded Topological Subgraphs. In: Proc. STOC 2012, pp. 173–192 (2012)
14. Grohe, M.: Fixed-point definability and polynomial time on graphs with excluded minors. In: Proc. LICS 2010, pp. 179–188 (2010)
15. Heath, M., Ng, E., Peyton, B.: Parallel algorithms for sparse linear systems. SIAM Review 33(3), 420–460 (1991)
16. Kleitman, D., West, D.: Spanning Trees with Many Leaves. SIAM J. Discrete Math. 4, 99–106 (1991)
17. Kratsch, S., Schweitzer, P.: Isomorphism for Graphs of Bounded Feedback Vertex Set Number. In: Kaplan, H. (ed.) SWAT 2010. LNCS, vol. 6139, pp. 81–92. Springer, Heidelberg (2010)
18. Lindell, S.: A logspace algorithm for tree canonization. In: Proc. STOC 1992, pp. 400–404 (1992)
19. Lovász, L.: Graph minor theory. Bulletin of the AMS 43(1), 75–86 (2006)
20. Luks, E.: Isomorphism of graphs of bounded valence can be tested in polynomial time. Journal of Computer and System Sciences (1982)
21. Manne, F.: An Algorithm for Computing an Elimination Tree of Minimum Height for a Tree. Tech. Rep. CS-91-59, University of Bergen, Norway (1992)
22. Miller, G.: Isomorphism testing for graphs of bounded genus. In: Proc. STOC 1980, pp. 225–235 (1980)
23. Nešetřil, J., Ossona de Mendez, P.: Sparsity: Graphs, Structures and Algorithms. Algorithms and Combinatorics, vol. 28. Springer (2012)
24. Nešetřil, J., Ossona de Mendez, P.: Tree-depth, subgraph coloring and homomorphism bounds. European Journal of Combinatorics 27(6), 1022–1041 (2006)

25. Ponomarenko, I.: The isomorphism problem for classes of graphs that are invariant with respect to contraction. Zap. Nauchn. Sem. Leningrad. Otdel. Mat. Inst. Steklov (LOMI) 174, 147–177 (1988) (Russian)
26. Robertson, N., Seymour, P.: Graph minors XX. Wagners conjecture. Journal of Combinatorial Theory, Series B 92, 325–357 (2004)
27. Toda, S.: Computing automorphism groups of chordal graphs whose simplicial components are of small size. IEICE Transactions on Information and Systems E89-D(8), 2388–2401 (2006)
28. Yamazaki, K., Bodlaender, H.L., de Fluiter, B., Thilikos, D.M.: Isomorphism for Graphs of Bounded Distance Width. Algorithmica 24(2), 105–127 (1999)

# Parameterized Algorithmics and Computational Experiments for Finding 2-Clubs

Sepp Hartung, Christian Komusiewicz, and André Nichterlein

Institut für Softwaretechnik und Theoretische Informatik, TU Berlin, Berlin, Germany
{sepp.hartung,christian.komusiewicz,andre.nichterlein}@tu-berlin.de

**Abstract.** Given an undirected graph $G = (V, E)$ and an integer $\ell \geq 1$, the NP-hard 2-CLUB problem asks for a vertex set $S \subseteq V$ of size at least $\ell$ such that the subgraph induced by $S$ has diameter at most two. In this work, we extend previous parameterized complexity studies for 2-CLUB. On the positive side, we give polynomial kernels for the parameters "feedback edge set size of $G$" and "size of a cluster editing set of $G$" and present a direct combinatorial algorithm for the parameter "treewidth of $G$". On the negative side, we first show that unless NP $\subseteq$ coNP/poly, 2-CLUB does not admit a polynomial kernel with respect to the "size of a vertex cover of $G$". Next, we show that, under the strong exponential time hypothesis, a previous $O^*(2^{|V|-\ell})$ search tree algorithm [Schäfer et al., Optim. Lett. 2012] cannot be improved and that, unless NP $\subseteq$ coNP/poly, there is no polynomial kernel for the dual parameter $|V| - \ell$. Finally, we show that, in spite of this lower bound, the search tree algorithm for the dual parameter $|V| - \ell$ can be tuned into an efficient exact algorithm for 2-CLUB that substantially outperforms previous implementations.

## 1 Introduction

Finding cohesive subnetworks in a network is an important task in graph-based data mining and social network analysis. The natural cohesiveness requirement is to demand that the subnetwork is a complete graph, that is, a clique. However, this requirement is often too restrictive and thus relaxed versions such as $s$-cliques [1], $s$-plexes [22], and $s$-clubs [17] have been proposed. An $s$-club is a graph with diameter at most $s$, and $s$-clubs are thus a distance-based relaxation of cliques (which are exactly the graphs of diameter 1). For a constant integer $s \geq 1$, the problem of finding large $s$-clubs is defined as follows.

$s$-CLUB
**Input:** An undirected graph $G = (V, E)$ and an integer $\ell \geq 1$.
**Question:** Is there a vertex set $S \subseteq V$ of size at least $\ell$ such that $G[S]$ has diameter at most $s$?

Clearly, 1-CLUB is equivalent to the well-known CLIQUE problem. In this work, we consider 2-CLUB, the most basic variant of $s$-CLUB that is different from

D.M. Thilikos and G.J. Woeginger (Eds.): IPEC 2012, LNCS 7535, pp. 231–241, 2012.

Clique. Furthermore, 2-Club is also an important special case concerning the applications: For biological networks, 2-clubs and 3-clubs have been identified as the most reasonable diameter-relaxations of Clique [19], further applications of 2-Club arise in the analysis of social networks [16]. Consequently, experimental evaluations concentrate on finding 2-clubs and 3-clubs [5, 6, 15].

*Related Work.* For all $s \geq 1$, $s$-Club is NP-complete on graphs of diameter $s+1$ [3]; 2-Club is NP-complete even on split graphs and, thus, also on chordal graphs [3]. For all $s \geq 1$, $s$-Club is NP-hard to approximate within a factor of $n^{1/2-\epsilon}$ [2]; a simple approximation algorithm obtains a factor of $n^{1/2}$ for even $s \geq 2$ and a factor of $n^{2/3}$ for odd $s \geq 3$ [2]. Several heuristics [5] and integer linear programming formulations [3, 5] for $s$-Club have been proposed and experimentally evaluated [15]. 1-Club is equivalent to Clique and thus W[1]-hard with respect to $\ell$. In contrast, for $s \geq 2$, $s$-Club is fixed-parameter tractable with respect to $\ell$ [6, 20, 21]. $s$-Club can be solved in $O^*(2^{n-\ell})$ time by a search tree algorithm [6, 20, 21][1]. $s$-Club can be formulated in monadic second order logic and thus is fixed-parameter tractable with respect to the treewidth of $G$ [21]. Moreover, $s$-Club does not admit a polynomial kernel with respect to $\ell$ (unless NP $\subseteq$ coNP/poly), but a so-called *Turing*-kernel with at most $k^2$ vertices for even $s$ and at most $k^3$ vertices for odd $s$ [20]. 2-Club is solvable in polynomial time on bipartite graphs, on trees, and on interval graphs [21].

*Our Contribution.* We make progress towards a systematic classification of the complexity of 2-Club with respect to structural parameters of the input graph. In Section 2, we give an $O(k^2)$-vertex kernel for the parameter "size of a cluster editing set" and an $O(k)$-size kernel for the parameter "feedback edge set size". The kernelization results for these rather large parameters are motivated by our negative results: We show that 2-Club does not admit a polynomial kernel with respect to the size of a vertex cover of the underlying graph, unless NP $\subseteq$ coNP/poly. This excludes polynomial kernels for many prominent structural parameters such as "feedback vertex set size", pathwidth, and treewidth. In Section 3, we give a direct combinatorial algorithm solving 2-Club in $2^{O(2^{\omega})}n^2$ time on graphs of treewidth $\omega$. Notably, up to a constant in the exponent, this is also the current best running time for the parameter vertex cover size (which we present in Theorem 4). In Section 4, we study $s$-Club, $s \geq 2$, parameterized by the dual parameter $k' := n - \ell$. We prove that unless the *Strong Exponential Time Hypothesis* (SETH)[2] fails, $s$-Club cannot be solved in $O^*((2-\epsilon)^{k'})$ time for all $\epsilon > 0$. This is evidence that the previous search tree algorithm [20] is

[1] Schäfer et al. [20] actually considered finding an $s$-club of size *exactly* $\ell$. The claimed fixed-parameter tractability with respect to $n-\ell$ however only holds for the problem of finding an $s$-club of size *at least* $\ell$. The other fixed-parameter tractability results hold for both variants.

[2] The SETH fails if the satisfiability problem for boolean formulas in conjunctive normal form, the so-called Cnf-Sat problem, is solvable in $O^*((2-\epsilon)^n)$ time for some $\epsilon > 0$ where $n$ denotes the number of variables; a recent survey on the (S)ETH is given by Lokshtanov et al. [14].

optimal with respect to the parameter $k'$. Moreover, the presented reduction also implies that $s$-CLUB does not admit a polynomial kernel with respect to $k'$ unless NP $\subseteq$ coNP/poly.

Having explored the limits of parameterized algorithmics for the dual parameter $k'$ on the theoretical side, in Section 5 we examine its usefulness for solving 2-CLUB in practice. To this end, we implemented the search tree strategy for the dual parameter together with data reduction rules that are partially deduced from our findings in Section 2. We show that our implementation outperforms all previously implemented exact algorithms for 2-CLUB on random and on large-scale real world graphs. Especially on large graphs the concept of Turing kernelization turns out to be the most efficient technique in our "parameterized toolbox".

*Preliminaries.* We only consider undirected and simple graphs $G = (V, E)$ where $n := |V|$ and $m := |E|$. For a vertex set $S \subseteq V$, let $G[S]$ denote the *subgraph induced by* $S$ and $G - S := G[V \setminus S]$. We use $\text{dist}_G(u, v)$ to denote the *distance between* $u$ *and* $v$ in $G$, that is, the length of a shortest path between $u$ and $v$. For a vertex $v \in V$ and an integer $t \geq 1$, denote by $N_t(v) := \{u \in V \setminus \{v\} \mid \text{dist}(u, v) \leq t\}$ the set of vertices within distance at most $t$ to $v$. Moreover, set $N_t[v] := N_t(v) \cup \{v\}$, $N[v] := N_1[v]$, and $N(v) := N_1(v)$. Two vertices $v$ and $w$ are *twins* if $N(v) \setminus \{w\} = N(w) \setminus \{v\}$. The following simple observation will be used throughout this work.

**Observation 1.** Let $S$ be an $s$-club in a graph $G = (V, E)$ and let $u, v \in V$ be twins. If $u \in S$ and $|S| > 1$, then $S \cup \{v\}$ is also an $s$-club in $G$.

For the relevant notions of parameterized complexity refer to [9, 10, 18]. A more recent concept not presented in these monographs is *Turing kernelization.* Roughly speaking, in Turing kernelization one creates polynomially many problem kernels instead of one problem kernel. Then, the solution to the parameterized problem can be computed by solving the problem separately on each of these problem kernels. Throughout this work, we assume that, unless stated otherwise, the structural parameter under consideration is provided as an additional input of the 2-CLUB instance. Due to the space restrictions, most of the proofs are deferred to a long version of this article.

## 2 Kernelization Algorithms and Lower Bounds

In this section, we provide polynomial-size problem kernels for 2-CLUB parameterized by "cluster editing set size" and "feedback edge set size", respectively. While these parameters can often be rather large, we show that for the (also relatively large) parameter "vertex cover size of $G$", there exists no polynomial-size problem kernel (unless NP $\subseteq$ coNP/poly).

*A Quadratic-Vertex Kernel for the Parameter Cluster Editing Set Size.* A cluster editing set of $G$ is a set of edge additions and deletions that transforms $G$ into a vertex-disjoint union of cliques. Let $G = (V, E)$, an integer $\ell$, and a cluster editing set $D$ be an instance of 2-CLUB; the parameter is $k := |D|$. Denote by $V(D)$ the set of all endpoints of the edges in $D$ and observe that $G - V(D)$ is a cluster graph. The following rules yield an $O(k^2)$-vertex kernel for 2-CLUB. The first reduction rule is obvious.

**Rule 1.** If there is a cluster $C$ in $G - V(D)$ with $|C| \geq \ell$, then reduce to a trivial yes-instance.

It follows that any 2-club of size at least $\ell$ has a nonempty intersection with $V(D)$, implying the correctness of the following data reduction rule.

**Rule 2.** If there is a cluster $C$ in $G - V(D)$ such that $N(v) \cap V(D) = \emptyset$ for all $v \in C$, then delete $C$.

After exhaustive application of Rule 2, at most $|V(D)| \leq 2k$ clusters remain in $G - V(D)$. Since Rule 1 has been applied, each cluster in $G - V(D)$ has size at most $\ell - 1$. Hence, if $\ell \leq 2k+1$, then there are at most $4k^2 + 2k$ vertices left and we are done. Now, consider the case where $\ell > 2k + 1$. To bound the size of the clusters in $G - V(D)$ we use the following observation. Its correctness follows from the fact that two vertices in different clusters of $G - V(D)$ are not adjacent and have no common neighbor.

**Observation 2.** For every 2-club $S$ in $G$ there is at most one cluster $C$ in $G - V(D)$ such that $S$ has a nonempty intersection with $C$.

Observation 2 implies that every 2-club of size at least $\ell$ contains at least $\ell - 2k$ vertices from exactly one cluster $C$ of $G - V(D)$. Since all vertices in $C$ are twins, Observation 1 now implies that in an inclusion-maximal 2-club either all or no vertices from $C$ or are contained. Hence, for $\ell > 2k + 1$ decreasing $\ell$ and the size of each cluster $C$ by $\ell - 2k - 1$ produces an equivalent instance. This leads to the following data reduction rule.

**Rule 3.** Delete $\ell - 2k - 1$ arbitrary vertices in each cluster $C$ of $G - V(D)$ and set $\ell := 2k + 1$.

Note that in case $|C| \leq l - 2k - 1$ we simply delete all vertices of $C$. After exhaustive application of Rule 3 for each cluster $C$ it holds that $1 \leq |C| < 2k + 1$. Thus, we arrive at the following.

**Theorem 1.** 2-CLUB *parameterized by the cluster editing set size $k$ admits an $O(k^2)$-vertex kernel that can be computed in $O(n + m)$ time.*

*A Linear Kernel for the Parameter Feedback Edge Set Size.* A *feedback edge set* of a graph is an edge set whose deletion leads to a forest. Let $F \subset E$ be a feedback edge set for $G = (V, E)$. Furthermore, let $T := (V, E \setminus F)$ be the forest obtained by deleting $F$ from $G$. The correctness of the first data reduction rule follows from the fact that for each vertex $v$ the set $N[v]$ is a 2-club.

**Rule 4.** If there is a vertex $v \in V$ with $|N[v]| \geq \ell$, then reduce to a trivial yes-instance.

In the following, we exploit that after application of Rule 4 all 2-clubs of size at least $\ell$ have to "use" feedback edges. The next rule removes all vertices that are not on paths between the endpoints between feedback edges. These vertices are defined as follows.

**Definition 1.** *For a feedback edge $\{u, v\} \in F$ the path $P_{\{u,v\}}$ between $u$ and $v$ in $T$ is called* feedback edge path. *If a vertex $w$ lies on the path $P_{\{u,v\}}$, then the edge $\{u, v\}$ is a* spanning feedback edge *of $w$.*

**Rule 5.** Let $(G, \ell)$ be reduced with respect to Rule 4. Then, delete all vertices that do not lie on any feedback edge path.

The final rule removes vertices that are too far away from feedback edges.

**Rule 6.** If there is a vertex $v$ that has in $G$ distance at least three to at least one endpoint of every spanning feedback edge, then remove $v$.

Applying these data reduction rules exhaustively results in a linear kernel:

**Theorem 2.** *The 2-*Club *problem parameterized by the size $k$ of a feedback edge set admits a kernel of size $6k$ that can be computed in $O(n^4)$ time.*

*A Kernelization Lower Bound for the Parameter Vertex Cover.* We next show that 2-Club does not admit a polynomial kernel with respect to the parameter vertex cover size. This result implies that 2-Club does not admit a polynomial kernel for many structural graph parameters such as feedback vertex set number or treewidth.

**Theorem 3.** *2-*Club *parameterized by vertex cover has no polynomial kernel unless NP $\subseteq$ coNP/poly.*

## 3 Fixed-Parameter Tractability with Respect to Treewidth

In this section, we show that 2-Club is fixed-parameter tractable when parameterized by treewidth. To demonstrate the principle idea behind the algorithm, we first describe a fixed-parameter algorithm for the parameter vertex cover.

**Theorem 4.** *$s$-*Club *is solvable in $O(2^k \cdot 2^{2^k} \cdot nm)$ time where $k$ denotes the size of a vertex cover.*

Extending the ideas behind Theorem 4, we now give a direct combinatorial algorithm for the parameter treewidth which uses the following lemma.

**Lemma 1.** *Let $G$ be a graph and let $S$ be a 2-club in $G$. Then, for any tree-decomposition of $G$ there is at least one vertex $v \in S$ such that there is a bag that contains $N[v] \cap S$.*

*Proof.* Let $T = (X_1 \cup \ldots \cup X_r, E)$ be a tree-decomposition of $G$. Fix an arbitrary vertex $u \in S$ and denote by $X^u$ any bag in $T$ that contains $u$. Now, choose a vertex $w \in S$ such that the length of the path from $X^u$ to the first bag that contains $w$ is maximum. Denote this bag by $X^w$. We show that $N(w) \cap S \subseteq X^w$. Suppose there is a neighbor $v \in N(w) \cap S$ that is not contained in $X^w$. Since $v$ and $w$ are adjacent and thus are together contained in at least one bag, $v$ is not contained in any bag on the path between $X^u$ and $X^w$. This implies that the path from $X^u$ to the first bag that contains $v$ is longer than the path between $X^u$ and $X^w$; a contradiction to the choice of $w$. □

**Theorem 5.** 2-Club *is solvable in* $2^{O(2^\omega)} \cdot n^2$ *time where* $\omega$ *denotes the treewidth.*

## 4 Optimality of Dual Parameter Algorithm

In this section, we prove algorithmic lower bounds for $s$-Club when parameterized by the dual parameter $k' := n - \ell$. We first show that there is a reduction from CNF-Sat to $s$-Club with certain properties that allows to infer these lower bounds.

**Lemma 2.** *There is a parameterized reduction from* CNF-Sat *to* $s$-Club *where the dual parameter in the constructed* $s$-Club *instance is equal to the number of variables in the boolean formula of the* CNF-Sat *instance.*

Denoting the number of variables in a boolean formula by $n$, the Strong Exponential Time Hypothesis (SETH) fails if CNF-Sat can be solved in $O^*((2-\epsilon)^n)$ time for some $\epsilon > 0$ [13]. Thus, by Lemma 2 an algorithm for $s$-Club running in $O^*((2-\epsilon)^{k'})$ time for some $\epsilon > 0$ would disprove the SETH. This bound is tight since $s$-Club can be solved in $O^*(2^{k'})$ time [20].

**Corollary 1.** *Unless the SETH fails,* $s$-Club *parameterized by the dual parameter* $k' := n - \ell$ *cannot be solved in* $O^*((2-\epsilon)^{k'})$ *time for all* $s \geq 2$.

Chen et al. [7] showed that CNF-Sat does not admit a polynomial kernel unless coNP ⊆ NP/poly. Since Lemma 2 provides a polynomial time and parameter transformation [4] this lower bound result transfers to 2-Club.

**Corollary 2.** $s$-Club *parameterized by the dual parameter* $k'$ *does not admit a polynomial problem kernel unless coNP* ⊆ *NP/poly.*

## 5 Implementation and Experiments

*Search Tree.* We implemented the following search tree strategy to find a maximum 2-club $S$ in a given graph $G = (V, E)$: If $G$ is not a 2-club, then find a vertex $v \in V$ such that $|N_2(v)|$ is minimum among all vertices. Then, branch into the cases to either delete $v$ from $G$ or to mark $v$ to be contained in $S$ and subsequently delete all vertices in $V \setminus N_2[v]$. During branching we maintain a lower bound, that is, the size $\ell'$ of a largest 2-club found so far; this lower bound is initialized by the maximum degree plus one. Branching is aborted if the current graph has less than $\ell'$ vertices. After exploring all branches, we output the current lower bound (along with a 2-club of this size).

The above search tree strategy was introduced by Bourjolly et al. [5] and has been already experimentally evaluated [6]. By simple recursion analysis the running time can be bounded by $O^*(\alpha^n)$ where $\alpha$ is the golden ratio with $\alpha \approx 1.62$ [6]. Schäfer et al. [20] showed that for the dual size parameter $k'$ this search tree strategy runs in $O^*(2^{k'})$ time (if branching is aborted if more than $k'$ vertices have been removed). Note that by Corollary 1, the search tree size measured by $k'$ cannot be improved unless the SETH fails.

*Turing Kernelization.* Before starting the search tree algorithm, we use the Turing kernelization introduced by Schäfer et al. [20]. That is, we compute the Turing kernels consisting of $N_2[v]$ for all vertices $v$ of the input graph.[3] We say $N_2[v]$ is the Turing kernel for vertex $v$. Then, as long as at least one Turing kernel is left, we apply the search tree algorithm to the smallest one, say the one for vertex $v$, to find the largest 2-club in the Turing kernel of $v$ that contains $v$. Afterwards, we delete $v$ in all other Turing kernels. The maximum 2-club found during this iteration is the output. Note that this is indeed equivalent to what the search tree algorithm does: In one case $v$ is contained in the maximum 2-club $S$ and thus $S \subseteq N_2[v]$, in the other case $v$ is not contained and can thus be deleted. This observation explains the effectiveness of the search tree algorithms on the considered real-world data from social network analysis: There, the smallest two-neighborhood in the graph is typically much smaller than the entire vertex set, ensuring that all except $n$ nodes in the search tree have limited size.

*Heuristic Speed-Up.* Our main tool for accelerating the search tree algorithm is an extensive application of the following data reduction rules in each branching step. We describe the rules in descending order of observed effectiveness. Herein, let $G = (V, E)$ be the graph of the current branching step.

I1 *Vertex Cover Rule:* Let $G' = (V, E')$ be the graph where two vertices are adjacent iff they have distance at least three in $G$. Clearly, if a minimum vertex cover of $G'$ has size at least $x$, then at least $x$ vertex deletions have to be performed in $G$ to obtain a 2-club. We compute a 2-approximate vertex cover $C$ for $G'$ that is disjoint to the marked vertices (as they may not be deleted). If $|V| - \lceil |C|/2 \rceil$ is less than the current lower bound, then abort this branch.

I2 *Cleaning conflicts with marked vertices:* If there is a vertex $v \in V$ that has distance at least three to a vertex that is marked to be contained in the 2-club, then delete $v$. If $v$ is marked, then abort this branch.

I3 *Common neighbors of marked vertices:* If there are two marked vertices with only one common neighbor $v$, then mark $v$.

I4 *Degree-one vertices:* Remove each vertex $v$ that has degree one. If $v$ is marked, then abort this branch.

[3] After applying Rule 4 in advance, $|N_2[v]| \leq \ell^2$.

**Table 1.** Experimental results on random instances. For each combination of density, $n$, $a$, $b$ the other values are the average over 50 instances.

| density | [a ; b] | n | m | max deg | avg deg | 2-club size | time(s) |
|---|---|---|---|---|---|---|---|
| 0.05 | [0.00 ; 0.10] | 160 | 633.60 | 18.30 | 7.40 | 19.30 | 0.18 |
| 0.10 | [0.05 ; 0.15] | 160 | 1279.54 | 28.54 | 15.46 | 29.54 | 2.43 |
| | [0.00 ; 0.20] | 160 | 1276.40 | 31.68 | 15.44 | 33.08 | 2.79 |
| 0.15 | [0.10 ; 0.20] | 150 | 1674.28 | 35.72 | 21.84 | 56.98 | 98.44 |
| | [0.10 ; 0.20] | 160 | 1899.78 | 38.15 | 23.26 | 63.44 | 372.52 |
| | [0.05 ; 0.25] | 160 | 1906.68 | 41.04 | 23.26 | 77.12 | 21.54 |
| | [0.00 ; 0.15] | 160 | 1894.08 | 44.52 | 23.20 | 88.60 | 1.80 |
| 0.20 | [0.10 ; 0.30] | 160 | 2544.66 | 50.28 | 31.36 | 143.16 | 0.04 |

The correctness of Rules I1–I3 is obvious. Rule I4 is correct since we initialized our lower bound by a 2-club formed by a maximum degree vertex and thus a larger 2-club cannot contain degree-one vertices (note that Rule I4 is a special case of Rule 5).

We ran all our experiments on an Intel(R) Core(TM) i3-2130 CPU 3.40GHz machine with 8GB memory under the Debian GNU/Linux 6.0 operating system. The program is implemented in Java and runs under Java 1.6.0.18. The source code is freely available from `http://fpt.akt.tu-berlin.de/two_club/`. We tested our program on random instances as well as on real world data from the 10th DIMACS challenge [8] and compare our running times with recent implementations [6, 15].

*Random Instances.* As in previous experimental evaluations [6, 15], we use the random graph generator proposed by Gendreau et al. [11] where the density of the resulting graphs is controlled by two parameters, $0 \leq a \leq b \leq 1$, and the expected density is $(a + b)/2$. Table 1 summarizes our findings. As first observed by Bourjolly et al. [5], density 0.15 produces the hardest instances. We solve instances of these types for $n = 150$ typically within 2min; previous implementations needed about 6min [6], or up to an hour [15] for these instances. We observed that the key point for the good behavior of our algorithm on these instances is the *Vertex Cover Rule* that allows quite frequently to prune the search tree.

*Real-World Networks.* We considered real-world data taken from the 2012 DIMACS challenge [8]. To investigate the usefulness of 2-Club as natural clique relaxation concept, we ran our algorithm on instances from the *clustering* category; to test our algorithm on large scale social network graphs we ran it on graphs from the *co-author and citation* category. These graphs were obtained by the co-author relationship or the citation relation among authors listed in the DBLP and Citeseer database. In addition to the DIMACS instances, we created a further DBLP coauthor graph, which is the largest instance in our experiments (dblp_thres_1). Table 2 shows the results.

**Table 2.** Experimental results on instances from the DIMACS implementation [8] taken from the clustering and the coauthor/citation category

| category | name | n | m | max deg | avg deg | 2-club size | time(s) |
|---|---|---|---|---|---|---|---|
| clustering | email | 1133 | 5451 | 71 | 9 | 72 | 3.27 |
| | hep-th | 8361 | 15751 | 50 | 3 | 51 | 4.18 |
| | PGPgiantcompo | 10680 | 24316 | 205 | 4 | 206 | 3.22 |
| | polblogs | 1490 | 16715 | 351 | 22 | 352 | 9.93 |
| | power | 4941 | 6594 | 19 | 2 | 20 | 2.53 |
| coauthor | citationCiteseer | 268495 | 1156647 | 1318 | 8 | 1319 | 429.83 |
| | coAuthorsCiteseer | 227320 | 814134 | 1372 | 7 | 1373 | 23.04 |
| | coAuthorsDBLP | 299067 | 977676 | 336 | 6 | 337 | 216.64 |
| | dblp_thres_01 | 715633 | 2511988 | 804 | 7 | 805 | 1742.57 |
| | dblp_thres_02 | 282831 | 640697 | 201 | 4 | 202 | 119.03 |

We observe that, since the average degree in real world graphs is small, the Turing kernelization typically produces small graphs for our search tree algorithm. We thus can solve all instances from the clustering category within 10s. This is a significant performance increase in comparison to [15] who needed up to 70min for these instances. Moreover, although the co-author/citation graphs are quite large (up to 715,000 vertices), Turing kernelization enabled us to handle them within roughly 30min.

We observed, however, the unexpected behavior that the largest 2-club is on a majority of the real-world instances "just" a maximum degree vertex together with its neighbors. Thus, the question arises whether the resulting community structures are meaningful. In a first step to examine this, we created from a DBLP coauthor graph subgraphs of the pattern dblp_thres_i where two authors are related by an edge if they coauthored *at least $i$ papers*. We expected that for moderate values of $i$, say 2 or 3, the resulting (2-club) communities would have a stronger meaning because there are no edges between authors that are only loosely related. Unfortunately, even for values up to $i = 6$ this seems not to be the case. We think the main reason for this is the large gap between the maximum degree vertex (around 1000) and the average degree (less than 10). Thus, there seem to be some authors that dominate the overall structure because of their large number of coauthors. Notably, there are only few of these "dominating" authors: less than 200 authors have more than 200 coauthors.[4]

## 6 Conclusion

On the theoretical side, we extended existing fixed-parameter tractability results for the 2-Club problem by providing polynomial-size kernels for the parameters cluster editing set size and feedback edge set size. We further gave a direct algorithm for the parameter treewidth of $G$. Complementing these positive results,

[4] This implies that the so-called $h$-index of the real-world instances is low and thus a promising parameter. In companion work, however, we showed that 2-Club is W[1]-hard with respect to the $h$-index of the input graph [12].

we showed lower bounds on the kernel size for parameter vertex cover and on the running time as well as on the kernel size for the dual parameter $k'$. On the practical side, we provide the currently best implementation for 2-CLUB which solves 2-CLUB in reasonable time even on large real-world graphs with more than 700,000 vertices.

Still, there are many open questions that deserve further investigations: Is there a substantially better algorithm for the parameter vertex cover than the one for treewidth? Concerning the parameter solution size $\ell$, can the, so far impractical, running time or the size of the Turing kernel be improved [20]? Are there stronger parameters than the ones considered here for which 2-CLUB admits polynomial-size problem kernels? Finally, it would be interesting to transfer our results to 3-CLUB which is also of interest in practice [15, 19].

## References

[1] Alba, R.: A graph-theoretic definition of a sociometric clique. J. Math. Sociol. 3(1), 113–126 (1973)
[2] Asahiro, Y., Miyano, E., Samizo, K.: Approximating Maximum Diameter-Bounded Subgraphs. In: López-Ortiz, A. (ed.) LATIN 2010. LNCS, vol. 6034, pp. 615–626. Springer, Heidelberg (2010)
[3] Balasundaram, B., Butenko, S., Trukhanovzu, S.: Novel approaches for analyzing biological networks. J. Comb. Optim. 10(1), 23–39 (2005)
[4] Bodlaender, H.L., Thomassé, S., Yeo, A.: Kernel bounds for disjoint cycles and disjoint paths. Theor. Comput. Sci. 412(35), 4570–4578 (2011)
[5] Bourjolly, J.-M., Laporte, G., Pesant, G.: An exact algorithm for the maximum $k$-club problem in an undirected graph. European J. Oper. Res. 138(1), 21–28 (2002)
[6] Chang, M.S., Hung, L.J., Lin, C.R., Su, P.C.: Finding large $k$-clubs in undirected graphs. In: Proc. 28th Workshop on Combinatorial Mathematics and Computation Theory (2011)
[7] Chen, Y., Flum, J., Müller, M.: Lower bounds for kernelizations and other preprocessing procedures. Theory Comput. Syst. 48(4), 803–839 (2011)
[8] DIMACS. Graph partitioning and graph clustering (2012), `http://www.cc.gatech.edu/dimacs10/` (accessed April 2012)
[9] Downey, R.G., Fellows, M.R.: Parameterized Complexity. Springer (1999)
[10] Flum, J., Grohe, M.: Parameterized Complexity Theory. Springer (2006)
[11] Gendreau, M., Soriano, P., Salvail, L.: Solving the maximum clique problem using a tabu search approach. Ann. Oper. Res. 41(4), 385–403 (1993)
[12] Hartung, S., Komusiewicz, C., Nichterlein, A.: On structural parameterizations for the 2-club problem (manuscript, June 2012)
[13] Impagliazzo, R., Paturi, R., Zane, F.: Which problems have strongly exponential complexity? J. Comput. System Sci. 63(4), 512–530 (2001)
[14] Lokshtanov, D., Marx, D., Saurabh, S.: Lower bounds based on the exponential time hypothesis. Bulletin of the EATCS 105, 41–72 (2011)
[15] Mahdavi, F., Balasundaram, B.: On inclusionwise maximal and maximum cardinality $k$-clubs in graphs. Discrete Optim. (to appear, 2012)

[16] Memon, N., Larsen, H.L.: Structural Analysis and Mathematical Methods for Destabilizing Terrorist Networks Using Investigative Data Mining. In: Li, X., Zaïane, O.R., Li, Z.-h. (eds.) ADMA 2006. LNCS (LNAI), vol. 4093, pp. 1037–1048. Springer, Heidelberg (2006)
[17] Mokken, R.J.: Cliques, Clubs and Clans. Quality and Quantity 13, 161–173 (1979)
[18] Niedermeier, R.: Invitation to Fixed-Parameter Algorithms. Oxford University Press (2006)
[19] Pasupuleti, S.: Detection of Protein Complexes in Protein Interaction Networks Using n-Clubs. In: Marchiori, E., Moore, J.H. (eds.) EvoBIO 2008. LNCS, vol. 4973, pp. 153–164. Springer, Heidelberg (2008)
[20] Schäfer, A., Komusiewicz, C., Moser, H., Niedermeier, R.: Parameterized computational complexity of finding small-diameter subgraphs. Optim. Lett. 6(5) (2012)
[21] Schäfer, A.: Exact algorithms for s-club finding and related problems. Diploma thesis, Friedrich-Schiller-Universität Jena (2009)
[22] Seidman, S.B., Foster, B.L.: A graph-theoretic generalization of the clique concept. J. Math. Sociol. 6, 139–154 (1978)

# Finding Dense Subgraphs of Sparse Graphs*

Christian Komusiewicz and Manuel Sorge**

Institut für Softwaretechnik und Theoretische Informatik, TU Berlin
{christian.komusiewicz,manuel.sorge}@tu-berlin.de

**Abstract.** We investigate the computational complexity of the DENSEST-$k$-SUBGRAPH (D$k$S) problem, where the input is an undirected graph $G = (V, E)$ and one wants to find a subgraph on exactly $k$ vertices with a maximum number of edges. We extend previous work on D$k$S by studying its parameterized complexity. On the positive side, we show that, when fixing some constant minimum density $\mu$ of the sought subgraph, D$k$S becomes fixed-parameter tractable with respect to either of the parameters maximum degree and $h$-index of $G$. Furthermore, we obtain a fixed-parameter algorithm for D$k$S with respect to the combined parameter "degeneracy of $G$ and $|V| - k$". On the negative side, we find that D$k$S is W[1]-hard with respect to the combined parameter "solution size $k$ and degeneracy of $G$". We furthermore strengthen a previous hardness result for D$k$S [Cai, Comput. J., 2008] by showing that for every fixed $\mu$, $0 < \mu < 1$, the problem of deciding whether $G$ contains a subgraph of density at least $\mu$ is W[1]-hard with respect to the parameter $|V| - k$.

## 1 Introduction

Identifying dense regions of graphs is a fundamental computational problem with many important applications, for instance in computational biology [19] and social network analysis [3]. There are many different definitions of what a dense subgraph is [11, 17] and for almost all of these formulations, the corresponding computational problems are NP-hard.

In this work, we study the problem of finding subgraphs with a fixed number $k$ of vertices and a maximum number of edges. This problem is known as DENSEST-$k$-SUBGRAPH. For fixed $k$, maximizing the number of edges is the same as maximizing the *density* of a graph $G = (V, E)$ which is defined as $2|E|/(|V|(|V| - 1))$. Using the notion of density, the NP-hard DENSEST-$k$-SUBGRAPH problem [11, 15] can be defined as follows.

DENSEST-$k$-SUBGRAPH (D$k$S) :
**Input:** A graph $G = (V, E)$, and a nonnegative integer $k$.
**Task:** Find a vertex set $S \subseteq V$ of size exactly $k$ such that $G[S]$ has maximum density.

---

* One result of this work (Thm. 4) is contained in the first author's dissertation [16].

** Supported by DFG projects PABI (NI 369/7-2) and DAPA (NI 369/12).

D.M. Thilikos and G.J. Woeginger (Eds.): IPEC 2012, LNCS 7535, pp. 242–251, 2012.

D$k$S is at least as hard as the well-studied CLIQUE problem which asks for finding a complete graph of order exactly $k$. In this work, our aim is to provide a better overview of when D$k$S becomes computationally hard or tractable, respectively. To this end, we consider how two types of parameters influence the complexity of D$k$S. The first type of parameters are the classical parameters "solution size $k$" and "parameterization by dual $|V| - k$". The second type of parameters measure the sparseness of the input graph $G$: maximum degree $\Delta$, $h$-index,[1] and degeneracy $d$. Bounded maximum degree means that *all* vertices have few neighbors, bounded $h$-index means that *most* vertices have few neighbors, and bounded degeneracy means that *there is always* a vertex with few neighbors. By definition, $\Delta \geq h\text{-index} \geq d$. The study of these three parameters is motivated by two facts: First, many real-world networks such as biological and social networks are relatively sparse since they contain many vertices of low degree and only few vertices of high degree (the network "hubs"). Second, the otherwise notoriously hard CLIQUE problem is much easier on sparse graphs. For example, all maximal cliques can be enumerated in $O(d^{3/d} \cdot d \cdot n)$ time on graphs with degeneracy $d$ [9].

We study the complexity of D$k$S mostly by considering the following problem which can be seen as a decision variant of D$k$S. Here, one asks whether there is a $k$-vertex subgraph with density *at least* $\mu$, where $0 \leq \mu \leq 1$ is some fixed constant. We call such subgraphs $\mu$-cliques, that is, a graph $G = (V, E)$ is a *$\mu$-clique* if the density of $G$ is at least $\mu$.

$\mu$-CLIQUE:
**Input:** A graph $G = (V, E)$, and a nonnegative integer $k$.
**Question:** Is there a vertex set $S \subseteq V$ of size at least $k$ such that $G[S]$ is a $\mu$-clique?

Throughout this work, we assume that $\mu$ is a fixed constant, in other words, that it is independent of $k$ and $n$. This assumption can be motivated by the fact that in many applications, the dense subgraphs that one wants to find should be almost complete graphs.

*Related Work.* CLIQUE, which asks to find a complete subgraph of order $k$ is W[1]-hard with respect to the parameter $k$ and fixed-parameter tractable with respect to the dual parameter $n - k$ [7]. Finding a densest subgraph of order exactly $k$ is NP-hard and W[1]-hard with respect to $k$, as it is a generalization of CLIQUE. Moreover, D$k$S is W[1]-hard with respect to the parameter $n - k$ [5]. It is, however, fixed-parameter tractable with respect to the combined parameter "maximum degree $\Delta$ and $k$" [6]. Holzapfel et al. [13] showed that D$k$S remains NP-hard, even when looking only for subgraphs with average degree at least $2 + \Omega(1/k^{1-\epsilon})$ for $0 < \epsilon < 2$. Finding $k$-vertex subgraphs of average degree at least $2 + O(1/k)$, however, can be done in polynomial time. Furthermore, D$k$S is NP-hard even in graphs with maximum degree three and degeneracy two [10].

[1] The structural graph parameter $h$-index was introduced by Eppstein and Spiro [8] in the context of triangle counting in dynamic graphs. For a definition, see Section 2.

**Table 1.** Summary of our results and previous results for $\mu$-Clique and D$k$S. Note that hardness transfers from $\mu$-Clique to D$k$S and tractability transfers in the reverse direction. For fixed-parameter tractability (FPT) results, we write a rough estimate of the exponential running time factor. Herein, $k$ denotes the order of the sought $\mu$-clique, $\ell = n - k$ is the number of vertices that have to be deleted in order to obtain a $\mu$-clique or a densest subgraph, and $d$ denotes the degeneracy of the input graph.

| parameter | $\mu$-Clique | D$k$S |
|---|---|---|
| max. degree $\Delta$ | FPT: $\Delta^{O(\Delta)}$ (Theorem 1), no poly. kernel (Theorem 7) | NP-hard for $\Delta = 3$ [10] |
| $h$-index $h$ | FPT: $h^{O(h)}$ (Theorem 2), no poly. kernel (Theorem 7) | |
| degeneracy $d$ | $\in$ XP (Lemma 1(iii)) | NP-hard for $d = 2$ [10] |
| $(k, d)$ | W[1]-hard (Theorem 6) | |
| $\ell$ | W[1]-hard (Theorem 4) | W[1]-hard [5] |
| $(\ell, d)$ | | FPT: $(\ell + d)^{O(\ell)}$ (Theorem 3) |

The "densest subgraph" in this reduction, however, has very low, non-constant density. For an overview of computational aspects of finding dense subgraphs, we refer to the survey of Kosub [17]. A related problem is Minimum Subgraph of Minimum Degree, where the task is to find a subgraph of order at most $k$ such that each vertex has a given minimum degree. Minimum Subgraph of Minimum Degree is W[1]-hard with respect to the parameter $k$ but becomes fixed-parameter tractable on graphs of bounded local treewidth and graphs with excluded minors [2]. A further related problem is to find a subgraph that has maximum average degree (without constraint on the order). This problem is polynomial-time solvable using network flow techniques [12].

*Our Results.* Table 1 gives an overview of our results; note that all negative results that were obtained for $\mu$-Clique immediately transfer to D$k$S. Our results can be summarized as follows. Finding dense subgraphs is significantly harder than finding cliques since $\mu$-Clique and D$k$S are W[1]-hard with respect to the parameter $(d, k)$. Furthermore, we show that the W[1]-hardness for D$k$S parameterized by $n - k$ [5] can also be generalized to hold for $\mu$-Clique for all $\mu$, $0 < \mu < 1$. Finally, we show that, in contrast to D$k$S, $\mu$-Clique is fixed-parameter tractable for the parameters maximum degree $\Delta$ and $h$-index $h$ of $G$. In particular, we show that the practically relevant case of finding subgraphs whose density $\mu$ deviates not too much from the maximum density (that is, $1/\mu$ is small) is still tractable for bounded $\Delta$ or $h$.

## 2 Preliminaries

We consider simple undirected graphs $G = (V, E)$ where $n := |V|$ and $m := |E|$. The *order* of a graph is the number of vertices. For a vertex set $S \subseteq V$ we denote by $N(S) := \bigcup_{v \in S} N(v) \setminus S$ the *neighborhood of* $S$, and by $\deg(v)$ the *degree* of $v$. We use $G[S]$ to denote the *subgraph induced by* $S$. The *degeneracy* of a graph $G$ is the smallest integer $d$ such that every induced subgraph of $G$ has at least one vertex with degree at most $d$. The *h-index* of a graph $G$ is the maximum integer $h$ such that $G$ contains $h$ vertices of degree at least $h$. The property of being a $\mu$-clique is not hereditary, but has a "nestedness" property [17]: Every $\mu$-clique $G = (V, E)$ has an induced subgraph $G'$ on $|V| - 1$ vertices that is also a $\mu$-clique. For the relevant notions from parameterized complexity, refer to [7].

## 3 Fixed-Parameter Algorithms

Here, we present fixed-parameter algorithms for the parameters maximum degree $\Delta$ of $G$, $h$-index of $G$ and the combined parameter that comprises $n - k$ and degeneracy of $g$. Before presenting these algorithms, we observe relationships between the order of $\mu$-cliques and the sparsity parameters under consideration. We also give an observation about the enumeration of certain subgraphs in degree bounded graphs, which yields a subroutine used in our algorithms.

*Preparations.* The relation between the order of a $\mu$-clique and its maximum degree, $h$-index and degeneracy is as follows.

**Lemma 1.** *A $\mu$-clique with*
*(i) maximum degree $\Delta$ has order at most $\Delta/\mu + 1$.*
*(ii) $h$-index $h$ has order at most $\frac{h \cdot (h-1) + 2 \cdot (n-h) \cdot h}{\mu \cdot (n-1)} < \frac{2 \cdot h}{\mu}$.*
*(iii) degeneracy $d$ has order less than $(4 \cdot d + \mu)/2 \cdot \mu$.*

The upper bound $\frac{h \cdot (h-1) + 2 \cdot (n-h) \cdot h}{\mu \cdot (n-1)}$ on the order of $\mu$-cliques is tight as a graph consisting of a clique of order $h$ and of $n - h$ further vertices that are an independent set but adjacent to all vertices of the clique has density exactly $\mu$ if $n$ is equal to the upper bound.

Continuing the preparation for our tractability results, a central observation is the following.

**Lemma 2.** *Let $G$ be a graph with maximum degree $\Delta$ and let $v$ be a vertex in $G$. There are at most $4^k \cdot (\Delta - 1)^k$ connected subgraphs of $G$ that contain $v$ and have order at most $k$. Furthermore, these subgraphs can be enumerated in $O(4^k \cdot (\Delta - 1)^k \cdot (n + m))$ time.*

*Proof.* We describe a search tree for enumerating these subgraphs. In each search tree node, we maintain two vertex sets $P$ (the "pivot set") and $N$, where $N$ is a subset of $P$ and the task is to enumerate all vertex sets $S$ such that 1) $G[S]$ is connected, 2) $P \subseteq S$, and 3) the vertices of $N$ have no neighbors in $S \setminus P$. Furthermore, in each of the search tree nodes, there will be a distinguished *active* vertex $v$ of $P \setminus N$. We will consider adding neighbors of the active vertex first. The details are as follows.

Assume that there is an arbitrary but fixed ordering of the vertices of $G$. Initialize the search by setting the pivot set $P := \{v\}$ and setting $N = \emptyset$ where $v$

is the vertex with lowest index in the fixed ordering. Furthermore, set $v$ as active vertex. Then, in each search tree node, do the following. First, report $G[P]$. Then, if $|P| = k$, abort this branch. Otherwise, if $|P| < k$ and there is no active vertex, then choose the vertex $v \in P \setminus N$ that has lowest index in the fixed ordering as new active vertex. Now, branch into the following cases to add neighbors of the active vertex $v$: First, for each neighbor $u$ of $v$ in $V \setminus P$ that is not adjacent to any vertex in $N$ create a search tree branch with $(P \cup \{u\}, N)$ that is, one branch for each possibility to add a neighbor of $v$ and keep $v$ as active vertex in these branches. Second, create one further search tree branch with $(P, N \cup \{v\})$, that is, a branch in which we assume that no further neighbors of $v$ may be added; in this branch $v$ will become *inactive*.

Since a vertex never leaves $P$ once it has been added and only neighbors of vertices in $P$ are added to $P$, clearly, the graph $G[P]$ is connected, contains $v$ and has order at most $k$ in each search tree node. Furthermore, each connected graph that contains $v$ and is of order at most $k$ is equal to $G[P]$ in some search tree node: for each vertex that is a neighbor of the current pivot set and not a neighbor of $N$, we branch at some point into the case that this vertex is added.

To bound the number of search tree nodes, observe that at most $k$ vertices can be added to $P$ and, hence, at most $k$ vertices can be added to $N$. Now, assume that we branch in advance into all the cases to either add a neighbor of an active vertex or move an active vertex to $N$, that is, we fix in advance that we add say $x_1$ neighbors of the first vertex, $x_2$ neighbors of the second active vertex and so on. The number of possible branchings is $2^{2k}$, since in the first branch, we add a vertex to $P$ and in the second branch, we add a vertex to $N$ and the cardinality of both sets is at most $k$. Now, assume that we branch for each such fixed case into the different cases to add a neighbor of the active vertex $v$. Then, this can be done by a search tree of depth at most $k$ and in each search tree node, we branch into at most $\Delta - 1$ cases, to add a vertex in $V \setminus P$ to $P$ (note that except for the first branching, every active vertex $v$ has at least one neighbor in $P$ since $G[P]$ is connected). Hence, the overall search tree size is $O(4^k \cdot (\Delta - 1)^k)$. Since the steps at each search tree node can be performed in $O(n + m)$ time, the search tree also gives a $O(4^k \cdot (\Delta - 1)^k \cdot (n + m))$ running time enumeration algorithm. □

Next, we present fixed-parameter algorithms for the parameters maximum degree $\Delta$ and $h$-index. We gradually develop the algorithms starting with the (easiest) case of finding connected $\mu$-cliques in graphs with maximum degree $\Delta$. Then, we present an algorithm for disconnected $\mu$-cliques in graphs with maximum degree $\Delta$. Finally, we describe an algorithm for the parameter $h$-index. Note that we can restrict ourselves to finding $\mu$-cliques of order exactly $k$ due to the nestedness property.

*Finding Connected $\mu$-cliques.* We use the enumeration algorithm described in Lemma 2. For every vertex $v$, we start an enumeration of all connected graphs of order at most $k$ that contain $v$ and instead of reporting the graphs, we check whether it is a $\mu$-clique and report it or not accordingly. Plugging in the bound for $k$ given by Lemma 1(i), we obtain the following.

**Proposition 1.** *All connected $\mu$-cliques in a graph $G$ with maximum degree $\Delta$ can be enumerated in $O(4^{\Delta/\mu+1} \cdot (\Delta-1)^{\Delta/\mu+1} \cdot n(n+m))$ time.*

*Finding Disconnected $\mu$-cliques.* The idea for finding disconnected $\mu$-cliques is to combine different connected subgraphs such that the sum of edges and vertices yields a graph with density at least $\mu$. In the process of combining these connected $\mu$-cliques, we have to ensure that we only combine these numbers for disjoint graphs. Otherwise, a dense subgraph might be counted twice. To this end, we use *color coding* [1] to obtain a randomized fixed-parameter algorithm with one-sided error. The algorithm can be derandomized using standard techniques with an additional running time factor of $2^{O(k)}$ [1]. Assume that the input graph contains a $\mu$-clique of order $k$, and let $S$ be the vertex set of this $\mu$-clique. The basic idea of color coding is to color the vertices of the input graph uniformly at random with a set $C$ of $k$ colors and to hope that $S$ is colorful, that is, for each color in $C$ there is exactly one vertex in $S$ that has received this color. Assuming the graph is colored this way, first use the enumeration algorithms for connected $\mu$-cliques described above, and then "combine" these connected graphs by applying dynamic programming. The color-coding/enumeration/dynamic programming routine is repeated sufficiently often to achieve constant error probability. The details are as follows.

After the coloring, first compute for every subset $C'$ of $C$ the densest connected subgraph that has color set $C'$. This can be easily achieved by adapting the above enumeration algorithm to only report colorful $\mu$-cliques. Using this enumeration, we fill a table $D$ where for each color set $C' \subseteq C$, the entry $D(C')$ contains the maximum number of edges in a connected $\mu$-clique in $G$ whose vertices have exactly the colors from $C'$. Afterwards, we find the maximum density of a colorful $\mu$-clique of order $k$ using another table $T$. Here, the entry in $T(C')$ for some color set $C' \subseteq C$ contains the number of edges of a (possibly disconnected) $\mu$-clique with maximum density in $G$ whose vertices have exactly the colors from $C'$. Observe that either $T(C') = D(C')$, or there is a partition of $C'$ into $C'_1, C'_2$ such that $T(C') = T(C'_1) + T(C'_2)$. Thus, we fill $T(C')$ by the following recurrence:

$$T(C') = \max\{D(C'), \max_{C'' \subset C'} \{T(C'') + T(C' \setminus C'')\}\}.$$

The maximum density of a colorful $\mu$ clique of order $k$ is then found in $T(C)$. The table $T$ can be filled in $O(3^k)$ time, since there are at most this many triples $(C', C'_1, C'_2)$ such that $C' \subset C$ and $(C'_1, C'_2)$ partitions $C'$ (each color in $C$ either has to be in $C'_1, C'_2$, or $C \setminus C'$); for each such triple there is only one table lookup. Adding the running time for filling $T$, we obtain a running time of $O(3^k + 4^k \cdot (\Delta-1)^k \cdot n(n+m)) = O(4^k \cdot (\Delta-1)^k \cdot n(n+m))$, where $\Delta$ is the maximum degree of $G$.

The error probability can be bounded as follows [1]. When coloring the vertices with $k$ colors uniformly at random, the probability of a $\mu$-clique $S$ being colorful is exactly $k!/k^k$, since there are $k^k$ distinct colorings of $S$ and $k!$ colorful ones. By Stirling's approximation, this probability is at least $e^{-k}$ and by repeating $e^k$ times the random coloring and the algorithm above, the probability of missing a feasible $\mu$-clique is at most $(1-e^{-k})^{e^k} \leq 1/e$. Using Lemma 1(i) we obtain the following.

**Theorem 1.** *$\mu$-Clique can be solved in time $O((4e \cdot (\Delta - 1))^{\Delta/\mu+1} n(n+m))$, reporting a yes-instance as a no-instance with probability at most $1/e$, where $\Delta$ is the maximum degree in the input graph.*

It has previously been shown that, using random separation, D$k$S can be solved in $2^{O(\Delta k)}$ time with one-sided error and constant error probability [6]. Our algorithm above applied to D$k$S runs in $2^{O(\log(\Delta)k)}$ time.

*Parameterization by h-index.* We now describe how to adapt the algorithm from Theorem 1 to obtain a fixed-parameter algorithm for the parameter $h$-index of the input graph. In many practical applications the $h$-index is much smaller than the maximum degree. For instance, social and biological networks have few so-called hubs, that is, vertices of very high degree, and many low-degree vertices. Hence, the $h$-index is much smaller than the maximum degree for these graphs.

The main idea of the algorithm is as follows. Let $H$ be the set of the at most $h$ vertices with degree at least $h$, and assume that $S$ is a vertex set of size $k$ such that $G[S]$ is a $\mu$-clique. First, by trying all $2^h$ partitions of $H$, guess the set $H_S$ of vertices that are in $S \cap H$. We annotate every vertex $v \in V \setminus H$ with the number of neighbors it has in $H_S$. Let the *weight* of a subgraph $G'$ of the input graph $G = (V, E)$ be the sum of the vertex annotations in $G'$ and the number of edges in $G'$. Now the task is to find a subgraph in $G[V \setminus H]$ of order at most $k - |H_S|$ that has maximum weight. If we have such subgraphs for all possible choices of $H_S$, we can compare them, also accounting for the edges in $G[H_S]$, to obtain a densest subgraph of order at most $k$. To find the maximum weight subgraphs in $G[V \setminus H]$, we proceed analogously to the algorithm given above for Theorem 1; we omit the details. Using the size bound for $k$ from Lemma 1(ii), we obtain the following running time.

**Theorem 2.** *$\mu$-Clique can be solved in time $O(2^h \cdot (4e \cdot (h-1))^{h/\mu+1} \cdot h \cdot n(n+m))$, reporting a yes-instance as no-instance with probability at most $1/e$ where $h$ is the $h$-index of the input graph.*

*Degeneracy and Dual Parameter.* In this section we show that D$k$S is fixed-parameter tractable with respect to the combined parameter degeneracy and $\ell := n - k$, where $n$ is the number of vertices in the input graph. Remember that in $\mu$-Clique we fix some constant minimum density $\mu$ of the sought graph. This is necessary to bound the maximum value of $k$ and, ultimately, obtain feasible running time bounds. For the combined parameter $(d, \ell)$ this constraint can be dropped. The algorithm is based on the following observation.

**Lemma 3.** *Let $G = (V, E)$ be a graph and let $S \subseteq V$ such that $G[S]$ is densest possible and $S$ has size $k$. Then, there is no vertex in $V \setminus S$ that has degree at least $\ell + d$, where $\ell = n - k$.*

*Proof.* Assume that there is a vertex $v$ of degree at least $\ell + d$ in $V \setminus S$. Since $v$ has at most $\ell - 1$ neighbors in $V \setminus S$, it has at least $d + 1$ neighbors in $S$. However, because $G$ is $d$-degenerate, there is a vertex $u$ of degree at most $d$ in $G[S]$. Thus, $G[(S \setminus \{u\}) \cup \{v\}]$ is a graph with at least one edge more than $G[S]$. This contradicts the fact that $G[S]$ is densest possible. □

Note that we can regard D$k$S as the problem of deleting a set of $\ell$ vertices whilst removing the least possible number of edges. Let us call such a vertex set *sparsest $\ell$-deletion set.* To exploit Lemma 3, first mark every vertex with degree at least $\ell + d$ undeletable. Then, find a sparsest $\ell$-deletion set in the degree-bounded graph induced by the deletable vertices. To find this deletion set, we employ, much in the spirit of the algorithms for maximum degree and $h$-index, color coding and use dynamic programming to first find connected sparsest $(\leq \ell)$-deletion sets and then combine them to an optimum one; we omit the details. By the same argument as for the algorithm for $\mu$-CLIQUE and maximum degree, it suffices to repeat $e^{\ell}$ times the random coloring and dynamic programming procedure to obtain an error probability of at most $1/e$. In summary, we have the following.

**Theorem 3.** DENSEST-$k$-SUBGRAPH *can be solved in* $O((4e \cdot (\ell + d - 1))^{\ell} n(n + m))$ *time, reporting a yes-instance as no-instance with probability at most* $1/e$*, where* $\ell = n - k$ *and* $d$ *is the degeneracy of the input graph.*

# 4 Hardness Results

In this section, we present two reductions that show the limits of the approach presented above.

## 4.1 W[1]-Hardness for Parameterization by Dual

First, we show that considering only the dual parameter $\ell$ leads to W[1]-hardness also in the case of $\mu$-CLIQUE.

**Theorem 4.** *For any fixed* $\mu$, $0 < \mu < 1$, $\mu$-CLIQUE *is W[1]-hard with respect to the parameter* $\ell = n - k$.

Somehow counter-intuitively, the reduction used in Theorem 4 suggests that in order to obtain a graph with density $\mu$ it might be of advantage to delete a clique from the input graph. Hence, one cannot expect that the set of removed vertices induces a sparse graph. From the above reduction, we also obtain a lower bound on the running time of algorithms for $\mu$-CLIQUE. This bound is based on the exponential-time hypothesis (ETH) which implies that 3SAT cannot be solved in $O^*(2^{o(n)})$ time [14, 18].

**Theorem 5.** $\mu$-CLIQUE *cannot be solved in time* $O^*(2^{o(\Delta/\mu)})$ *for every* $0 < \mu \leq 1$ *unless the exponential time hypothesis (ETH) fails. Here,* $h$ *is the* $h$*-index of the input instance.*

Clearly, Theorem 5 also excludes algorithms with running time $O^*(2^{o(h/\mu)})$.

### 4.2 W[1]-Hardness for Parameterization by Degeneracy and Solution Size

Next, we show that the parameter $h$-index cannot be replaced by the smaller parameter degeneracy.

**Theorem 6.** *For any fixed $\mu$, $0 < \mu < 1$, $\mu$-CLIQUE is W[1]-hard parameterized by $(d, k)$, where $d$ denotes the degeneracy of the input graph.*

We can use the reduction behind Theorem 6 to also exclude polynomial-size problem kernels for the parameters maximum degree and $h$-index.

**Theorem 7.** *$\mu$-CLIQUE does not admit polynomial-size problem kernels with respect to either maximum degree or $h$-index unless NP $\subseteq$ coNP/poly.*

*Proof.* It suffices to prove the statement for the larger maximum degree parameter. For this, we observe that the reduction used in Theorem 6 implies a cross-composition [4] from CLIQUE into $\mu$-CLIQUE parameterized by maximum degree. A cross-composition from a language $L \subseteq \Sigma^*$ into a parameterized problem $P$ is an algorithm that, given $t$ strings $x_1, x_2, \ldots, x_t \in \Sigma^*$, computes an instance $x^*$ of $P$ with parameter value $k$ such that its running time is bounded by a polynomial in $\sum_{i=1}^{t} |x_i|$, $k$ is bounded by a polynomial in $\max_{i=1}^{t} |x_i|$, and $x^* \in P$ if and only if $x_i \in L$ for some $1 \leq i \leq t$.[2] If a parameterized problem that has a cross-composition from an NP-hard language also admits a polynomial-size problem kernel, then NP $\subseteq$ coNP/poly [4].

Let a number of instances of CLIQUE be given and without loss of generality, assume that each instance asks for a clique of order $k'$.[3] Merge the instances into one instance of CLIQUE by taking the disjoint union of the graphs. It is clear that this graph contains a clique of given order if and only if one of its connected components does. Then, apply the reduction used in Theorem 6 to the resulting graph. To obtain that this procedure is a cross-composition, it remains to show that the maximum degree in the created instance is bounded by a polynomial in the maximum size of the input instances. This follows since the reduction used for Theorem 6 does not merge any connected components and the introduced gadget graph has size polynomial in $k'$. Thus, there is cross-composition from CLIQUE into $\mu$-CLIQUE parameterized by the maximum degree. □

## 5 Outlook

Several research tasks remain. First, it would be interesting to improve the presented algorithms. We conjecture, however, that it is not possible to achieve a running time of $O^*(2^{o((\Delta/\mu)\log \Delta)})$, but have no proof for this at the moment. Furthermore, it would be interesting to obtain nontrivial Turing kernels for $\mu$-CLIQUE and any of the considered parameters. Also, is there a better

---

[2] For readability, we simplified the more general definition of cross-composition here.

[3] If an instance asks for a smaller clique, simply add a new vertex and connect it to all other vertices of this instance.

polynomial-time algorithm for $\mu$-CLIQUE on planar graphs than the XP-algorithm for degeneracy? Finally, a further restriction that can be made in the area of community detection is to bound the size of the neighborhood of the $\mu$-cliques. Efficient algorithms exploiting such bounds would be interesting and also practically relevant.

## References


1. Alon, N., Yuster, R., Zwick, U.: Color-coding. J. ACM 42(4), 844–856 (1995)
2. Amini, O., Sau, I., Saurabh, S.: Parameterized Complexity of the Smallest Degree-Constrained Subgraph Problem. In: Grohe, M., Niedermeier, R. (eds.) IWPEC 2008. LNCS, vol. 5018, pp. 13–29. Springer, Heidelberg (2008)
3. Balasundaram, B., Butenko, S., Hicks, I.V.: Clique relaxations in social network analysis: The maximum $k$-plex problem. Oper. Res. 59(1), 133–142 (2011)
4. Bodlaender, H.L., Jansen, B.M.P., Kratsch, S.: Cross-composition: A new technique for kernelization lower bounds. In: Proc. 28th STACS. LIPIcs, vol. 9, pp. 165–176. Schloss Dagstuhl–Leibniz-Zentrum fuer Informatik (2011)
5. Cai, L.: Parameterized complexity of cardinality constrained optimization problems. Comput. J. 51(1), 102–121 (2008)
6. Cai, L., Chan, S.M., Chan, S.O.: Random Separation: A New Method for Solving Fixed-Cardinality Optimization Problems. In: Bodlaender, H.L., Langston, M.A. (eds.) IWPEC 2006. LNCS, vol. 4169, pp. 239–250. Springer, Heidelberg (2006)
7. Downey, R.G., Fellows, M.R.: Parameterized Complexity. Springer (1999)
8. Eppstein, D., Spiro, E.S.: The $h$-Index of a Graph and Its Application to Dynamic Subgraph Statistics. In: Dehne, F., Gavrilova, M., Sack, J.-R., Tóth, C.D. (eds.) WADS 2009. LNCS, vol. 5664, pp. 278–289. Springer, Heidelberg (2009)
9. Eppstein, D., Löffler, M., Strash, D.: Listing All Maximal Cliques in Sparse Graphs in Near-Optimal Time. In: Cheong, O., Chwa, K.-Y., Park, K. (eds.) ISAAC 2010, Part I. LNCS, vol. 6506, pp. 403–414. Springer, Heidelberg (2010)
10. Feige, U., Seltser, M.: On the densest k-subgraph problem. Technical report, The Weizmann Institute, Department of Applied Math. and Computer Science (1997)
11. Feige, U., Peleg, D., Kortsarz, G.: The dense $k$-subgraph problem. Algorithmica 29(3), 410–421 (2001)
12. Gallo, G., Grigoriadis, M.D., Tarjan, R.E.: A fast parametric maximum flow algorithm and applications. SIAM J. Comput. 18(1), 30–55 (1989)
13. Holzapfel, K., Kosub, S., Maaß, M.G., Täubig, H.: The complexity of detecting fixed-density clusters. Discrete Appl. Math. 154(11), 1547–1562 (2006)
14. Impagliazzo, R., Paturi, R., Zane, F.: Which problems have strongly exponential complexity? J. Comput. Syst. Sci. 63(4), 512–530 (2001)
15. Khuller, S., Saha, B.: On Finding Dense Subgraphs. In: Albers, S., Marchetti-Spaccamela, A., Matias, Y., Nikoletseas, S., Thomas, W. (eds.) ICALP 2009, Part I. LNCS, vol. 5555, pp. 597–608. Springer, Heidelberg (2009)
16. Komusiewicz, C.: Parameterized Algorithmics for Network Analysis: Clustering & Querying. PhD thesis, Technische Universität Berlin, Berlin, Germany (2011)
17. Kosub, S.: Local Density. In: Brandes, U., Erlebach, T. (eds.) Network Analysis. LNCS, vol. 3418, pp. 112–142. Springer, Heidelberg (2005)
18. Lokshtanov, D., Marx, D., Saurabh, S.: Lower bounds based on the exponential time hypothesis. Bulletin of the EATCS 105, 41–72 (2011)
19. Saha, B., Hoch, A., Khuller, S., Raschid, L., Zhang, X.-N.: Dense Subgraphs with Restrictions and Applications to Gene Annotation Graphs. In: Berger, B. (ed.) RECOMB 2010. LNCS, vol. 6044, pp. 456–472. Springer, Heidelberg (2010)

# Enumerating Neighbour and Closest Strings*

Naomi Nishimura and Narges Simjour

Cheriton School of Computer Science, University of Waterloo,
Waterloo, Ontario, Canada
{nishi,nsimjour}@uwaterloo.ca

**Abstract.** We present the first parameterized enumeration algorithm for the *neighbour string problem*, where a neighbour string of $n$ input strings, each of length $\ell$, is a string that differs from input $s_i$ in no more than $d_i$ positions. The problem is NP-complete even when the $d_i$'s are equal; this is the well-known *closest string problem.*

Our new approach gives us the ability to tune the running time to optimize the algorithm for varying relative values of $n$ and $d = \max_i d_i$. For strings over an alphabet $\Sigma$, we can choose a tuning constant $\lambda$ to obtain an algorithm that runs in time $O(n\ell + (nd)^{f(\lambda)}(|\Sigma| - 1)^d 5^{(1+\lambda)d})$, where $f$ is a function that decreases with increasing $\lambda$. When $\Sigma = \{0, 1\}$, the dependency on $d$ is an asymptotic improvement over the previous best parameterized time bound of $O(n\ell + nd^3 6.7308^d)$ for finding a single solution.

## 1 Introduction

In both the neighbour string problem [15] and the well-studied closest string problem [7,16,2,13,8,10,6], the goal is to determine a string that is not too different from any of the $n$ length-$\ell$ input strings. In the latter case, the solution cannot differ from any string $s_i$ in more than $d$ positions; in the former case, for each $s_i$ a bound $d_i$ is specified as part of the input, and the solution cannot differ from $s_i$ in more than $d_i$ positions. Aside from the application in coding theory [7,8], to remove errors from sequences originating from a single sequence, these problems have several applications in computational biology [16,2,11].

As the closest string problem is NP-complete even for binary strings [7], much of the work has focused on finding approximate solutions [2,8,11]. Polynomial-time approximation schemes (PTAS) [13,1,15], the most recent with running time $O(\ell(n\ell)^{O(\epsilon^{-2})})$, have been proposed for the problem.

Gramm et al. [9] gave an integer programming formulation of the closest string problem in $(n-1)B(n)$ variables, where the Bell number $B(n)$ is at most $n!$; since integer programs can be solved in time polynomial in the size of the problem and number of variables [12], as a consequence the closest string problem is fixed-parameter tractable (in FPT) when parameterized by $n$, albeit with a huge function on $n$. Further developments have included solutions for the special cases of $n = 3$ [9] and, for binary strings, $n = 4$ [3].

* Supported by the Natural Sciences and Engineering Research Council of Canada (NSERC).

D.M. Thilikos and G.J. Woeginger (Eds.): IPEC 2012, LNCS 7535, pp. 252–263, 2012.

Solutions to the neighbour string problem have been built on a search tree algorithm of Gramm et al. [9] for the closest string problem (which was used to prove the fixed-parameter tractability with parameter $d$), and share the key ideas of careful selection of two strings $s_x$ and $s_y$ and branching on all acceptable substrings in the region $R$ of positions on which $s_x$ and $s_y$ differ. The StringSearch algorithm of Ma and Sun [15], a modification of the closest string algorithm, computes a neighbour string in time $O(n\ell + nd\binom{d+b}{b}(|\Sigma|-1)^b 4^b)$ for instances having $\min_i d_i \leq b$ and $\max_i d_i \leq d$. This algorithm generalized that of Gramm et al. [9] by considering all substrings of $R$ in a single round rather than position by position. A series of further papers culminated in a running time of $O(n\ell + nd(\sqrt{2}|\Sigma| + \sqrt[4]{8}(\sqrt{2}+1)(1+\sqrt{|\Sigma|-1}) - 2\sqrt{2}))^d)$ [17,18,5]; these papers made refinements in the choice of the strings $s_x$ and $s_y$ to aid in the analysis. Lokshtanov et al. [14] showed that there does not exist any $O(2^{o(d \log |\Sigma|)} \cdot (n\ell)^{O(1)})$ time algorithm, unless the Exponential Time Hypothesis (ETH) fails.

Recently, Chen et al. [6] added a third input string to the computations, and obtained an $O(n\ell + nd^3 6.7308^d)$-time algorithm for binary strings and an $O(n\ell + nd1.612^d(|\Sigma| + \beta^2 + \beta - 2)^d)$-time algorithm 3-String for arbitrary alphabet $\Sigma$, where $\beta = \alpha^2 + 1 - 2\alpha^{-1} + \alpha^{-2}$ with $\alpha = \sqrt[3]{\sqrt{|\Sigma|-1}+1}$. For small $|\Sigma|$'s, this time bound is an improvement over the previous best running time, due to Chen and Wang [5]. The algorithm and analysis are both considerably more complicated than those for the approaches using two strings; as the number of strings increases, so does the number of different kinds of difference regions.

The reduction used to prove the NP-hardness of the closest string problem can also be used to prove that the corresponding counting problem is #P-hard [4]. The integer programming formulation of Gramm et al. can also be used to enumerate all the solutions [4], thus proving the fixed-parameter tractability of the enumeration problem with parameter $n$. Obtaining a parameterized enumeration algorithm with parameter $d$ is not necessarily possible unless all $d_i$'s are *minimal*, in the sense that there does not exist a neighbour string with all the distances strictly less than the required distances. When there are non-minimal $d_i$'s, there can be as many as $\binom{\ell}{\max_i d_i}$ solutions for such instances. The simplest example is an instance consisting of a single string $s_1$ of length $\ell$ and an arbitrary number as $d_1$. Given this constraint, we will consider an algorithm to be an enumeration algorithm if it produces all solutions when the $d_i$'s are minimal, and produces at least one solution otherwise.

We observe that in its original form, the $O(n\ell + nd(d+1)^d)$-time search tree algorithm of Gramm et al. [9,10] does not always enumerate all solutions. Starting with an input string $s_x$, the algorithm tries to find an input string $s_y$ such that $H(s_x, s_y) > d$. If there is no such string, then $s_x$ is produced as the sole solution. When the set of solutions is a superset of the input strings, even if the algorithm were modified to take the union of solutions found from starting at each of the input strings, the algorithm would fail to produce any solution that is not an input. In the example of the strings $\{0010, 0100, 1001, 1111\}$ and the value $d = 3$, there are eight solutions outside the set of inputs, such as 0001 and 0011, which are not found by this algorithm.

**Our Contributions.** We propose a new enumeration algorithm, CROSSOVER-SEARCH, for the neighbour string problem. As in previous algorithms, we construct a region $R$ on which two strings differ. In order to take advantage of all $n$ of the input strings instead of just two or three, we take different approaches to solutions that are "close" to $R$ and "far" from $R$. For the strings that are "close", we form $n+1$ representative strings such that each "close" solution is very close to one of the representatives, allowing us to handle all such solutions in $n+1$ recursive calls. The removal of the "close" strings allows us to exploit the additional structure that results on the remaining "far" strings, allowing us to improve bounds on the reduced instance.

Our analysis can be viewed as simulating the idea of considering a third string as in 3-STRING, and, further, of extending it to the consideration of all input strings. In doing so, we avoid the detailed case analysis that was required to handle all the different subregions formed in 3-STRING.

Another contribution of our work is to introduce tuning constants $\epsilon$ and $\delta$ that allow us to optimize the running time of our algorithm for values of $n$ and $d$ that are related in different ways. This results in an overall running time of $O(n\ell + ndN_\delta(d, b, \epsilon d, n))$, where $N_\delta(d, b, t_0, n)$ is the maximum number of leaves in the search trees (and hence the number of solutions) of input instances having $n$ strings of $\max_i d_i \leq d$ and $\min_i d_i \leq b$, called with threshold values $\delta$ and $t_0$. Furthermore, we prove that

$$N_\delta(d, d, \epsilon d, n) \leq ((n+1)(d+1))^{\lceil \log_{(1-\frac{\delta}{2})} \epsilon \rceil} (|\Sigma| - 1)^d 5^{d(1+\epsilon+\delta)}.$$

The measure $k_{min}$, used in defining "close" and "far", can also be used to advantage in the analysis of our modification of STRINGSEARCH, ENUM-STRINGSEARCH, designed to enumerate all neighbour strings. We prove a bound on the maximum number of leaves in the search tree of ENUMSTRINGSEARCH for input instances having $n$ strings of $\max_i d_i \leq d$ and $d_1 \leq b$, denoted by $M(d, b, n)$. In particular, we prove that

$$M(d, d, n) \leq 2(n(d+1))^{\lceil \lg \frac{1}{\epsilon} \rceil} (|\Sigma| - 1)^d 6^{d(1+\epsilon)},$$

for any $0 \leq \epsilon \leq 1$, and that the running time is in $O(n\ell + ndM(d, d, n))$.

## 2 Definitions

Throughout this article, we use $\Sigma^\ell$ to denote the set of all strings of length $\ell$ over the alphabet $\Sigma$. For any $s \in \Sigma^\ell$, the character in position $p$ in $s$, $1 \leq p \leq \ell$, is denoted by $s[p]$. For a set of positions or *region* $R \subseteq \{1, 2, \ldots, \ell\}$, we use $\overline{R}$ to denote $\{1, 2, \ldots, \ell\} - R$, the positions not in the region $R$. We define $s|R$ as the string formed by removing the characters in $\overline{R}$, and making the remaining characters consecutive. For the example $s = 01101$ and $R = \{1, 3, 4\}$, $s|R$ is the string 010. By extension, for a set of strings $S \subseteq \Sigma^\ell$, $S|R$ is defined as $\{s|R : s \in S\}$. For convenience, we call $s|R$ ($S|R$) the *restriction of $s$ ($S$) to $R$*. For strings $s_1$ and $s_2$ of lengths $|R|$ and $\ell - |R|$, we define $s_1 \oplus_R s_2$ as the string

$s$ of length $\ell$ such that $s|R = s_1$ and $s|\overline{R} = s_2$, and $s_1 \oplus_R S$ as the set of strings $\{s_1 \oplus_R s_2 : s_2 \in S\}$.

We use a distance measure to describe the degree to which strings differ from each other. For two strings $s_1, s_2 \in \Sigma^\ell$, $P(s_1, s_2)$ is defined as the set $\{p : s_1[p] \neq s_2[p]\}$ of positions on which the strings differ. The *Hamming distance* between $s_1$ and $s_2$, denoted by $H(s_1, s_2)$, is the number of positions in $P(s_1, s_2)$; more informally, we say that $s_1$ is *at distance* $H(s_1, s_2)$ *from* $s_2$. For any $d \geq H(s_1, s_2)$, we say that $s_1$ ($s_2$) is *within distance* $d$ of $s_2$ ($s_1$). Informally, we can think of $s_1$ and $s_2$ as being far if $H(s_1, s_2)$ is larger than some threshold value and as being close otherwise.

Each solution to the neighbour string problem is close to all input strings, where for each input a threshold value is supplied. Given an instance $\mathcal{I} = \{(s_1, d_1), (s_2, d_2), \ldots, (s_n, d_n)\}$, where $s_i \in \Sigma^\ell$ and $d_i$ is a non-negative integer (the *distance allotment*) for $1 \leq i \leq n$, a *neighbour string* for $\mathcal{I}$ is a string $\sigma$ such that for all $i$, $H(\sigma, s_i) \leq d_i$. We use $\mathcal{I}_s$ to denote the set of strings in instance $\mathcal{I}$, that is, $\mathcal{I}_s = \{s_1, \ldots, s_n\}$, and $NB(\mathcal{I})$ to denote the set of all neighbour strings for $\mathcal{I}$.

## 3 Enumerating Neighbour Strings

Our algorithm CrossoverSearch makes use of new ideas in both the algorithm and its analysis. Before discussing the algorithm, we introduce some of our analysis in a simpler example, namely our minor modification EnumStringSearch of the StringSearch algorithm of Ma and Sun [15].

### 3.1 Analysis of EnumStringSearch

The algorithm EnumStringSearch showcases a simplified form of our new analysis technique; the algorithm itself is a minor modification of Ma and Sun's StringSearch that forms an enumerative algorithm. The algorithm proceeds by first choosing $s_1$ as the starting string $s_x$ and deciding if it is a solution (line 3). A region $R$ is defined as the set of positions in which $s_x$ differs from a second string, $s_y$. In Example 1 below, $s_2$ is chosen as $s_y$, as it satisfies the condition (line 5) that $s_x$ and $s_y$ differ in at least $d_2 = 3$ positions. The algorithm tries each possible assignment of characters to the positions in $R$, and for each such assignment $w$, forms solutions by recursively solving the problem on a reduced instance consisting of the substrings formed by removing all positions in $R$. In lines 7 and 8, values of $w$ are grouped by $k$, where $k = H(w, s_x|R)$, the distance between $w$ and the restriction of $s_x$ to $R$. The distance allotment $d_i'$ for the reduced instance is determined by subtracting from $d_i$ the number of positions that differ between $s_i|R$ and $w$ (line 9). We observe that since $s_x|\overline{R} = s_y|\overline{R}$, we can set $d_x'$ to the minimum of $d_x'$ and $d_y'$ (line 10).

*Example 1.* For $\mathcal{I} = \langle(00000000, 3), (11111000, 3), (10011110, 3), (01111001, 3)\rangle$, $s_x = s_1$, $s_y = s_2$, $R = \{1, 2, 3, 4, 5\}$, and $w = 00011$, the $d_i'$'s are determined as $d_2' = 3-3 = 0$, $d_3' = 3-1 = 2$, $d_4' = 3-2 = 1$, and $d_x' = d_1' = \min\{3-2, d_2'\} = 0$.

**Algorithm 1.** CrossoverSearch

**Require** : An instance $\mathcal{I} = \langle (s_1, d_1), (s_2, d_2), \ldots, (s_n, d_n) \rangle$, a real number $\delta$, an integer $t_0$, and $(s_x, d_x)$

**Assume** : $\nexists s \forall 1 \leq i \leq n H(s, s_i) < d_i$

1 **if** $d_x \leq t_0$ **then return** EnumStringSearch($\langle (s_x, d_x), (s_1, d_1), (s_2, d_2), \ldots, (s_n, d_n) \rangle$);

2 Choose $s_y \in \mathcal{I}_s$ such that $H(s_x, s_y) \geq d_y$;

3 $R \leftarrow P(s_x, s_y)$;

4 $t \leftarrow \lfloor (1 - \frac{\delta}{2}) d_x \rfloor + 1$;

5 $NB \leftarrow$ CrossoverSearch($\mathcal{I}, \delta, t_0, (s_x, t-1)$);

6 **foreach** $1 \leq i \leq n$ **do**

7 $\quad \hat{s} = s_i | R \oplus_R s_x | \overline{R}$;

8 $\quad NB \leftarrow NB \cup$ CrossoverSearch($\mathcal{I}, \delta, t_0, (\hat{s}, t-1)$);

9 **end**

10 **foreach** $\max\{|R| - d_y, t - \frac{d_x + d_y - |R|}{2}\} \leq k \leq d_x$ **do**

11 $\quad$ **foreach** $w \in \Sigma^{|R|}$ *with* $H(w, s_x|R) = k$ **do**

12 $\quad\quad$ **foreach** $1 \leq i \leq n$ **do** $d'_i \leftarrow d_i - H(w, s_i|R)$;

13 $\quad\quad$ $d'_x \leftarrow \min\{d'_x, d'_y\}$;

14 $\quad\quad$ **if** $\min_i H(w, s_i|R) + d'_x < t$ **then** produce a fake leaf and continue to line 11;

15 $\quad\quad$ $NB \leftarrow NB \cup (w \oplus_R$ CrossoverSearch($\langle (s_1|\overline{R}, d'_1), \ldots, (s_n|\overline{R}, d'_n) \rangle, \delta, t_0, (s_x|\overline{R}, d'_x)$));

16 $\quad$ **end**

17 **end**

18 **return** $NB$;

The algorithm shown in Algorithm 2 differs from StringSearch only in lines 5 (where the original $>$ is replaced by $\geq$) and 11. The importance of the change to line 5 is that the search will continue even when a neighbour string is found. If the $d_i$'s are not minimal, the algorithm will find at least one solution (if any exist), but is not guaranteed to find all solutions. To be precise, the algorithm will terminate early at line 5 if $s_x$ is a neighbour string that is closer than $d_i$ for each $s_i$.

Lemma 1 below shows that, despite the change to line 5, the bounds obtained for StringSearch still hold. The proof, omitted due to lack of space, follows from the analogue in the analysis of StringSearch and the fact that $H(s_x, s_y)$ can equal $d_y$ has no impact on those proofs. The first item generalizes the fact that if $|R| > d_x + d_y$, no neighbour string exists for $\mathcal{I}$; the closer $|R|$ is to $d_x + d_y$, the smaller the value of $d'_x$, and thus the easier the reduced instance. The second item shows that at each recursive call, the size of $d_x$ is at most half that at the previous level; we will see how to generalize this result in CrossoverSearch. The bound given by the third item will prove useful in improving the results of each algorithm, as in Theorem 1, where this result is used as a way to take an early exit from a recursive analysis.

**Algorithm 2.** EnumStringSearch

**Require** : An instance $\mathcal{I} = \langle (s_1, d_1), (s_2, d_2), \ldots, (s_n, d_n) \rangle$
**Assume** : $\nexists s \forall 1 \le i \le n H(s, s_i) < d_i$

$NB \leftarrow \emptyset$; $x \leftarrow 1$;
**if** $d_x < 0$ **then return** $NB$;
**if** $H(s_x, s_i) \le d_i$ *for all* $1 \le i \le n$ **then** $NB \leftarrow \{s_x\}$;
**if** $d_x = 0$ **then return** $NB$;
Choose $s_y \in \mathcal{I}_s$ such that $H(s_x, s_y) \ge d_y$;
$R \leftarrow P(s_x, s_y)$;
**foreach** $|R| - d_y \le k \le d_x$ **do**
  **foreach** $w \in \Sigma^{|R|}$ *with* $H(w, s_x|R) = k$ **do**
    **foreach** $1 \le i \le n$ **do** $d'_i \leftarrow d_i - H(w, s_i|R)$;
    $d'_x \leftarrow \min\{d'_x, d'_y\}$;
    $NB \leftarrow NB \cup (w \oplus_R$ EnumStringSearch$(\langle (s_1|\overline{R}, d'_1), (s_2|\overline{R}, d'_2), \ldots, (s_n|\overline{R}, d'_n) \rangle))$;
  **end**
**end**
**return** $NB$;

**Lemma 1.** 1. $d'_x \le \frac{d_x + d_y - |R|}{2}$ *(or, equivalently,* $|R| \le d_x + d_y - 2d'_x$*)*
2. $d'_x \le \frac{d_x}{2}$
3. *For all* $0 \le b \le d$, $M(d, b, n) \le \binom{d+b}{b}(|\Sigma| - 1)^b 4^b$.

Using a new analysis, we derive a new recursive bound for $M(d, b, n)$ in Lemma 2. Critical to our analysis is the value $k_{min} = \min_i H(w, s_i|R)$ (it does not appear explicitly in the algorithm), defined for a particular value of $w$. Unlike in the original analysis of StringSearch, where a bound was based on $k = H(w, s_x|R)$, we instead categorize different branches depending on $k_{min}$. In Example 1, $k_{min} = H(w, s_3|R) = 1$ and $k = H(w, s_x|R) = 2$.

We will use a combination of the recursive bounds of Lemma 2 and Lemma 1, item 3, to prove the complexity of EnumStringSearch in Theorem 1 and Corollary 1. Due to space limitations, only a high-level idea of the proof is provided.

**Theorem 1.** *Given an instance* $\mathcal{I} = \langle (s_1, d_1), (s_2, d_2), \ldots, (s_n, d_n) \rangle$ *for which there does not exist* $s$ *with* $H(s, s_i) < d_i$ *for all* $1 \le i \le n$, EnumStringSearch$(\mathcal{I})$ *returns* $NB(\mathcal{I})$ *in* $O(n\ell + n \max_i d_i \cdot M(\max_i d_i, d_1, n))$ *time, where*

$$M(d, b, n) \le (n(b+1))^{\lceil \lg \frac{b}{\epsilon d} \rceil} (|\Sigma| - 1)^b 2^b g(d(1+\epsilon), \min\{\lceil \frac{2d(1+\epsilon)}{3} \rceil, b\}),$$

*for all* $0 \le b \le d$, $0 \le \epsilon \le 1$, *and for* $g(x, y) = \binom{x}{y} 2^y$. *Furthermore,* $|NB(\mathcal{I})| \le M(\max_i d_i, d_1, n)$.

Stirling's inequality will simplify the bound for the closest string problem, where all $d_i$'s are equal to $d$, and thus $b = d$.

**Corollary 1.** *Given an instance* $\mathcal{I} = \langle (s_1, d), (s_2, d), \ldots, (s_n, d) \rangle$ *for which there does not exist* $s$ *with* $H(s, s_i) < d$ *for all* $1 \leq i \leq n$, ENUMSTRINGSEARCH($\mathcal{I}$) *returns* $NB(\mathcal{I})$ *in* $O(n\ell + nd \cdot M(d, d, n))$ *time, where*

$$M(d, d, n) \leq 2(n(d+1))^{\lceil \lg \frac{1}{\epsilon} \rceil} (|\Sigma| - 1)^d 6^{d(1+\epsilon)},$$

*for all* $0 \leq \epsilon \leq 1$. *Furthermore,* $|NB(\mathcal{I})| \leq M(d, d, n)$.

At the heart of the proof of Theorem 1 is the recursive bound for $M(d, b, n)$ given in Lemma 2. The formula is generated by summing over every string $w$ produced at line 8 the number of search tree leaves in the branch for $w$. For any $w$, we will use $v(w)$ to denote the value of variable $v$ in the body of the loop processing that $w$. In particular, $k_{min}(w) = \min_i H(w, s_i|R)$.

The $n$ factor, not present in previous bounds [9,15], results from the fact that although the number of $w$'s at distance $c$ from $s_x$ is $\binom{|R|}{c}(|\Sigma| - 1)^c$, there can be as many as $n\binom{|R|}{c}(|\Sigma| - 1)^c$ $w$'s at minimum distance $c$ from the $s_i$'s.

**Lemma 2.** *For all* $0 \leq b \leq d$,

$$M(d, b, n) \leq n(b+1) \cdot \max_{0 \leq i \leq \frac{b}{2}, 0 \leq j \leq b-i} \binom{d+b-2i}{j} (|\Sigma| - 1)^j M(d-j, i, n).$$

*Proof.* We consider an input $\mathcal{I} = \langle (s_1, d_1), \ldots, (s_n, d_n) \rangle$, with $\max_i d_i \leq d$ and $d_1 \leq b$, maximizing the number of leaves in the search tree. The lemma holds for $b \leq 0$, as the algorithm stops at line 2 or 4 producing only one node in the search tree.

For $b > 0$, the algorithm will branch on every $w$ produced at line 8. For a specific $j$, we consider all $w$'s such that $k_{min}(w) = j$. We can view each such $w$ as being formed by choosing an input string $s_i$, choosing $j$ positions in $R$, and choosing symbols for those positions that differ from the corresponding symbols in $s_i$; in total the number of such $w$'s is thus at most $n\binom{|R|}{j}(|\Sigma| - 1)^j$. We claim that the number of search tree leaves in the branch for any such $w$ is at most

$$A(j) = \max_{0 \leq i \leq \min\{b-j, \frac{d_y+b-|R|}{2}\}} M(d-j, i, n). \qquad (1)$$

An upper bound on the total number of nodes in the search tree is then found by summing over all values of $j$ the product of $A(j)$ and the number of $w$'s with $k_{min}(w) = j$, or

$$\sum_{0 \leq j \leq b} n \binom{|R|}{j} (|\Sigma| - 1)^j A(j),$$

which can be shown by straightforward mathematical manipulations to be at most

$$n(b+1) \cdot \max_{0 \leq i \leq \frac{b}{2}, 0 \leq j \leq b-i} \binom{d+b-2i}{j} (|\Sigma| - 1)^j M(d-j, i, n), \qquad (2)$$

as needed to complete the proof. We also used the fact that the range of $i$ in (2) is a superset of its range in (1) as $\frac{d_y+b-|R|}{2} \leq \frac{b}{2}$.

All that remains is to prove the claim. We consider an arbitrary $w$ produced at line 8. Assuming that $j = k_{min}(w)$, we show that the number of leaves produced in the loop is no more than $A(j)$.

In the case in which $d'_x(w)$ is non-negative, the number of leaves produced at the function call at line 11 is at most $M(\max_i d'_i(w), d'_x(w), n)$. Since $\max_i d'_i(w) \leq \max_i\{d_i - H(w, s_i|R)\} \leq \max_i d_i - \min_i H(w, s_i|R)$, we have $\max_i d'_i(w) \leq d - j$. There are thus at most $M(d-j, d'_x(w), n)$ leaves produced in the loop. We now wish to show that $M(d-j, d'_x(w), n)$ is bounded above by $A(j)$. We complete the proof by showing that $d'_x(w)$ is in the range $[0, \min\{b-j, \frac{d_y+b-|R|}{2}\}]$. By definition, $d'_x(w) \leq d_x - H(w, s_x|R) \leq b - k_{min}(w) = b - j$, covering the first case in the minimum. By Lemma 1, item 1, $d'_x(w) \leq \frac{d_y+d_x-|R|}{2} \leq \frac{d+b-|R|}{2}$.

For a negative $d'_x(w)$, the recursive call terminates at the second line, yielding a single node. Here we show that the number of leaves produced in the loop, i.e. 1, is bounded above by $A(j)$ by demonstrating that $1 \leq M(d-j, i, n)$ for some $i$ in the range $[0, \min\{b-j, \frac{d+b-|R|}{2}\}]$. In fact, $M(d-j, i, n) = 1$ for $i = 0$, as needed to complete the proof. □

The challenge in proving Theorem 1 by induction stems from the fact that the relative values of $i$ and $\lceil \frac{2(d-j)(1+\epsilon)}{3} \rceil$ are not known for the $i$ and $j$ maximizing the Lemma 2 recursive bound. Setting $u$ to 2 in Lemma 3 below, given without proof, allows us to correlate the relative values of $b$ and $\lceil \frac{2d(1+\epsilon)}{3} \rceil$ with the relative values of $i$ and $\lceil \frac{2(d-j)(1+\epsilon)}{3} \rceil$.

**Lemma 3.** *For any integer $u$ and for all $i \in [0, \frac{b}{2}]$ and $j \in [0, b-i]$, and $d \geq b$, if $i \geq \lceil \frac{u(d-j)}{u+1} \rceil$, then $b \geq \frac{2ud}{2u+1}$ and $b \geq \lceil \frac{ud}{u+1} \rceil$.*

Another challenge is that every time the recursive bound of Lemma 2 is applied, an $n(b+1)$ factor is generated. However, $\log b$ applications of the bound are needed before a constant $b$ is reached. To reduce the exponent of $n(b+1)$ to a constant, we stop using the recursive function of Lemma 2 after $b$ becomes smaller than $\epsilon d$, for a tuning constant $\epsilon$, and then use the bound $M(d, b, n) \leq \binom{d+b}{b}(|\Sigma|-1)^b 4^b$ of Ma and Sun (Lemma 1, item 3). As $\epsilon$ increases, the recursive depth (and hence the exponent on $n(d+1)$) decreases, as the ending condition is met sooner. Since the optimal choice will depend on the relative values of $n$ and $b$, the $\epsilon$ in Theorem 1 can be set as best for each circumstance.

### 3.2 Overview of CrossoverSearch

In this section, we highlight the ideas in CrossoverSearch, shown in Algorithm 1. Following previous algorithms, we begin by finding a difference region $R$ (lines 2–3). The new approach introduced in this algorithm is the classification of all solutions into two types: an *R-close solution* is "close" to the restriction of some input string to $R$, and an $\overline{R}$-*close solution* is "close" to restrictions of all input strings to $\overline{R}$. For the appropriate definition of "close", each solution must be of one of these types, since if a solution is not $R$-close, then $H(\sigma|R, s_i|R)$ is large, and hence $H(\sigma|\overline{R}, s_i|\overline{R})$ cannot be very big, since $H(\sigma|R, s_i|R) + H(\sigma|\overline{R}, s_i|\overline{R})$ must be at most $d_i$. The analysis of $\overline{R}$-close solutions uses the measure $k_{min}$ in a manner analogous to the analysis of EnumStringSearch in Section 3.1. The fake leaves produced at line 14 are for technical reasons only; they make the running time of the algorithm a nice function of the number of leaves, as otherwise, some of the time-consuming branches do not produce many leaves.

Each $R$-close solution will be close to at least one of the $n+1$ strings $s_i|R \oplus_R s_x|\overline{R}$, $s_i \in \mathcal{I}_s \cup \{s_x\}$. The algorithm will find all such neighbour strings through the $n+1$ recursive calls at lines 5 and 8, each of which uses one of the $n+1$ strings as $s_x$ along with a small distance allotment $t-1$. It is the threshold $t$, then, that defines "close" to distinguish $R$-close and $\overline{R}$-close solutions. The smaller this threshold is, the more distances are considered "far", and hence the more solutions are $\overline{R}$-close. In Example 1, if $t = 3$, the solution $\sigma = 00011000$ is considered $R$-close since $H(s_3|R, \sigma|R) = 1 \leq t - 1$. Consequently, $\sigma$ will be found efficiently through the function call at line 8 for $\hat{s} = s_3|R \oplus_R s_x|\overline{R}$. After line 9, the search is confined to solutions whose restrictions to $R$ are at distance at least 3 from all the $s_i$'s.

The roles of lines 1 and 4 are related to the use of tuning constants for the analysis. In EnumStringSearch a single tuning constant $\epsilon$ was confined to the analysis, used to determine when to stop the recursive calls; here we add a second tuning constant $\delta$ and, unlike in EnumStringSearch, introduce both constants into the algorithm itself. Here, $t_0$ plays the same role as $\epsilon$. The additional constant $\delta$ plays a role in choosing the threshold $t$.

We can think of the analysis as occurring in two stages. In the first stage, we recursively reduce our bound on $d_x$ (line 4); this is similar to the reduction by halving in EnumStringSearch (Lemma 1, item 2). Here instead of using $\frac{1}{2}$, we use $1 - \frac{\delta}{2}$; the bigger the value of $\delta$, the smaller the value of $t$, which plays a role in defining the new $d_x$. The first stage ends when $d_x \leq t_0$ (line 1). In the second stage, i.e. within the instance of EnumStringSearch called at line 1, the bound is halved at each recursive call, and Lemma 1, item 3 is invoked to complete the analysis.

The inputs include not only the $n$ strings and distance allotments and the tuning constants $\delta$ and $t_0$, but also a specified string and distance allotment pair $(s_x, d_x)$, where $s_x$ (as formed in line 7) is not required to be one of the input strings. To avoid an increase in the number of strings in each invocation of the algorithm, $s_x$ is not merged into $\mathcal{I}$ at line 8; also, whenever a constructed string $\hat{s}$ is expected to be closer than $s_x$ to the solutions, $\hat{s}$ will replace the current value of $s_x$ (line 8).

### 3.3 Analysis of CrossoverSearch

We prove the complexity of CrossoverSearch in Theorem 2. The ideas in the proof of Theorem 1, such as the use of $k_{min}$ and the use of the recursive bound of Ma and Sun (Lemma 1, item 3) for small values of $b$ are also used in the proof of Theorem 2. In comparison, the parameter $k_{min}$ appears in the algorithm (line 14), and a more complicated condition is needed at line 10. Determining a condition that satisfies both the correctness and the required complexity is a challenge. Due to space constraints, we omit the proof, and only mention high level ideas.

The ratio of $t_0$ to $d$ has the same role as the tuning constant $\epsilon$ in ENUMSTRINGSEARCH. The proof is optimized for values of $\delta$ smaller than 0.75, since these are the only values of interest. Larger values of $\delta$ will produce $5^{d(1+\epsilon+0.75)}$ factors in the bound, already worse than the $16^d$ previous bound [15], also mentioned in Lemma 1, item 3.

**Theorem 2.** *Given an instance* $\mathcal{I} = \langle(s_1, d_1), (s_2, d_2), \ldots, (s_n, d_n)\rangle$, *for which there does not exist* $s$ *with* $H(s, s_i) < d_i$ *for all* $1 \leq i \leq n$, *and for any* $0 < \delta \leq 0.75$ *and* $0 \leq \epsilon \leq 1$, CROSSOVERSEARCH$(\mathcal{I}, \delta, \epsilon d, (s_1, d_1))$ *returns* $NB(\mathcal{I})$ *in time* $O(n\ell + n \max_i d_i \cdot N_\delta(\max_i d_i, d_1, \epsilon d, n))$, *where*

$$N_\delta(d, b, \epsilon d, n) \leq ((n+1)(b+1))^{\lceil \log_{(1-\frac{\delta}{2})} \frac{\epsilon d}{b} \rceil} (|\Sigma| - 1)^b f(d', \min\{\lceil \frac{4d'}{5} \rceil, b\}),$$

*for any* $0 \leq b \leq d$, *and for* $f(x, y) = \binom{x}{y} 4^y$ *and* $d' = d + \epsilon d + \delta b$. *Furthermore,* $|NB(\mathcal{I})| \leq N_\delta(\max_i d_i, d_1, \epsilon d, n)$.

Again, Stirling's inequality simplifies the bound for the special case of the closest string problem, where $b = d$.

**Corollary 2.** *Given an instance* $\mathcal{I} = \langle(s_1, d), (s_2, d), \ldots, (s_n, d)\rangle$, *for which there does not exist* $s$ *with* $H(s, s_i) < d$ *for all* $1 \leq i \leq n$, *and for any* $0 < \delta \leq 0.75$ *and* $0 \leq \epsilon \leq 1$, CROSSOVERSEARCH$(\mathcal{I}, \delta, \epsilon d, (s_1, d))$ *returns* $NB(\mathcal{I})$ *in time* $O(n\ell + nd \cdot N_\delta(d, d, \epsilon d, n))$, *where*

$$N_\delta(d, d, \epsilon d, n) \leq ((n+1)(d+1))^{\lceil \log_{(1-\frac{\delta}{2})} \epsilon \rceil} (|\Sigma| - 1)^d 5^{d(1+\epsilon+\delta)}.$$

*Furthermore,* $|NB(\mathcal{I})| \leq N_\delta(d, d, \epsilon d, n)$.

The bound is mainly derived from the recursive functions in Observation 1 and Lemma 4. If $b \leq t_0$, the function call at line 1 makes the algorithm run ENUMSTRINGSEARCH with the additional string $s_x$, thus producing no more than $M(d, t_0, n+1)$ leaves.

**Observation 1.** *For all* $0 \leq t_0 \leq d, 0 \leq b \leq t_0$, $N_\delta(d, b, t_0, n) \leq M(d, b, n+1)$.

The proof for larger values of $b$ will use much of the analysis appearing in the proof of Lemma 2.

**Lemma 4.** *For every* $0 \leq t_0 \leq d$, $t = \lfloor (1 - \frac{\delta}{2})b \rfloor$, *and all* $t_0 < b \leq d$, $N_\delta(d, b, t_0, n)$ *is less than or equal to*

$$\max \begin{cases} (n+1)\left(N_\delta(d, t-1, t_0, n) + b \cdot \max_{0 \leq i \leq \frac{b}{2}, t-i \leq j \leq b-i} \binom{d+b-2i}{j} (|\Sigma| - 1)^j N_\delta(d-j, i, t_0, n)\right) \\ \max_{\hat{d} \leq d, \hat{b} \leq b} N_\delta(\hat{d}, \hat{b}, t_0, n) \end{cases}.$$

*Proof.* We use strong induction on $b$, considering the instance $\mathcal{I}$ and values $(s_x, d_x)$ that maximize the number of leaves in the search tree. When $d_x < b$, the number of leaves is bounded by $N_\delta(\max_i d_i, d_x, t_0, n)$, covered by the second line in the recursive formula.

From now on, we can assume that $d_x = b$, and set $d' = \max_i d_i$. We let $s_y$ be the string chosen at line 2; $R$ will be $P(s_x, s_y)$. The algorithm will make $n+1$ function calls at lines 5 and 8, each producing at most $N_\delta(d', t-1, t_0, n)$ nodes. It will then branch on every $w$ produced at line 11. We claim that for any produced $w$ there exist $0 \leq j \leq b$ and $s \in \mathcal{I}_s \cup \{s_x\}$ such that $H(w, s|R) = j$ and the number of leaves reached from $w$ is at most

$$A(j) = \max_{\max\{t-j,0\} \leq i \leq \min\{b-j, \frac{d_y+b-|R|}{2}\}} N_\delta(d'-j, i, t_0, n).$$

In total, the number of $w$'s mapped to a certain $j$ and $s$ is at most $\binom{|R|}{j}(|\Sigma|-1)^j$.

An upper bound on the total number of leaves in the search tree is then found by the counts for lines 5, 8, and 15, for a total of

$$(n+1) \cdot N_\delta(d', t-1, t_0, n) + \sum_{0 \leq j \leq b} (n+1) \binom{|R|}{j} (|\Sigma|-1)^j A(j),$$

which by straightforward mathematical manipulations is at most

$$(n+1)\left( N_\delta(d', t-1, t_0, n) + b \cdot \max_{0 \leq i \leq \frac{b}{2}, t-i \leq j \leq b-i} \binom{d_y+b-2i}{j} (|\Sigma|-1)^j N_\delta(d'-j, i, t_0, n) \right)$$
$$\leq (n+1)\left( N_\delta(d, t-1, t_0, n) + b \cdot \max_{0 \leq i \leq \frac{b}{2}, t-i \leq j \leq b-i} \binom{d+b-2i}{j} (|\Sigma|-1)^j N_\delta(d-j, i, t_0, n) \right).$$

All that remains is to prove the claim. We omit details due to space limitations. We consider an arbitrary $w$ produced at line 11.

In the case in which $d'_x(w)$ is non-negative and $\min_i H(w, s_i|R) + d'_x(w) \geq t$, the idea is to demonstrate that the number of leaves produced in the loop, which is no more than $N_\delta(\max_i d'_i(w), d'_x(w), t_0, n)$ in this case, is at most $A(j)$ for $j = k_{min}(w)$.

For a negative $d'_x$ or for $\min_i H(w, s_i|R) + d'_x$ smaller than $t$, the algorithm will produce a single node. In this case, we show that the number of leaves produced in the loop, i.e. 1, is bounded above by $A(j)$ for $j = k$, i.e. $H(w, s_x|R)$. □

## 4 Conclusions

Aside from theoretical improvements, the new bounds are indications of the effectiveness of our approach. In particular, we expect the ideas in CROSSOVER-SEARCH to be useful in practice, for cases $\frac{\min_i d_i}{d} > \frac{2}{3}$, which are considered hard for previous algorithms.

**Acknowledgements.** The authors are grateful to anonymous reviewers for their feedback and to Bin Ma for helpful discussions and mention of an instance for which previous algorithms failed to find all closest strings.

## References

1. Andoni, A., Indyk, P., Patrascu, M.: On the optimality of the dimensionality reduction method. In: Proc. 47th Annu. IEEE Symp. Foundations of Computer Science, pp. 449–458. IEEE Computer Society, Washington (2006)
2. Ben-Dor, A., Lancia, G., Perone, J., Ravi, R.: Banishing Bias from Consensus Sequences. In: Apostolico, A., Hein, J. (eds.) CPM 1997. LNCS, vol. 1264, pp. 247–261. Springer, Heidelberg (1997)
3. Boucher, C., Brown, D.G., Durocher, S.: On the Structure of Small Motif Recognition Instances. In: Amir, A., Turpin, A., Moffat, A. (eds.) SPIRE 2008. LNCS, vol. 5280, pp. 269–281. Springer, Heidelberg (2008)
4. Boucher, C., Omar, M.: On the Hardness of Counting and Sampling Center Strings. In: Chavez, E., Lonardi, S. (eds.) SPIRE 2010. LNCS, vol. 6393, pp. 127–134. Springer, Heidelberg (2010)
5. Chen, Z., Wang, L.: Fast exact algorithms for the closest string and substring problems with application to the planted $(l, d)$-motif model. IEEE/ACM Trans. Comput. Biol. Bioinformatics 8, 1400–1410 (2011)
6. Chen, Z.Z., Ma, B., Wang, L.: A three-string approach to the closest string problem. J. Comput. Syst. Sci. Int. 78(1), 164–178 (2012)
7. Frances, M., Litman, A.: On covering problems of codes. Theory Comput. Syst. 30, 113–119 (1997)
8. Gąsieniec, L., Jansson, J., Lingas, A.: Efficient approximation algorithms for the hamming center problem. In: Proc. 10th Annu. ACM SIAM Symp. Discrete Algorithms, pp. 905–906. SIAM, Philadelphia (1999)
9. Gramm, J., Niedermeier, R., Rossmanith, P.: Exact Solutions for CLOSEST STRING and Related Problems. In: Eades, P., Takaoka, T. (eds.) ISAAC 2001. LNCS, vol. 2223, pp. 441–453. Springer, Heidelberg (2001)
10. Gramm, J., Niedermeier, R., Rossmanith, P.: Fixed-parameter algorithms for closest string and related problems. Algorithmica 37(1), 25–42 (2003)
11. Lanctot, J.K., Li, M., Ma, B., Wang, S., Zhang, L.: Distinguishing string selection problems. Inform. and Comput. 185(1), 41–55 (2003)
12. Lenstra, H.W.: Integer programming with a fixed number of variables. Math. Oper. Res. 8(4), 538–548 (1983)
13. Li, M., Ma, B., Wang, L.: Finding similar regions in many strings. In: Proc. 31st Annu. ACM Symp. Theory of Computing, pp. 473–482. ACM, New York (1999)
14. Lokshtanov, D., Marx, D., Saurabh, S.: Slightly superexponential parameterized problems. In: Proc. 22nd Annu. ACM-SIAM Symp. Discrete Algorithms, pp. 760–776. SIAM, San Francisco (2011)
15. Ma, B., Sun, X.: More efficient algorithms for closest string and substring problems. SIAM J. Comput. 39(4), 1432–1443 (2009)
16. Stojanovic, N., Berman, P., Gumucio, D., Hardison, R., Miller, W.: A Linear-time Algorithm for the 1-Mismatch Problem. In: Dehne, F., Rau-Chaplin, A., Sack, J., Tamassia, R. (eds.) WADS 1997. LNCS, vol. 1272, pp. 126–135. Springer, Heidelberg (1997)
17. Wang, L., Zhu, B.: Efficient Algorithms for the Closest String and Distinguishing String Selection Problems. In: Deng, X., Hopcroft, J.E., Xue, J. (eds.) FAW 2009. LNCS, vol. 5598, pp. 261–270. Springer, Heidelberg (2009)
18. Zhao, R., Zhang, N.: A more efficient closest string algorithm. In: Proc. 2nd Int. Conf. Bioinformatics and Computational Biology, pp. 210–215 (2010)

# An Improved Kernel for the Undirected Planar Feedback Vertex Set Problem

Faisal N. Abu-Khzam[1] and Mazen Bou Khuzam[2]

[1] Department of Computer Science and Mathematics
Lebanese American University
Beirut, Lebanon
faisal.abukhzam@lau.edu.lb

[2] Department of Mathematics, Kuwait University
Al Safat, Kuwait
mazen@sci.kuniv.edu.kw

**Abstract.** We consider the parameterized Feedback Vertex Set problem on unweighted, undirected planar graphs. We present a kernelization algorithm that takes a planar graph $G$ and an integer $k$ as input and either decides that $(G, k)$ is a no instance or produces an equivalent (kernel) instance $(G', k')$ such that $k' \leq k$ and $|V(G')| < 97k$. In addition to the improved kernel bound (from $112k$ to $97k$), our algorithm features simple linear-time reduction procedures that can be applied to the general Feedback Vertex Set problem.

## 1 Introduction

For a given undirected graph $G$, a feedback vertex set is a set of vertices whose removal yields an induced forest. The problem of finding a feedback vertex set of minimum size has applications in several domains including (for example) constraint processing and Bayesian inference [1,7]. Formally, the Feedback Vertex Set problem, henceforth FVS, is defined as follows:

Given: A graph $G$ and a positive integer $k$.
Question: Does $G$ have a feedback vertex set of cardinality $k$ or less?

FVS is NP-complete in general [11], and remains NP-complete when the input is restricted to planar graphs [15]. When parameterized by the solution size, FVS is fixed parameter tractable [9]. This has motivated many successful efforts to design parameterized FVS algorithms [6,8,12,13]. In particular, kernelization algorithms for FVS received some attention lately [2,5,14].

The Feedback Vertex Set problem has a kernelization algorithm that guarantees a quadratic-order kernel in general [14]. Linear kernels have been obtained on graphs of bounded genus and $H$-minor free graphs [3,10]. On planar graphs, a kernel of order $112k$ has been presented in [4]. In this paper, we present a linear-time kernelization algorithm for planar graphs that delivers kernels of order $97k-203$ (in the worst-case). The restriction to planar graphs allows us to use structural constraints that bound the order (and size) of our reduced instances. However, all the reduction procedures apply to general problem instances.

D.M. Thilikos and G.J. Woeginger (Eds.): IPEC 2012, LNCS 7535, pp. 264–273, 2012.

## 2 Preliminaries

Throughout this paper, the pair $(G, k)$ denotes a given planar instance of Feedback Vertex Set. $(G, k)$ will undergo a sequence of reduction operations that are based on a set of rules. Each reduction operation transforms $(G, k)$ into a potentially smaller instance, $(G', k')$, such that $k' \leq k$ and $G$ has a feedback vertex set of size $k$ if and only if $(G', k')$ has a solution of size $k'$.

We shall use common graph theoretic terminologies. In particular, we denote by $N_H(v)$ the set of neighbors of a vertex $v$ in a graph (or subgraph) $H$. For $A \subset V$, $G[A]$ denotes the subgraph of $G$ induced by $A$, and $N_H(A) = \cup_{v \in A} N_H(v) - A$. We denote by $V(G)$ and $E(G)$ the sets of vertices and edges of the graph $G$, respectively. A graph $H$ is bipartite if $V(H) = A \cup B$ so that each of the subsets $A$ and $B$ induces an edge-less subgraph. We shall refer to such graph by $H = (A, B)$. The complete bipartite graph is denoted by $K_{a,b} = (A, B)$ where $a = |A|$ and $b = |B|$.

We assume the graph $G$ is given with a particular planar embedding. We denote the number of vertices, edges and faces of $G$ by $n$, $e$ and $f$ respectively. These three values are related via the invariance formula of Euler for connected planar graphs, which states that $n - e + f = 2$. Euler's formula leads to upper bounds on the number of edges for many sub-families of planar graphs. In particular, a simple bipartite planar graph has at most $2n - 4$ edges. In general, a planar graph satisfies: $e \leq 3n - 6$. These two well known formulae are used to compute the upper bound on our kernel size.

During the reduction process, it may be possible to determine that a vertex belongs to a solution of $(G, k)$, if one exists. Such vertex will be deleted and placed in a set $S$ that, henceforth, denotes a potential solution of $(G, k)$. Moreover, the value of $k$ will be decremented by one. This action is justified by the rather trivial fact that deleting a vertex $v$ deletes all cycles containing it. Thus, for $v \in S$, $S \setminus \{v\}$ is a solution of $(G - v, k - 1)$.

Multiple edges between adjacent vertices are possible, as they may arise due to some operations performed on $G$. When this occurs, we assume that $G$ is drawn (or re-drawn) in the plane in such a way that any cycle of length two is the boundary of a face (i.e., it has an empty interior). Intuitively speaking, this is always possible since a multi-edge can be envisioned as obtained from *thickening* an edge and (then) splitting it. With this in mind, we can always assume that a pair of vertices share at most two edges, since additional edges do not introduce more cycles. When a pair, $\{u, v\}$, of adjacent vertices share two edges, we say that edge $uv$ is a *double-edge*. Otherwise it is a *single-edge*.

We shall also assume that $G$ has no self loops. The existence of a self loop yields an easy detection of an element of $S$ (the single vertex that forms the cycle), which results in deleting the corresponding vertex. Vertex deletion, on the other hand, are applied frequently by our algorithm and could result in disconnecting the graph in question. When this happens, we process the resulting connected components separately, in a sequential manner. So we may always assume that $G$ is connected.

For a vertex $v \in V(G)$, we denote by $C(v)$ the set of all cycles of $G$ that contain $v$. We extend this notation to subsets of $V(G)$ by setting $C(A) = \cup_{v \in A} C(v)$, where $A \subset V(G)$. We denote by $P_i$ any (sub)graph isomorphic to the path of length $i$. As usual, the path length is the number of edges forming the path. Moreover, the set of interior vertices of a path $P$ shall be denoted by $I(P)$.

Finally, the concept of cycle domination plays a role in simplifying our arguments. For $\{u, v\} \subset V$, we say that $u$ is dominated by $v$ when $C(u) \subset C(v)$ (every cycle passing through $u$ passes also in $v$). In other words, a solution that contains $v$ is at most as large as any solution that contains $u$. Detecting such cycle domination is easy: for each pair $\{u, v\} \subset V$, delete $v$ and check whether $u$ belongs to a cycle of the remaining graph. This checking can be performed via depth-first search in linear time since $G$ has a linear number of edges.

## 3 Reduction Rules

In this section, we present the reduction rules that are adopted by our kernelization algorithm. Every such rule is stated as a lemma that describes a pair (*condition*, *action*) and asserts that the *action* can be performed if the corresponding *condition* holds. A reduction rule is safe (or sound) if its action transforms $(G, k)$ into an equivalent instance, while updating the potential solution $S$. We shall assume that, for each $i$, rules 1 through $i$ are applied exhaustively before applying rule $i + 1$. Some reductions include the addition of edges or double-edges. While this seems contrary to the concept of data reduction, it will be an intermediate step to further reduction in the number of vertices (see Rules 5 and 7 below).

Some of the following rules are folkloric, and are left without proof. In particular, Rules 1-3 have appeared in several FVS algorithms (see [5,8]). We also introduce a few rules that work for general FVS instances and allow us to use planarity to achieve a linear-bound on the kernel size.

**Reduction Rule 1.** Vertices whose degree is less than two can be deleted.

**Reduction Rule 2.** If a vertex $u$ has degree two, with incident single-edges $uv$ and $uw$, then we can delete $u$ and add the edge $vw$.

**Reduction Rule 2*.** If a vertex $u$ has two neighbors $v$ and $w$ such that $uv$ is a single-edge and $uw$ is a double-edge, then we can place $w$ in $S$, decrement $k$ by one, and delete $u$ and $w$.

*Proof.* At least one of $u$ and $w$ is in $S$ and $C(u) \subset C(w)$. So it is safe to place $w$ in $S$ and decrement $k$ by one.

Note that Rules 2 and 2* are special cases of the following general reduction rule: if $u$ and $v$ are adjacent vertices and $C(u) \subset C(v)$, then edge $uv$ can be contracted. This rule, however, is not needed to obtain the linear bound presented in this paper.

**Reduction Rule 3.** If there are more than three edges between two vertices, then we can safely remove all but two of these edges.

**Reduction Rule 4.** If a vertex $u$ has exactly two neighbors $v$ and $w$ such that $uv$, $uw$ and $vw$ are double-edges, then we can add both $v$ and $w$ to $S$, decrement $k$ by two, and delete $u$, $v$ and $w$.

**Reduction Rule 5.** Let $uv$ be a double-edge in $G$ and let $w$ be a degree-three vertex whose neighbors are $u, v$ and $w'$. Then deleting $w$ and adding the two edges $uw'$ and $vw'$ results in an equivalent instance (See figure 1). This action can be performed even if $w'$ is a neighbor of $u$ or $v$ (or both).

*Proof.* At least one element of $\{u, v\}$ belongs to $S$. W.l.o.g., assume $u \in S$. Then the instance $(G-u, k-1)$, which results from deleting $u$ is equivalent to $(G, k)$. In this resulting instance, $w$ can be bypassed, by Rule 2 ($w$ is deleted and the edge $vw'$ is added). Observe also that adding the two edges $uw'$ and $vw'$ introduces only one additional cycle, namely $(u, v, w')$. If $(G, k)$ is a yes instance, then the new cycle will be covered by at least one element of $\{u, v\}$. Q.E.D.

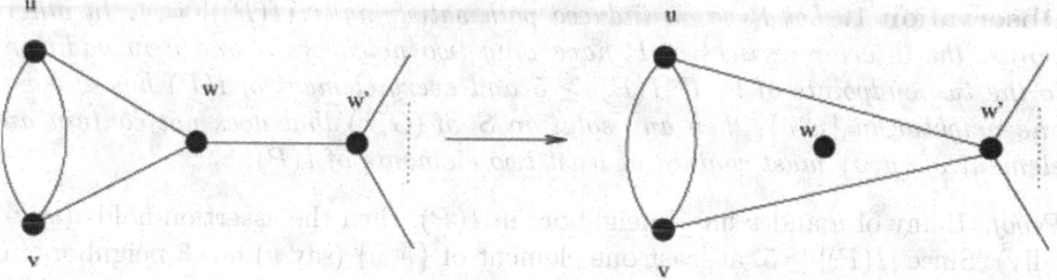


**Fig. 1.** Rule 5

**Reduction Rule 6.** Let $P$ be an induced path of endpoints $u$ and $v$ such that $N(P \setminus \{u, v\}) = \{w\}$. If $P$ has more than two interior vertices, then an equivalent instance can be obtained by placing $w$ in $S$ and deleting it. This action results in replacing $P$ by the edge $uv$ due to (the re-application of) Rule 2.

*Proof.* First note that $S$ must cover the local cycles of $G[P \cup \{w\}]$ and any (global) cycles that pass through the endpoints of $P$. So it is possible that $S$ contains an element of $I(P)$. If $P$ has interior vertices $w_1, w_2, \cdots, w_s$ such that $s \geq 3$, then each $w_i$ is dominated by $\{w, u\}$ (or $\{w, v\}$). If $s > 3$, and $w$ does not belong to $S$, then at least two interior vertices of $P$ are in $S$. So it is safe to add $w$ to $S$. If $s = 3$ and the solution $S$ has only one element from $I(P)$, then $w_2 \in S$ since neither $w_1$ nor $w_3$ covers the cycles through $w$ in $G[P \cup \{w\}]$. However, deleting $w_2$ will not result in disconnecting $G[P \cup \{w\}]$. Therefore we can replace $w_2$ by $w$ and obtain an equivalent (if not better) solution.

**Reduction Rule 7.** Let $(A, B) = K_{2,b}$ be a subgraph of $G$. If $b \geq 3$ and every vertex of $B$ is of degree three, then we can add a double-edge between the two elements of $A$. By Rule 5, and (possibly) Rule 4, this leads to deleting all elements of $B$.

*Proof.* Note that every element of $B$ is dominated by $A$ (this being true since every element of $B$ has only one neighbor outside $A$). Let $A = \{v, v'\}$ and $B = \{v_1, v_2, \cdots, v_b\}$. If none of the elements of $A$ is in $S$, then at least two elements of $B$ are needed to cover the cycles of the $K_{2,b}$-subgraph induced by $A \cup B$. It follows that adding the double-edge between the two elements of $A$ is sound. This addition does not affect the planarity of $G$ for the following reason: once the double-edge $vv'$ is introduced, Rule 5 applies and every common neighbor of degree three of $v$ and $v'$ is deleted, except when two elements of $B$ are adjacent, in which case we use Rule 5 to delete one of them and (then) Rule 4 to delete the other.

The last reduction rule deals with induced paths whose interior vertices have only two neighbors in addition to their endpoints. At this stage, we assume every vertex has degree three or more. We observe the following.

**Observation 1.** *Let $P$ be an induced path satisfying $|N(I(P))| = 4$. In other words, the interior vertices of $P$ have only two neighbors $u$ and $v$ in addition to the two endpoints of $P$. If $|I(P)| \geq 5$ and every element of $I(P)$ has at least one neighbor in $\{u, v\}$, then any solution $S$ of $(G, k)$ that does not contain an element of $\{u, v\}$ must contain at least two elements of $I(P)$.*

*Proof.* If any of $u$ and $v$ has 4 neighbors in $I(P)$, then the assertion holds (trivially). Since $|I(P)| \geq 5$, at least one element of $\{u, v\}$ (say $v$) has 3 neighbors in $I(P)$, while the other has at least two neighbors. Assume one element $x \in I(P)$ covers all cycles of $G[P \cup \{u, v\}]$. Obviously, $x$ must be a neighbor of $v$ that is interior to the path connecting $v$'s neighbors in $I(P)$. Moreover, $x$ must also be interior to the path connecting the neighbors of $u$ in $I(P)$ (even when $u$ has only two such neighbors, since in that case $u$ and $v$ cannot have common neighbors). If $x$ is deleted, the remaining neighbors of $u$ and $v$ (which "surround" $x$ in $P$) would be connected via both $u$ and $v$, forming a cycle of length 6, which contradicts the assumption.

**Reduction Rule 8.** Let $P$ be an induced path satisfying $|N(I(P))| = 4$. As in the above Observation, the interior vertices of $P$ have only two neighbors $u$ and $v$ in addition to the two endpoints of $P$. If $|I(P)| = 6$, and $|N_{I(P)}(v)| \geq |N_{I(P)}(u)|$, then a solution of $(G, k)$ exists if and only if $(G, k)$ has a solution that contains $v$. So, in this case, we delete $v$ and decrement $k$ by one.

*Proof.* Let $V(P) = \{w_0, w_1, w_2, \cdots, w_s\}$ where $w_i w_{i+1}$ is an edge of $G[P]$ $(0 \leq i < s;\ s \geq 8)$. Any solution of $(G, k)$ must cover all the local cycles and possibly disconnect the two endpoints of $P$ in $G[P \cup \{u, v\}]$. We show that two elements of $I(P)$ are either not enough to cover all local cycles or one of them can be replaced by $v$ (safely). To see this, consider first the case where $u$ has at most two neighbors in $I(P)$ and $v$ has at least 4 such neighbors. If $v$ is not in the solution, at least two elements of $I(P)$ are needed to disconnect the local cycles through $v$. An equivalent (if not better) solution would include $v$ and one of the neighbors of $u$ (to disconnect $P$ and cover the local cycle through $u$, if any).

Now consider the case where each of $u$ and $v$ has more than two neighbors in $I(P)$. Assume there is a solution that contains only two elements $w_i$ and $w_j$ ($0 < i < j < s$) and disconnets any local path between $w_0$ and $w_s$ in $G[P \cup \{u, v\}]$. Then $\{w_i, w_j\}$ is a feedback vertex set for the graph $H$ obtained from $G[P \cup \{u, v\}]$ by adding the edge $w_0 w_s$. Let $H'$ be the acyclic graph obtained by deleting $w_i$ and $w_j$. The total number of edges incident on $u$ and $v$ in $H'$ is at least 4 (since $|I(P)| \geq 6$), while the number of edges that are deleted from the cycle $\{w_0, w_1, \cdots, w_s\}$ is at most 4. It follows that $H'$ has at least $s + 1$ edges and exactly $s + 1$ vertices, which contradicts the assumtpion that $H'$ is acyclic. Therefore, any solution that contains only two elements of $I(P)$ can result in either not covering all the local cycles or not disconnecting all local paths between $w_0$ and $w_s$ (in $G[P \cup \{u, v\}]$). In such case, it would be safe to take both $u$ and $v$ into the solution.

The instance $(G, k)$ is said to be reduced with respect to the above reduction rules if none of them applies to $(G, k)$. We shall prove that such a reduced instance has a solution only if $|V(G)| < 97k$.

## 4 A Linear Kernel

Our reduction process consists of applying the rules of the previous section exhaustively until none of them applies. Applying a rule starts by a search for a graph structure that makes its condition holds. If this search is successful, it will be followed by the action described by the rule. The resulting instance, $(G, k)$, is called a reduced instance.

Note that every reduction rule can be applied in linear time. In fact, it should be clear that applying each of the Rules 1-4 takes linear time, especially if an updated degrees-array is used. Moreover, it takes linear-time to check if the condition of Rule 5 holds (select a degree-three vertex and check whether two of its neighbors share a double-edge). Each of the conditions of the remaining rules can be checked via depth-first search (DFS). Starting with any vertex of $G$, DFS takes $O(E(G))$ (thus, linear-time) to find a path or a $K_{2,b}$ that satisfies the required condition. Note that we require the "successive" application of all the rules until they cannot be applied. The process of applying the rules is repeated whenever $k$ is decremented. It follows that the total reduction time is in $O(kn)$.

Throughout the rest of this section we assume that $(G, k)$ is a reduced **yes** instance. Again, let $S$ be a corresponding (complete) solution, and assume $G$ is a plane graph. By Rule 2, the degree of every vertex of $G$ is bounded below by three.

The complement of $S$ in $V(G)$ induces a forest $F$. Let $L$ be the set of tree-leaves in $F$. Each such leaf must have at least two neighbors in $S$. This is obvious since each leaf has only one neighbor in $F$ and at least three neighbors in $G$. Note the use of Rule 2* here: the unique edge that is incident on a leaf of $F$ is a single-edge. So a leaf node with only one neighbor in $S$ will be deleted, even if its degree is three. Our objective is to compute an upper bound on the cardinality of $L$, and use it to bound the number of elements in $F$. The following observation is needed in the sequal.

**Observation 2.** *Let $u$ and $v$ be non-adjacent vertices of a planar graph $H$. If $u$ and $v$ have a common neighbor $w$ whose degree is two in $H$, then the graph obtained from $H$ by adding the edge $uv$ is also planar.*

*Proof.* The following operations preserve planarity: contract edge $uw$, duplicate the (resulting) edge $uv$, then subdivide one of the edges that connect $u$ and $v$ (to get $w$ back).

Our main result depends on some counting arguments that make use of the following lemmas. Hereafter, a $(c, 2)$-chain of a graph $G$ is an induced path on $c$ vertices (and length $c - 1 \geq 0$), each of which is of degree two in $G$. When the length of the induced path is not important (or arbitrary), such a path is dubbed a degree-two chain. A degree-two chain is maximal if it is not contained in a larger degree-two chain.

Consider a tree $T$ and let $C$ be a collection of disjoint $(c, 2)$-chains of interior vertices in $T$. We say that $C$ is maximal if $T[V(T) \setminus V(C)]$ has no $(c, 2)$-chains.

**Lemma 1.** *Let $T$ be a tree with at most $l$ leaves. If, for $c \geq 1$, every maximal collection of disjoint $(c, 2)$-chains has at most $n_c$ internal vertices, then $T$ has at most $n_c + 2cl - 2c + 1$ vertices.*

*Proof.* The number of degree-three vertices of $T$ is less than the number of its leaves. Let $T'$ be the tree obtained from $T$ by replacing every maximal degree-two chain by a single edge (by a sequence of edge-contractions). Then $T'$ has at most $l - 1$ internal vertices and at most $2l - 2$ edges. Consequently, the number of maximal degree-two chains in $T$ is bounded above by $2l - 2$. The length of any degree-two chain is either smaller than $c$ or consists of (or can be decomposed into) one or more $(c, 2)$-chains and at most one $(b, 2)$-chain, $b \leq c-1$. It follows that the number of internal degree-two vertices is bounded above by $n_c + (c-1)(2l-2)$. Q.E.D.

We now show that $n_6$ is bounded by a linear function of $k$ and that $F$ has a linear number of leaves.

**Lemma 2.** *Let $H = (A, B)$ be a simple bipartite planar graph. If $|A| = k$ and every vertex of $B$ has at least three neighbors in $A$, then $|B| \leq 2k - 4$.*

*Proof.* Recall that, due to a corollary of Euler's formula, $H$ has at most $2n_h - 4$ edges, where $n_h = |V(H)| = k + |B|$. Since every vertex of $B$ is of degree at least three, we have $2n_h - 4 \geq |E(H)| \geq 3|B|$. Replacing $n_h$ by $k + |B|$ gives the desired inequality.

**Lemma 3.** *If $C$ is a maximal collection of $(6,2)$-chains of internal vertices in $F$, then $|C| \leq 2k - 4$.*

*Proof.* Let $P$ be a $(6,2)$-chain of internal vertices in $F$. By Rules 6 and 8, $N(P)$ has at least three elements from $S$ (since $P$ is $I(P')$ for some path $P'$ of length 8 in $F$). Let $C'$ be the set of vertices obtained by contracting every chain in $C$ (thus replacing every $(6,2)$-chain by a single vertex of degree $\geq 3$). Ignoring the edges between elements of $S$, and replacing multiple edges by simple edges, we consider the resulting simple bipartite planar graph $(S, C')$. The result now follows from Lemma 2.

**Lemma 4.** *Let $L_2$ be the subset of $L$ consisting of leaves that have exactly two neighbors in $S$. Then $|L_2| \leq 6k - 12$.*

*Proof.* For each pair $\{u, v\} \subset S$, the number of elements of $L_2$ that are common neighbors of $u$ and $v$ is at most 2. This follows easily from Rule 7. Consider the planar graph $H$ induced by $S \cup L_2$. And let $H'$ be a graph obtained from $H$ by adding an edge between non-adjacent elements of $S$ that share a common neighbor. Then, by Observation 2, $H'$ is also planar. Moreover, the number of pairs of $S$ that can share a common neighbor (from $L$) is bounded above by the number of edges in $H'$, which is at most $3k - 6$. The proof is now complete (knowing that each such pair has at most 2 common neighbors in $L$).

We now consider the leaves of $F$, if any, that have at least three neighbors in $S$.

**Lemma 5.** *Let $L_3 = L \backslash L_2$ be the set of leaves of $F$ that have more than two neighbors in $S$. Then $|L_3| \leq 2k - 4$.*

*Proof.* Consider the planar bipartite subgraph $H$ whose vertex set is $S \cup L_3$ and whose edges are elements of $E(G)$ that connect elements of $S$ to those of $L_3$. The proof follows from Lemma 2 by letting $l$ be the number of leaves in $L_3$ and $n_h$ the number of vertices in $H$.

The above two lemmas guarantee that our reduced instance has a solution $S$ of size $k$ only if the corresponding induced forest $F$ has at most $8k - 16$ leaves and at most $2k - 4$ $(6,2)$-chains. In fact, the number of elements of $F$ that have three neighbors in $S$ and the number of $(6,2)$-chains of $F$ (which is $n_6/6$) sum up to at most $2k - 4$. This can be obtained by considering (again) a bipartite plane graph $H = (A, B)$ where $A = S$ and $B$ consists of "contracted" $(6,2)$-chains and all elements of $F$ that have three neighbiors in $S$. Therefore $\frac{n_6}{6} + |L_3| \leq 2k - 4$ (or $n_6 \leq 12k - 6|L_3| - 24$). Moreover, $l \leq 6k - 12 + |L_3|$ (again $l$ is the total number of leaves in $F$).

It follows, by Lemmas 1, 4 and 5 that $|F|$ is bounded above by: $n_6 + 2(6)l - 2(6) + 1 \leq (12k - 6|L_3| - 24) + 12(6k + |L_3| - 12) - 11 = 84k + 6|L_3| - 179 \leq 84k + 6(2k - 4) - 179 = 96k - 203$.

This proves our claimed linear kernel bound, which we state in the following theorem.

**Theorem 1.** *There is a linear-time algorithm that, given an arbitrary planar instance $(G, k)$ of Feedback Vertex Set, either decides that no solution exists or produces an equivalent instance whose size is bounded above by* $97k - 203$.

## 5 Conclusion

We showed how to reduce any planar instance of Feedback Vertex Set into one whose order is bounded by $97k - 203$, where $k$ is the input parameter. Our reduction rules apply to general FVS instances and might be useful on their own, as preprocessing steps.

Our kernel bound is obtained by counting arguments that made heavy use of planarity. This bound can be improved further via reductions that consider induced paths as in Rule 8, and by more detailed counting arguments. We continue to investigate the potential use of our techniques to obtain a much smaller kernel bound without loosing the simplicity and efficiency of the corresponding reduction procedures.

## References


1. Bar-Yehuda, R., Geiger, D., Naor, J.(S.), Roth, R.M.: Approximation algorithms for the vertex feedback set problem with applications to constraint satisfaction and bayesian inference. In: SODA 1994: Proceedings of the Fifth Annual ACM-SIAM Symposium on Discrete Algorithms, Philadelphia, PA, USA, pp. 344–354. Society for Industrial and Applied Mathematics (1994)
2. Bodlaender, H.L.: A Cubic Kernel for Feedback Vertex Set. In: Thomas, W., Weil, P. (eds.) STACS 2007. LNCS, vol. 4393, pp. 320–331. Springer, Heidelberg (2007)
3. Bodlaender, H.L., Fomin, F.V., Lokshtanov, D., Penninkx, E., Saurabh, S., Thilikos, D.M.: (meta) kernelization. In: Proceedings of the 2009 50th Annual IEEE Symposium on Foundations of Computer Science, FOCS 2009, pp. 629–638. IEEE Computer Society, Washington, DC (2009)
4. Bodlaender, H.L., Penninkx, E.: A Linear Kernel for Planar Feedback Vertex Set. In: Grohe, M., Niedermeier, R. (eds.) IWPEC 2008. LNCS, vol. 5018, pp. 160–171. Springer, Heidelberg (2008)
5. Burrage, K., Estivill-Castro, V., Fellows, M.R., Langston, M.A., Mac, S., Rosamond, F.A.: The Undirected Feedback Vertex Set Problem Has a Poly($k$) Kernel. In: Bodlaender, H.L., Langston, M.A. (eds.) IWPEC 2006. LNCS, vol. 4169, pp. 192–202. Springer, Heidelberg (2006)
6. Chen, J., Fomin, F.V., Liu, Y., Lu, S., Villanger, Y.: Improved Algorithms for the Feedback Vertex Set Problems. In: Dehne, F., Sack, J.-R., Zeh, N. (eds.) WADS 2007. LNCS, vol. 4619, pp. 422–433. Springer, Heidelberg (2007)

7. Dechter, R.: Enhancement schemes for constraint processing: backjumping, learning, and cutset decomposition. Artificial Intelligence 41(3), 273–312 (1990)
8. Dehne, F., Fellows, M.R., Langston, M.A., Rosamond, F.A., Stevens, K.: An $o^*(2^{O(k)})$ FPT Algorithm for the Undirected Feedback Vertex Set Problem. In: Wang, L. (ed.) COCOON 2005. LNCS, vol. 3595, pp. 859–869. Springer, Heidelberg (2005)
9. Downey, R.G., Fellows, M.R.: Parameterized Complexity. Springer (1999)
10. Fomin, F.V., Lokshtanov, D., Saurabh, S., Thilikos, D.M.: Bidimensionality and kernels. In: Proceedings of the Twenty-First Annual ACM-SIAM Symposium on Discrete Algorithms, SODA 2010, Philadelphia, PA, USA, pp. 503–510. Society for Industrial and Applied Mathematics (2010)
11. Garey, M.R., Johnson, D.S.: Computers and Intractability. W. H. Freeman, New York (1979)
12. Guo, J., Gramm, J., Hüffner, F., Niedermeier, R., Wernicke, S.: Improved Fixed-Parameter Algorithms for Two Feedback Set Problems. In: Dehne, F., López-Ortiz, A., Sack, J.-R. (eds.) WADS 2005. LNCS, vol. 3608, pp. 158–168. Springer, Heidelberg (2005)
13. Raman, V., Saurabh, S., Subramanian, C.R.: Faster fixed parameter tractable algorithms for finding feedback vertex sets. ACM Trans. Algorithms 2(3), 403–415 (2006)
14. Thomassé, S.: A $4k^2$ kernel for feedback vertex set. ACM Transactions on Algorithms TALG 6(2), 1–8 (2010)
15. Yannakakis, M.: Node-and edge-deletion np-complete problems. In: STOC 1978: Proceedings of the Tenth Annual ACM Symposium on Theory of Computing, pp. 253–264. ACM, New York (1978)

# Author Index